HYDRAULIC CONTROL SYSTEMS

HYDRAULIC CONTROL SYSTEMS

Second Edition

Noah D. Manring
Roger C. Fales
University of Missouri–Columbia

Registered Offices
John Wiley & Sons, Inc., 111 River Street, Hoboken, NJ 07030, USA
John Wiley & Sons Ltd, The Atrium, Southern Gate, Chichester, West Sussex, PO19 8SQ, UK

Editorial Office
The Atrium, Southern Gate, Chichester, West Sussex, PO19 8SQ, UK

For details of our global editorial offices, customer services, and more information about Wiley products visit us at www.wiley.com.
Wiley also publishes its books in a variety of electronic formats and by print-on-demand. Some content that appears in standard print versions of this book may not be available in other formats.

Library of Congress Cataloging-in-Publication Data

Names: Manring, Noah, author. | Fales, Roger, author.
Title: Hydraulic control systems / Noah D Manring, University of Missouri,
    MO, US, Roger C Fales, University of Missouri, MO, US.
Description: Second edition. | Hoboken : Wiley, 2020. | Includes
    bibliographical references and index.
Identifiers: LCCN 2019023875 (print) | LCCN 2019023876 (ebook) | ISBN
    9781119416470 (cloth) | ISBN 9781119416487 (adobe pdf) | ISBN
    9781119416494 (epub)
Subjects: LCSH: Hydraulic control.
Classification: LCC TJ843 .M2185 2020 (print) | LCC TJ843 (ebook) | DDC
    629.8/042—dc23
LC record available at https://lccn.loc.gov/2019023875
LC ebook record available at https://lccn.loc.gov/2019023876

Cover Design: Wiley
Cover Image: © Songdech Kothmongkol/Shutterstock

Set in 10/12pt and TimesTenLTStd by SPiGlobal, Chennai, India

V10015689_111819

To our parents, who inspired us from the beginning!

CONTENTS

PREFACE TO THE SECOND EDITION

It has been over 50 years since Herbert Merritt wrote and published his classical work, *Hydraulic Control Systems*, and almost 15 years since Noah Manring published a new work by the same title. In this second edition, several additions have been made to the previous work, written in 2005 – to add information about modern methods, improve the usability of the text, and add practical information needed to design a complete hydraulic system. Major additions have been made to Chapter 3, "Dynamic Systems and Controls," and a new chapter has been added, Chapter 7, "Auxiliary Components." In Chapter 3, new subsections have been added about block diagrams of dynamic systems, frequency domain control design, digital control, state feedback control, and state estimation. The additions to Chapter 3 will be used by the reader to help develop many types of hydraulic control systems with consideration of issues like discrete sampling in microcontrollers, complicated dynamics, and demanding performance requirements. In the new Chapter 7, auxiliary components are presented that must also be considered for building practical hydraulic control systems that operate smoothly for a long duration. These auxiliary components include accumulators, conduits, reservoirs, coolers, and filters. Other improvements have also been made throughout the text, including a helpful list of unit conversions.

N.D. Manring and R.C Fales 2015

PREFACE TO THE
FIRST EDITION

It has been 38 years since Herbert Merritt wrote and published his classical work, *Hydraulic Control Systems*. To say that Merritt's work is classical is not an overstatement, as almost every research paper that has been published within the last 30 years references this text, and certainly everyone in the hydraulic control industry knows Merritt to be the authority in this field. Since Merritt first published his book in 1967, other texts have appeared in the literature, but none have managed to displace his volume as it sets forth general principles and a more rigorous development of the fluid power science. Even so, Merritt's text has carried with it a significant deficiency that has made its use in the academic world somewhat cumbersome; viz., the classic text contains no pedagogical resources which ought to include homework and example problems. Furthermore, 38 years of textual stagnation has dated the book and modern updates have been needed for some time. For several years now, scholars in this field have discussed collaborative projects in which a replacement for Merritt's text has been proposed. Though our intentions have been well-meaning, these projects have never seemed to get off the ground, and for that reason the current author has gone it alone to see what can be done to improve the textbook situation. It is with great appreciation for Herbert Merritt that this updated text is being offered, and if this current text makes any improvements over his work, it is only because the current author has had the privilege to stand on the shoulders of a giant.

As previously noted, this present work has been written to remedy the pedagogical deficiencies of Merritt's text and to make updates where they are needed. To address the teaching needs, a preliminary section has been added to the book in which fundamental concepts of fluid mechanics and system dynamics are reviewed. Furthermore, each chapter has been concluded with a section of homework problems that may be used to exercise the student and to assist the instructor. Scattered throughout the chapters are example problems noted as Case Studies. These case studies are aimed at illustrating points within the text, and at showing the student how to implement and use the equations. Technical updates have also been added as research over the past 30 years has advanced the state of the art in this field. In particular, a more thorough analysis of transient fluid flow-forces has been added to the section on valves; and poppet valves, which were omitted from Merritt's text, have now been included. In the area of pumps, a discussion of flow ripple has been introduced for both gear pumps and axial-piston pumps. Also, an updated analysis of the pump control problem has been added for swash-plate type machines. For hydraulic system design, a methodology suggested by Richard Burton at the University of Saskatchewan has been used throughout the text. This methodology begins the analysis at the load point of the system and works its way backward to the ultimate power source. For each hydraulic control system that is discussed in the text, control objectives of position, velocity, and effort are illustrated using reduced order models and PID controllers. Though the efforts of this current work are well-intended, undoubtedly readers will find deficiencies that will need to be addressed at some time. With a bit of history in mind, the author welcomes future updates of this work and only pleads with the critic not to wait for 38 years.

N.D. Manring 2004

INTRODUCTION

Hydraulic control systems have been in use at least as long as paddle wheels have been used to turn the grindstone at the mill. However, if we consider the more modern era in which we live, we can see that current day hydraulic systems had their beginnings before the end of the 19th century, when the invention of efficient prime movers, like the Watt's steam engine and the development of the factory system, spurred the need to develop a method for transmitting power from one point to another. At that time, the transmission of power could be handled in a mechanical way with belts and drive trains, but this proved to be difficult to control and accuracy of effort was often sacrificed. The solution to this problem was sought in transmitting power using a fluid under pressure, rather than using a traditional mechanical system. During the Industrial Revolution, the development of fluid power was greatly emphasized and industrial countries like England developed large hydraulic circuits using high pressure water pipes and steam engine driven pumps which conveyed power to the mill where it was needed. To make these systems practical, numerous of hydraulic accumulators, control valves, and actuators were developed without much regard to theory; most power distribution needs were steady in nature and required only static considerations for design. Although these advances in fluid power were quite impressive, the birth of the modern hydraulic control system was soon to experience a major setback. At the beginning of the 20th century, the electrical power industry underwent phenomenal growth. Within a very short period of time, it was recognized that electricity could be transmitted much more economically over a larger distance than power that was produced hydraulically. This obvious fact, coupled with the nature of the engineering needs at the time, caused the fluid power industry to nearly die for several decades.

Near the time of World War II, applications for power transmission characterized by high effort and fast response were needed. These applications manifested themselves in the wartime machinery that was required on warship gun turrets, aircraft controls, and land roving vehicles. In these cases, electrical power was limited due to the fact that ferromagnetic materials saturate at a low flux density and therefore the torque output per unit mass of iron in a motor armature is relatively low. This means that the transmission of a large amount of torque using an electrical device is sluggish, and that fast responses for high torque devices had to be sought elsewhere. During World War II, hydraulically transmitted power was the obvious solution to the problem, as it offered a tremendous torque to inertia ratio; and, interestingly enough, it is still the high torque to inertia ratio that gives fluid power its niche in the marketplace today. In order to advance the understanding of hydraulic control systems, the U.S. government funded research programs at the Massachusetts Institute of Technology (MIT), exposing the basic science behind the technology of transmitting power hydraulically. This work proved to be foundational and eventually generated a textbook entitled, *Fluid Power Control* by Blackburn et al., which was published jointly by the Technology Press of MIT and John Wiley & Sons (1960). Thus, the wartime environment in which we live spawned great advances in the field of hydraulic control systems, and the work done at MIT proved to be instrumental for reviving the industry which now flourishes under the guiding hand of many practical engineers.

Indeed, during the post World War II era and up to this present day, the hydraulic control industry (as opposed to the university) has become the steward of technology for the fluid power world. Most notably, it was the fluid power industry that produced Herbert Merritt's text entitled, *Hydraulic Control Systems*, which was published by John Wiley & Sons (1967). Herbert Merritt was an engineer and section head at the Cincinnati Milling Machine Company, and his industrially produced text has remained the worldwide authority in fluid power for more than 50 years. Except for the strong presence of fluid power research throughout Europe and parts of Asia, the subject of hydraulic control systems nearly vanished from the face of the university for almost 40 years, with the result of most technology developments being proprietary and specific rather than open and general.

In the early 2000s, a resurgence of interest within the university regarding the science and technology of fluid power control systems appeared. Undoubtedly much of this interest pertains to the growing presence of high-speed microprocessors and the evermore popular field of Mechatronics – a field of control involving the advanced interface of the computer with mechanical systems. Alongside the field of control, the harsh environment in which hydraulic control systems operate has proved rich territory for machine design research as well. As previously mentioned, the primary advantage of hydraulic power transmission has been dynamical in nature; and while responsive power transmission remains a baseline expectation for hydraulic systems, today's emphasized needs have expanded to include optimized continuously variable control, reduced size and weight of hydraulic components, maximized effort at zero power conditions, hybrid systems of mechanical and electrical components, higher efficiency across a wide range of operating conditions, energy storage within hydraulic systems, and other issues associated with performance and cost. With the increased demand of the engineering environment comes the increased need for greater understanding of hydraulic control systems. It is with this awareness that industry has turned to her academic colleagues for developing stronger partnerships in the activities of both education and research. This resurgence of collaborative effort has been made evident in the United States by the recent academic outreach initiated by the National Fluid Power Association (NFPA); by the collaborative research activities of the fluid power industry and university researchers, made possible by the Center for Compact and Efficient Fluid Power (CCEFP); and by the endowed chair and funded laboratory activities that have been most recently pursued.

Historically, it has been the university that has provided an environment for generalizing advances in technology – even as MIT provided the initial catalyst for the development of what we call modern hydraulic control systems. As we turn toward an advancing stage of technology in this field, it is the hope of the author that this current text will be used to assist in the reintroduction of fluid power into the university education and research curriculum. Furthermore, this text is being offered to industry as a concise summary and resource for describing the state of the art in hydraulic control systems. With these goals in mind, this book has been written to include both instructional and informative content. While the instructional content will be of most interest to the college professor who is organizing and teaching a senior elective or a first-year graduate course in hydraulic control systems, the informative content will be of interest to the practicing engineer or the graduate student who is developing a future application.

The outline for the text is given in the Contents; the book provides a review of fundamental engineering concepts but is largely constructed with an eye toward supporting current day discussions on hydraulic components and control systems. This text heavily emphasizes the system dynamics aspects of hydraulic control systems with a focus on the design and development of the plant to be controlled. The control aspects of this text include classical methods carried out in the time and frequency domains as well as some more modern methods such as digital control, state feedback, and state estimation. The practicing engineer will find the control aspects of this text to be of great use as a first exercise in developing automated control structures. Students who are introduced to this text should have had a first course in Fluid Mechanics, Dynamic Systems, and Classical Control Theory. The first part of the book is aimed at reviewing these subjects, and instructors are encouraged to spend time reviewing this material with their class before proceeding toward the last two parts of the book.

I

FUNDAMENTALS

1

FUNDAMENTALS

1

FLUID PROPERTIES

1.1 INTRODUCTION

This text addresses the science and technology of controllably transmitting power using a pressurized fluid. As such, it is entirely appropriate to begin by considering the fundamental properties that characterize fluids that are typically used within hydraulic control systems. In this chapter, the fluid properties that describe the condition of a liquid are presented. Since imparting velocity to a pressurized fluid is the means for transmitting power hydraulically, it is important to consider the physical mechanisms that describe the pressurization of the fluid and the fluid's resistance to flow. In the presentation that follows, the mass density of a liquid is discussed with the bulk coefficients that characterize the equation of state. These coefficients are expanded on in subsequent sections dealing with the fluid bulk modulus of elasticity and the coefficient of thermal expansion. Another important fluid property of consideration is the fluid viscosity. This parameter is discussed as it provides insight to the impedance of fluid flow and the ultimate generation of heat associated with shearing hydraulic fluid. Other fluid property topics of chemical composition, thermal conductivity, and fluid vapor pressure are discussed. This chapter concludes with a summary of fluid types and offers practical suggestions for the selection of various hydraulic fluids.

1.2 FLUID MASS DENSITY

1.2.1 Equation of State

The equation of state for any substance is used to describe the mass density of that substance as it varies with exposed conditions, such as pressure and temperature. For many substances, the equation of state is extremely complex and difficult to describe exactly. An example of a relatively simple equation of state is that of an ideal gas. In this case, the ideal gas law is used to relate the pressure and temperature of the gas to the mass density using a determined gas constant. This result is given by

$$\rho = \frac{P}{RT}, \tag{1.1}$$

where ρ is the mass density of the gas, T is the absolute temperature of the gas (expressed in Kelvin or Rankine), P is the absolute gas pressure, and R is the universal gas constant that has been determined for the specific gas in question. Unfortunately, in the case of liquids, the equation of state is not so simple. However, since liquids are fairly incompressible it may be assumed that the mass density of a liquid will not change significantly with the exposed conditions of temperature and pressure. In this case, a first-order Taylor series approximation may be written to describe the small variations in density that occur due to changes in pressure and temperature.[1] This result is given by

$$\rho = \rho_o + \left.\frac{\partial \rho}{\partial P}\right|_o (P - P_o) + \left.\frac{\partial \rho}{\partial T}\right|_o (T - T_o), \tag{1.2}$$

where ρ_o, P_o, and T_o represent a reference density, pressure, and temperature, respectively. The equation of state may be more meaningfully expressed by making the definitions,

$$\frac{1}{\beta} \equiv \frac{1}{\rho}\frac{\partial \rho}{\partial P}, \quad \alpha \equiv -\frac{1}{\rho}\frac{\partial \rho}{\partial T}, \tag{1.3}$$

where β is the isothermal fluid bulk modulus and α is the isobar fluid coefficient of thermal expansion. Using Equation (1.3) with Equation (1.2) yields the following equation of state for a liquid:

$$\rho = \rho_o \left[1 + \frac{1}{\beta_o}(P - P_o) - \alpha_o(T - T_o)\right]. \tag{1.4}$$

If the nearly incompressible assumption of this equation is correct, one may infer that the fluid bulk modulus β is large and the thermal coefficient of expansion α is small. Indeed, this is the case for hydraulic fluids, as shown in subsequent sections of this chapter.

1.2.2 Density-Volume Relationship

To evaluate the relationship between fluid mass density and fluid volume, consider a fluid element of mass M. This mass may be described as

$$M = \rho V, \tag{1.5}$$

where ρ is the fluid mass density and V is the volume of the fluid element. An infinitesimal change in this mass would be described by

$$dM = \rho\, dV + V\, d\rho. \tag{1.6}$$

Since the mass itself cannot be diminished or increased the left-hand side of Equation (1.6) must be zero. Therefore, the following differential relationship between the fluid density and the fluid volume may be expressed as

$$-\frac{1}{\rho}\, d\rho = \frac{1}{V}\, dV. \tag{1.7}$$

Integrating both sides of this equation between say, condition 1 and condition 2, yields the following relationship between the fluid mass density and the fluid volume:

$$\frac{\rho_1}{\rho_2} = \frac{V_2}{V_1}. \tag{1.8}$$

---

[1] The Taylor series is discussed in Subsection 3.3.2 of Chapter 3.

With Equation (1.5) it can be seen that this result simply states that the mass at condition 1 must equal the mass at condition 2. These density and volume relationships will be useful when considering the discussion of the fluid bulk modulus.

1.3 FLUID BULK MODULUS

1.3.1 Definitions

The isothermal fluid bulk modulus describes the elasticity, or "stretchiness," of the fluid at a constant temperature. This property is experimentally determined using a stress-strain test in which the volume of fluid is decreased while keeping the mass constant. During this process, the stress of the fluid is measured by measuring the fluid pressure. A plot of the fluid pressure versus the fluid strain is then generated, and the slope of this plot is used to describe the elasticity of the fluid. This slope is generally referred to as the fluid bulk modulus. Figure 1-1 shows a plot of the stress-strain curve for a liquid. As shown in this figure, the stress strain curve is not linear as its slope is shown to vary in magnitude with pressure. In Figure 1-1, both the secant bulk modulus K and the tangent bulk modulus β are shown.

Using Figure 1-1, it can be seen that the secant bulk modulus is defined as

$$K = \frac{\Delta P}{\Delta \varepsilon},\qquad(1.9)$$

where P is the fluid pressure and ε is the fluid strain. For the secant bulk modulus, the fluid strain is defined by

$$\varepsilon \equiv \frac{\Delta V}{V_o} = \frac{V_o - V}{V_o},\qquad(1.10)$$

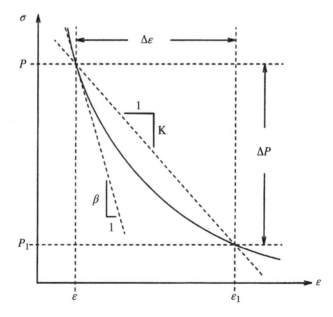

Figure 1-1. The stress-strain curve for a liquid showing the secant bulk modulus and the tangent bulk modulus.

where V_o is the fluid volume at atmospheric pressure and V is the fluid volume at another point of interest. From Figure 1-1 and Equation (1.10) it may be shown that

$$\Delta P = P - P_1 \quad \text{and} \quad \Delta \varepsilon = \frac{V_1 - V}{V_o}. \tag{1.11}$$

Substituting Equation (1.11) into Equation (1.9) yields the following result for the secant bulk modulus:

$$K = \frac{V_o(P - P_1)}{V_1 - V}. \tag{1.12}$$

By convention, the reference pressure and volume for the calculation of the secant bulk modulus is given by $P_1 = 0$ and $V_1 = V_o$. Notice that with this convention, gauge pressures are assumed. Substituting these reference values into Equation (1.12) yields the following conventional expression for the secant fluid bulk modulus:

$$K = \frac{V_o P}{V_o - V} = \frac{\rho P}{\rho - \rho_o}. \tag{1.13}$$

In this result, we have used the relationship between the fluid volume V and the fluid density ρ as shown in Equation (1.8).

The tangent fluid bulk modulus is defined by the slope of a line that is anywhere tangent to the stress-strain curve shown in Figure 1-1. This quantity is expressed mathematically by the following limit:

$$\beta = \lim_{\Delta \varepsilon \to 0} \frac{\Delta P}{\Delta \varepsilon} = \frac{dP}{d\varepsilon}. \tag{1.14}$$

The fluid strain for the calculation of the tangent bulk modulus is defined by

$$\varepsilon \equiv -\ln \left(\frac{V}{V_o} \right). \tag{1.15}$$

At this point, the reader may note that the fluid strain for the tangent bulk modulus has been defined differently than it was for the secant bulk modulus. See Equation (1.10). In fact, one might argue that the strain definition for the secant bulk modulus is more typical and expected than the strain definition presented in Equation (1.15). This, of course, would be true especially from a perspective of solid mechanics where the definition of material strain is very similar in form to that of the secant bulk modulus. The use of Equation (1.15) for describing the strain of the tangent bulk modulus is more an issue of convention rather than principle; however, it may give the reader comfort to recognize that for values of $V/V_o \approx 1$, Equation (1.15) may be linearly approximated as the secant bulk modulus just as it has been presented in Equation (1.10). Since liquid is fairly incompressible, this approximation is easily justified and the two strain definitions can be viewed as essentially the same. Using Equation (1.15), it may be shown that

$$d\varepsilon = -\frac{1}{V} dV. \tag{1.16}$$

Therefore, the tangent fluid bulk modulus may be more explicitly expressed as

$$\beta = -V \frac{dP}{dV} = \rho \frac{dP}{d\rho}. \tag{1.17}$$

Note: In this result we have used the relationship between the fluid volume V and the fluid density ρ, as shown in Equation (1.7).

1.3.2 Effective Bulk Modulus

General Equations. The fluid bulk modulus has been used to describe the elasticity of the fluid as it undergoes a volumetric deformation. This elasticity describes a spring effect that is often attributed to high-frequency resonance within hydraulic control systems. High-frequency resonance can create irritating noise problems and premature failures of vibrating parts; however, it usually does not present a control problem since the resonance occurs at a much higher frequency than the dominant natural frequency of the typical device being controlled. If on the other hand the effective spring rate of the hydraulic system becomes soft due to entrained air within the system or a large volume of compressed fluid or an overly compliant fluid container, the resonating frequency of the hydraulic system will become much lower and a potential for control difficulties will exist.

Figure 1-2 shows a schematic of a flexible container filled with a fluid mixture of liquid and air.

A piston is moved to the left to compress the fluid while also expanding the structural volume of the container in a radial direction. For the analysis that follows container deflection in the axial direction is neglected. The total volume of the chamber is given by

$$V = V_o + V_\delta - Ax = V_l + V_a, \tag{1.18}$$

where V_o is the initial volume of the container, V_δ is an additional volume that results from expanding the chamber, A is the cross-sectional area of the piston, x is the piston displacement, V_l is the volume of liquid within the chamber, and V_a is the volume of air within the chamber. By subtracting the deflection volume V_δ from each side of Equation (1.18), an effective volume may be calculated as

$$V_e = V_o - Ax = V_l + V_a - V_\delta. \tag{1.19}$$

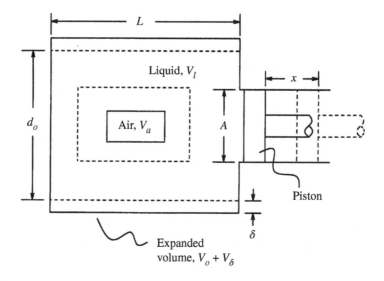

Figure 1-2. A pressurized flexible container filled with a fluid mixture of liquid and air.

This definition for the effective volume is useful since it represents a quantity that may be easily calculated without knowing the deformation characteristics of the chamber. The change in the effective volume is differentially expressed as

$$dV_e = dV_l + dV_a - dV_\delta. \tag{1.20}$$

Using the general form of Equation (1.17), the effective fluid bulk modulus for the device shown in Figure 1-2 may be expressed as

$$\frac{1}{\beta_e} = -\frac{1}{V_e}\frac{dV_e}{dP}, \tag{1.21}$$

where β_e is the effective fluid bulk modulus, V_e is the effective volume that undergoes deformation, and P is the fluid pressure within the hydraulic system. Substituting Equation (1.20) into Equation (1.21) produces the following result for the effective fluid bulk modulus:

$$\frac{1}{\beta_e} = \frac{V_l}{V_e}\left(-\frac{1}{V_l}\frac{dV_l}{dP}\right) + \frac{V_a}{V_e}\left(-\frac{1}{V_a}\frac{dV_a}{dP}\right) + \frac{1}{V_e}\frac{dV_\delta}{dP}. \tag{1.22}$$

By definition, the bulk modulus for liquid and air are given respectively as

$$\frac{1}{\beta_l} = -\frac{1}{V_l}\frac{dV_l}{dP}, \quad \frac{1}{\beta_a} = -\frac{1}{V_a}\frac{dV_a}{dP}. \tag{1.23}$$

It is also useful to define the bulk modulus of the container with respect to the effective fluid volume as

$$\frac{1}{\beta_c} = \frac{1}{V_e}\frac{dV_\delta}{dP}. \tag{1.24}$$

Notice: this definition is a different form than that of Equation (1.23) since the volume and differential volume terms are based on different volume quantities; however, the details of these differences are addressed later. Substituting Equations (1.23) and (1.24) into Equation (1.22) yields the following result for the effective fluid bulk modulus of the system shown in Figure 1-2:

$$\frac{1}{\beta_e} = \frac{V_l}{V_e}\frac{1}{\beta_l} + \frac{V_a}{V_e}\frac{1}{\beta_a} + \frac{1}{\beta_c}. \tag{1.25}$$

The volumetric ratios within this expression describe the fractional volume content of liquid and air. Equation (1.19) may be used with Equation (1.25) to show that

$$\frac{1}{\beta_e} = \left(1 + \frac{V_\delta}{V_e}\right)\frac{1}{\beta_l} + \left(1 - \frac{\beta_a}{\beta_l}\right)\frac{V_a}{V_e}\frac{1}{\beta_a} + \frac{1}{\beta_c}. \tag{1.26}$$

Recognizing that $V_e \gg V_\delta$ and $\beta_l \gg \beta_a$, Equation (1.26) may be closely approximated as

$$\frac{1}{\beta_e} = \frac{1}{\beta_l} + \frac{V_a}{V_e}\frac{1}{\beta_a} + \frac{1}{\beta_c}. \tag{1.27}$$

Equation (1.27) is a useful expression for describing the effective fluid bulk modulus within the flexible container shown in Figure 1-2; however, there are several unknowns in this equation that must be discussed further. In particular, useful expressions for the bulk moduli must be developed and a consideration of the fractional content of air must be discussed.

An Equivalent Spring System.   Figure 1-3 shows a spring system that is equivalent to the pressurized chamber shown in Figure 1-2. The equivalent spring system is useful for showing the effective spring rate of the hydraulic system and for lending insight into the reduction of natural frequencies of oscillation. In the top portion of Figure 1-3 a cylinder is shown with internal pistons that are separated by springs. The right-hand spring is intended to model the spring rate of the air within the system, the middle spring is used to model the spring rate of the liquid, and the left-hand spring is used to simulate the spring rate of the container. The pistons are free to slide within the cylinder and may change their location depending on the input force F.

From an overall system view, the input force may be described as

$$F = x\, k_e,\tag{1.28}$$

where x is the displacement of the first piston and k_e is the effective spring rate of the overall system. From a static analysis of the spring system, it may be shown that the input force is also described by

$$F = (x - x_1)\, k_a = (x_1 - x_2)\, k_l = x_2\, k_c,\tag{1.29}$$

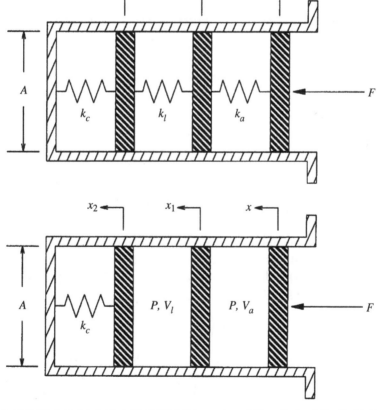

Figure 1-3. An equivalent spring system illustrating the compressibility effects of the liquid, the air, and the container.

where k_a, k_l, and k_c are the spring rates of the air, the liquid, and the container, respectively. From Equation (1.29) it may be shown that

$$x_2 = \frac{F}{k_c},$$

$$x_1 = x_2 + \frac{F}{k_l} = \frac{F}{k_c} + \frac{F}{k_l},$$

$$x = x_1 + \frac{F}{k_a} = \frac{F}{k_c} + \frac{F}{k_l} + \frac{F}{k_a}. \tag{1.30}$$

From Equation (1.28) it can be seen that $x = F/k_e$. Substituting this expression into the bottom result of Equation (1.30) yields the following equation for the effective spring rate of the system:

$$\frac{1}{k_e} = \frac{1}{k_c} + \frac{1}{k_l} + \frac{1}{k_a}. \tag{1.31}$$

This is the classical expression that is used to describe the effective spring rate for a group of springs that are placed in series with respect to one another.

The bottom portion of Figure 1-3 is shown with certain springs removed and now the spaces between pistons have been filled with liquid and air. The spring associated with the container remains, as this is the best model for the effect of deforming a solid material. From the bottom portion of Figure 1-3 it can be seen that the input force is statically equivalent to the pressure of the air (or liquid) times the cross sectional area of a single piston. This force is simply expressed by

$$F = PA. \tag{1.32}$$

From the definition of the effective fluid bulk modulus given in Equation (1.21) it can be seen that the pressure within the system is described by the differential expression

$$dP = -\beta_e \frac{1}{V_e} dV_e, \tag{1.33}$$

where β_e is the effective fluid bulk modulus and V_e is the effective volume of the fluid. Solving Equation (1.33) produces the following result for the fluid pressure:

$$P = \beta_e \ln\left(\frac{V_o}{V_e}\right) \approx \beta_e\left(\frac{V_o}{V_e} - 1\right), \tag{1.34}$$

where V_o is the volume of the fluid when the pressure is zero. *Note*: The right-hand side of Equation (1.34) assumes that $V_o \approx V_e$, which means that the changes in the fluid volume are small. Using Equation (1.34) with Equation (1.32) yields the following result for the input force to the equivalent spring system:

$$F = \beta_e\left(\frac{V_o}{V_e} - 1\right)A. \tag{1.35}$$

Setting Equation (1.35) equal to Equation (1.28) yields the following result for the effective spring rate of the system:

$$\frac{1}{k_e} = \frac{1}{\beta_e}\frac{V_e}{A^2}. \tag{1.36}$$

In this result, it has been recognized from Equation (1.19) that $V_e = V_o - A\ x$. Since the definitions of the effective fluid bulk modulus and the bulk modulus for air and liquid are similar in form (compare Equations (1.21) and (1.23)), a rerun of the previous analysis may be done for the columns of air and liquid shown in Figure 1-3. This analysis produces the following results for the spring rate of air and the spring rate of liquid:

$$\frac{1}{k_a} = \frac{1}{\beta_a}\frac{V_a}{A^2}, \quad \frac{1}{k_l} = \frac{1}{\beta_l}\frac{V_l}{A^2}. \tag{1.37}$$

From the static analysis of Figure 1-3, it may be shown that

$$k_c\, x_2 = P\, A. \tag{1.38}$$

Equation (1.24) presents the definition of the container bulk modulus and can be used to develop an equivalent expression for the fluid pressure within the system. This expression is determined by rearranging Equation (1.24) as:

$$dP = \beta_c\, \frac{1}{V_e}\, dV_\delta. \tag{1.39}$$

Solving this equation yields

$$P = \beta_c\, \frac{1}{V_e}\, V_\delta = \beta_c\, \frac{1}{V_e}\, Ax_2, \tag{1.40}$$

where it has been recognized that $V_\delta = Ax_2$. Substituting Equation (1.40) into Equation (1.38) yields the following result for the spring rate of the container:

$$\frac{1}{k_c} = \frac{1}{\beta_c}\frac{V_e}{A^2}. \tag{1.41}$$

By substituting the results of Equations (1.36), (1.37), and (1.41) into Equation (1.31), the effective fluid bulk modulus for the system may be expressed as

$$\frac{1}{\beta_e} = \frac{V_l}{V_e}\frac{1}{\beta_l} + \frac{V_a}{V_e}\frac{1}{\beta_a} + \frac{1}{\beta_c}, \tag{1.42}$$

which is the exact same expression presented in Equation (1.25). *Note*: Making the appropriate simplifications this result may be further reduced to the form of Equation (1.27).

All of this discussion shows that the compressibility effects within the hydraulic system may be considered as a series of springs, which describe the stiffness of the liquid, the air, and the container itself. Since these springs are arranged in series, the overall stiffness of the system will never exceed the stiffness of any one spring. The spring rate of each substance is dependent on geometry and the bulk modulus property. In the paragraphs that follow the bulk modulus of the liquid, the air, and the container, will be considered in their turn.

Bulk Modulus of Liquid.   From experiments it has been determined that over a limited pressure range the secant bulk modulus of all liquids increases linearly with pressure. That is to say

$$K = K_o + mP, \tag{1.43}$$

where K_o is the secant bulk modulus of the liquid at zero gauge pressure and m is the slope of increase. For any one liquid the value of m is practically the same at all temperatures; however, the value of K_o carries a temperature dependency with it. Equation (1.43) is valid for mineral oils and most other hydraulic fluids up to about 800 bar. This same equation is valid for water up to about 3000 bar. Consequently, if the appropriate values for K_o and m are known for any liquid the secant bulk modulus for that liquid can be easily calculated using Equation (1.43). By setting Equation (1.43) equal to Equation (1.13) it may be shown that

$$V = V_o \left(1 - \frac{P}{K_o + mP}\right), \quad \frac{dP}{dV} = -\frac{(K_o + mP)^2}{K_o V_o}. \tag{1.44}$$

Substituting this result into Equation (1.17) produces the following expression that may be used to evaluate the tangent bulk modulus of a liquid as a function of the liquid parameters K_o and m:

$$\beta_l = K_o \left(1 + \frac{(m-1)P}{K_o}\right)\left(1 + \frac{mP}{K_o}\right). \tag{1.45}$$

Table 1-1 presents typical fluid properties for liquids that are commonly used within hydraulic systems.

Bulk Modulus of Air.   There are two methods for determining the fluid bulk modulus of air: one method assumes that the temperature of the air remains constant (isothermal), the other method assumes that no heat transfer occurs in or out of the volume of air during the expansion and compression of the fluid (adiabatic). While the isothermal assumption is more consistent with our definition of the fluid bulk modulus, the adiabatic assumption is more commonly used and recommended [1].

To develop an expression for the isothermal bulk modulus of air we use the ideal gas law of Equation (1.1) and enforce the assumption of a constant temperature T. A convenient representation of this equation is given by

$$\frac{\rho}{P} = \frac{1}{RT} = \text{constant}, \tag{1.46}$$

where ρ is the density of air, P is the air pressure, R is the gas constant, and T is the constant air temperature. Taking the derivative of Equation (1.46) yields the following result:

$$\frac{1}{P} d\rho - \frac{\rho}{P^2} dP = 0. \tag{1.47}$$

Table 1-1. Fluid bulk modulus properties K_o and m, values for K_o are in kbar

Temperature °C	Mineral oil	Water	Water glycol	Water in oil emulsion	Phosphate ester*
0	20.7	19.7	32.0	20.8	29.7
10	19.8	20.9	31.8	20.2	28.1
20	19.0	21.8	31.5	19.6	26.5
30	18.1	22.4	31.1	19.0	25.0
40	17.3	22.6	30.5	18.4	23.6
50	16.4	22.7	29.9	17.8	22.3
60	15.6	22.5	29.1	17.2	21.1
70	14.7	22.2	28.2	16.6	19.9
80	13.9	21.6	27.2	16.0	18.8
90	13.0	21.1	26.0	15.4	17.8
100	12.2	20.4	24.8	14.8	16.9
m (for all temperatures)	5.6	3.4	4.5	5.0	5.5

*Viscosity greater than 50 cSt at 22°C

Rearranging this equation yields an expression for the isothermal bulk modulus of air. This result is given by

$$\frac{1}{\beta_a} = \frac{1}{\rho}\frac{d\rho}{dP} = \frac{1}{P}. \tag{1.48}$$

In other words, the isothermal bulk modulus of air is simply equal to the fluid pressure itself.

The adiabatic bulk modulus of air may be determined by assuming that no heat transfer occurs between the air and the surrounding liquid or container material. Using the first law of thermodynamics and the ideal gas law it can be shown that

$$PV_a^\gamma = \text{constant}, \tag{1.49}$$

where P is the air pressure, V_a is the air volume, and γ is the ratio of the constant pressure specific heat to the constant volume specific heat. *Note*: $\gamma = 1.4$ for air. Taking the derivative of Equation (1.49) it may be shown that

$$V_a^\gamma\, dP + P\gamma V_a^{(\gamma-1)}\, dV_a = 0. \tag{1.50}$$

Dividing this expression through by V_a^γ and rearranging terms produces the following expression for the bulk modulus of entrained air within the hydraulic system.

$$\frac{1}{\beta_a} = -\frac{1}{V_a}\frac{dV_a}{dP} = \frac{1}{P\gamma}. \tag{1.51}$$

A comparison of Equations (1.48) and (1.51) shows that the adiabatic bulk modulus of air differs from the isothermal bulk modulus of air by a factor of 1.4. Though the isothermal bulk modulus is more compatible with the equation of state, which has been used to define the bulk modulus, the adiabatic bulk modulus is more often used and will therefore by applied in the examples that follow.

Bulk Modulus of the Container.   To consider the bulk modulus of a container, it will be instructive to examine the cylindrical container shown in Figure 1-2. The volume of this container is given by

$$V_c = \frac{\pi}{4}d^2 L, \tag{1.52}$$

where d is the diameter of the container and L is the container length. If it is assumed that the container expands only in the radial direction then the expanded volume of the container may be expressed as

$$V_c = \frac{\pi}{4}(d_o + 2\delta)^2 L = \frac{\pi}{4}d_o^2\left[1 + 4\left(\frac{\delta}{d_o}\right) + 4\left(\frac{\delta}{d_o}\right)^2\right]L, \tag{1.53}$$

where d_o is the original diameter of the container and δ is the radial deflection, as shown in Figure 1-2. If it is assumed that $\delta/d_o \ll 1$, then the container volume may be closely approximated as

$$V_c = V_o + V_\delta, \tag{1.54}$$

where the original volume and the deformed volume are given respectively by

$$V_o = \frac{\pi}{4}d_o^2 L \quad\text{and}\quad V_\delta = \pi d_o\,\delta L. \tag{1.55}$$

Note: These are more explicit expressions for the volume terms than were used in Equation (1.18) for describing the total volume of the container. From a strength of materials textbook [2] we learn that the inside radial deflection of a thick-walled cylinder (without capped ends) is given by

$$\delta = \frac{d_o}{2} \frac{P}{E} \left(\frac{D_o{}^2 + d_o{}^2}{D_o{}^2 - d_o{}^2} + v \right),$$ (1.56)

where P is the internal pressure, E is the tensile modulus of elasticity for the cylinder material, v is Poisson's ratio, D_o is the original outside diameter of the cylinder, and d_o is the original inside diameter of the cylinder. *Note*: The result of Equation (1.56) is valid for both thick- and thin-walled cylinders. Using Equations (1.55) and (1.56) together, it may be shown that the deformed volume of the cylinder is given by

$$V_\delta = \frac{\pi}{2} d_o{}^2 \frac{P}{E} \left(\frac{D_o{}^2 + d_o{}^2}{D_o{}^2 - d_o{}^2} + v \right) L.$$ (1.57)

The derivative of this expression with respect to the pressure P is then

$$\frac{dV_\delta}{dP} = \frac{\pi}{2} \frac{d_o{}^2 L}{E} \left(\frac{D_o{}^2 + d_o{}^2}{D_o{}^2 - d_o{}^2} + v \right).$$ (1.58)

If we assume that the ratio of the displaced volume in the chamber to the original volume is much less than unity; that is, $Ax/V_o \ll 1$, then Equation (1.19) may be used to show that the effective volume of the chamber is given by

$$V_e \approx V_o = \frac{\pi}{4} d_o{}^2 L.$$ (1.59)

Equations (1.58) and (1.59) may now be substituted into Equation (1.24) to express the container bulk modulus as

$$\frac{1}{\beta_c} = \frac{2}{E} \left(\frac{D_o{}^2 + d_o{}^2}{D_o{}^2 - d_o{}^2} + v \right).$$ (1.60)

Recognizing that the inside diameter of the container is given by $d_o = D_o - 2t$, where t is the container wall thickness, an equivalent expression for the container bulk modulus may be written as

$$\frac{1}{\beta_c} = \frac{2}{E} \left[\frac{D_o}{2t} \left(1 + \frac{t}{D_o - t} \right) - (1 - v) \right].$$ (1.61)

Two special cases of these results are instructive. For very thick walls it may be assumed that $D_o/d_o \gg 1$. In this case Equation (1.60) may be approximated as

$$\frac{1}{\beta_c} = \frac{2(1 + v)}{E}.$$ (1.62)

On the other hand, for a thin-walled cylinder it may be assumed that $D_o/t \gg 1$. In this case Equation (1.61) may be used to show that

$$\frac{1}{\beta_c} = \frac{D_o/t}{E}.$$ (1.63)

Equations (1.62) and (1.63) may be used to describe the range for the bulk modulus of the container as

$$\frac{2(1+v)}{E} < \frac{1}{\beta_c} < \frac{D_o/t}{E}. \tag{1.64}$$

Table 1-2 presents the modulus of elasticity and Poisson's ratio for common materials that are used to construct hydraulic containers.

Case Study

It is instructive to consider typical calculations of the effective bulk modulus for the purposes of illustrating the wide variation that may be expected for this parameter. Let us consider a petroleum-based fluid (mineral oil) that is pressurized to 20 MPa (0.20 kbar) at a temperature of 70°C. Furthermore, let us consider a case where the fluid is contained within a cylindrical container with an outside diameter equal to six times the wall thickness ($D_o/t = 6$). From Equation (1.45) and Table 1-1 it may be shown that the tangent bulk modulus for this liquid is given by 15.82 kbar. From Equation (1.51) the bulk modulus of the air is given by 0.28 kbar. Calculate the effective bulk modulus of this system for the following two cases:

1. When the container is made of steel and the liquid has no entrained air.
2. When the container is made of high-pressure hose material and the liquid has 1% entrained air by volume ($V_a/V_e = 0.01$).

For the first case, the bulk modulus of the container may be calculated using Equation (1.61) and the material properties for steel that are shown in Table 1-2. This calculation shows that $\beta_c = 504.63$ kbar for the steel container. Using this result with Equation (1.27) the effective bulk modulus may be calculated as

$$\beta_e = \left(\frac{1}{\beta_l} + \frac{V_a}{V_e}\frac{1}{\beta_a} + \frac{1}{\beta_c} \right)^{-1} = \left(\frac{1}{15.82 \text{ kbar}} + 0 \times \frac{1}{0.28 \text{ kbar}} + \frac{1}{504.63 \text{ kbar}} \right)^{-1}$$
$$= 15.82 \text{ kbar}.$$

Similarly, for the second case the bulk modulus of the container may be calculated using Equation (1.61) and the material properties for high-pressure hose that are shown in Table 1-2. This calculation shows that $\beta_c = 13.82$ kbar for the hose. Using this result with Equation (1.27) the effective bulk modulus may be calculated as

$$\beta_e = \left(\frac{1}{\beta_l} + \frac{V_a}{V_e}\frac{1}{\beta_a} + \frac{1}{\beta_c} \right)^{-1} = \left(\frac{1}{15.82 \text{ kbar}} + 0.01 \times \frac{1}{0.28 \text{ kbar}} + \frac{1}{13.82 \text{ kbar}} \right)^{-1}$$
$$= 5.84 \text{ kbar}.$$

Table 1-2. Material properties for common hydraulic containers, modulus of elasticity E is reported in kbar

Property	Steel	Ductile cast iron	Copper	Brass	Aluminum	High-pressure hose*	TPE**
E	2069	1655	1103	1034	724	59	0.0393
v	0.30	0.28	0.30	0.34	0.33	0.47	0.47

*Flexible and reinforced with stainless steel braids (consult manufacturers for more accurate numbers)
**Thermoplastic elastomer (melt-processible rubber)

A comparison of these two results reveals a 63% difference in the effective bulk modulus for fluids that are presumably very similar and much of this difference is a result of simply putting the fluid in a different container.

Summary. The preceding work has shown that the effective bulk modulus is a result of various physical considerations. These considerations include the liquid properties, the amount of entrained air within the liquid, and the flexibility of the device that is used to contain the fluid. In summary, it may be noted that a wide variation in the effective bulk modulus can be expected from one hydraulic system to the next. As previously mentioned, this variation in the effective bulk modulus is dependent on the air content of the fluid, which may also be dependent on the circulation system for the fluid. For instance, one may suspect that an open circuit system where fluid is circulated through a ventilated reservoir may contain more entrained air than a closed circuit system where most of the fluid is contained within the working components of the system. Other issues related to the reservoir design may impact the turbulent mixing that occurs within the reservoir, which may also impact the amount of entrained air that is introduced by each application. Since the uncertainty of the fluid bulk modulus is significant from one application to the next, it is worth considering techniques that may be used to measure the fluid bulk modulus within an actual system. This is the topic of our next subsection.

1.3.3 Measuring the Fluid Bulk Modulus

Figure 1-4 shows a closed container with a piston that is supported by a fluid generally comprised of a mixture of liquid and air. The piston has a cross-sectional area given by A. When the fluid pressure is at zero gauge pressure, the height of the fluid column is given by the dimension l_o. When the force F is applied to the piston the fluid column is decreased in height by the dimension x. Obviously, to support the applied force, the fluid pressure within the container must increase.

From the definition of the fluid bulk modulus, the fluid pressure within the closed container may be differentially expressed as

$$dP = -\beta \frac{1}{V} dV, \tag{1.65}$$

where β is the effective fluid bulk modulus and V is the effective fluid volume. By treating β as a constant Equation (1.65) may be solved and rearranged to show that the effective fluid bulk modulus is given by

$$\beta = \frac{P}{\ln(V_o/V)}, \tag{1.66}$$

where V_o is the fluid volume in the chamber when the pressure is zero. This volume would be given more explicitly as Al_o. From geometry, the effective fluid volume after compression takes place may be written as

$$V = V_o - A x = A (l_o - x). \tag{1.67}$$

Substituting this result into Equation (1.66) yields the following result for the effective fluid bulk modulus of the device shown in Figure 1-4:

$$\beta = \frac{P}{\ln[l_o/(l_o - x)]}. \tag{1.68}$$

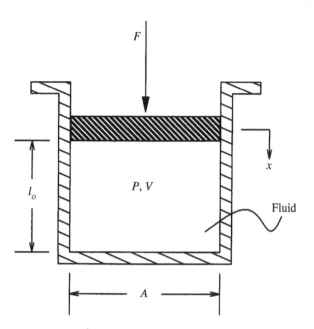

Figure 1-4. Geometry of a test device used to measure the fluid bulk modulus within a closed container.

From this equation it can be seen that the parameters that must be measured for determining the fluid bulk modulus are the fluid pressure P, the original height of the fluid column l_o, and the displaced distance of the piston x.

As all experimentalists know, the measurements of the parameters shown in Equation (1.68) will always deviate slightly from the actual physical values. If we denote these measured parameters using primed notation, the measured value for the fluid bulk modulus may be written as

$$\beta' = \frac{P'}{\ln[l_o'/(l_o' - x')]}. \tag{1.69}$$

Using this expression a dimensionless uncertainty in our measurements may be defined as

$$\varepsilon = \frac{\beta - \beta'}{\beta}. \tag{1.70}$$

Using Equation (1.68) the dimensionless uncertainty may be written more explicitly as

$$\varepsilon = 1 - \beta' \frac{\ln[l_o/(l_o - x)]}{P}, \tag{1.71}$$

where the unprimed values of P, l_o, and x are the unknown true values of these parameters. If it is assumed that the measurement uncertainty is small, a first-order Taylor series expansion of Equation (1.71) may be written as

$$\varepsilon = \left(\frac{\partial \varepsilon}{\partial P}\right)'(P - P') + \left(\frac{\partial \varepsilon}{\partial l_o}\right)'(l_o - l_o') + \left(\frac{\partial \varepsilon}{\partial x}\right)'(x - x'). \tag{1.72}$$

Using Equation (1.71) and assuming that $x'/l'_o \ll 1$ it can be shown that

$$\left(\frac{\partial \varepsilon}{\partial P}\right)' = \frac{1}{P'}, \quad \left(\frac{\partial \varepsilon}{\partial l_o}\right)' \approx \frac{1}{l'_o}, \quad \left(\frac{\partial \varepsilon}{\partial x}\right)' \approx -\frac{1}{x'}. \tag{1.73}$$

Substituting these results back into Equation (1.72) yields the following result for the uncertainty associated with measuring the fluid bulk modulus in the closed container:

$$\varepsilon = \frac{P}{P'} + \frac{l_o}{l'_o} - \frac{x}{x'} - 1. \tag{1.74}$$

The reader will recall that the primed variables are the measured values while the unprimed variables are the true values. The accuracies for instruments that are used for measuring pressures and displacements are typically specified in terms of some percentage of the full-scale capabilities of the instrument itself. This means that the true values of the measured parameters are somewhere within a range that has been identified by the manufacturer of the instrument. Mathematically this maximum range of uncertainty for each measurement may be described as

$$(P - P') = \pm \xi_P \; P_{max}, \quad (l_o - l'_o) = \pm \xi_l \; l_{max}, \quad (x - x') = \pm \xi_x \; x_{max}, \tag{1.75}$$

where the values of ξ_P, ξ_l, and ξ_x are supplied by the instrument manufacturers and P_{max}, l_{max}, and x_{max} are the maximum measurement ranges for these instruments. Using the results of Equations (1.72), (1.73), and (1.75) an expression for the maximum measurement uncertainty may be given as

$$\varepsilon_{max} = \pm \left\{ \xi_P \frac{P_{max}}{P'} + \xi_l \frac{l_{max}}{l'_o} + \xi_x \frac{x_{max}}{x'} \right\}. \tag{1.76}$$

Equation (1.76) describes the maximum range of uncertainty associated with the measurement of the fluid bulk modulus for the mechanical device shown in Figure 1-4. If a strain-gauged pressured transducer is used, it is common to be able to measure pressures accurately within $\pm 1.5\%$ of the full-scale reading. This means that $\xi_P = 0.015$. For measuring linear dimensions such as l_o the method used may vary from a highly accurate coordinate-measuring machine (CMM) to a simple visual inspection of a calibrated scale. Therefore, the range of accuracies for making the linear measurement of the original height of the fluid column may be anywhere between $\pm 1 \times 10^{-4}$ % and $\pm 1\%$ of the full-scale reading [3]. This means that $1 \times 10^{-6} < \xi_l < 0.01$. In practice, it is quite common to use a linear variable-differential-transformer (LVDT) for making linear displacement measurements. These devices are typically accurate within $\pm 0.5\%$ of full-scale readings and therefore $\xi_x = 0.005$.

Case Study

The test device shown in Figure 1-4 is used to measure the effective fluid bulk modulus. While taking this measurement each instrument is being used at half of its maximum capacity and the instrument accuracies are given by $\xi_P = 0.015, \xi_l = 0.005$, and $\xi_x = 0.005$. Calculate the uncertainty in the measurement of the bulk modulus.

To calculate the uncertainty in this measurement Equation (1.76) will be used. This calculation is given by

$$\varepsilon_{\max} = \pm \left\{ \xi_P \frac{P_{\max}}{P'} + \xi_l \frac{l_{\max}}{l_o'} + \xi_x \frac{x_{\max}}{x'} \right\} = \pm(0.015 \times 2 + 0.005 \times 2 + 0.005 \times 2)$$

$$= \pm 0.05.$$

This result shows a measurement uncertainty of plus or minus 5%.

As shown by the previous case study, measurements of the effective fluid bulk modulus may be taken with the device of Figure 1-4 while maintaining a fair degree of confidence in the results. Of course, the uncertainty in this measurement can get better or worse depending on the quality of the instruments, their operating range, and the operating point of the measurement. In any event the range of expected uncertainty should be checked before conducting the experiment to see whether the measurement is worthwhile, or to design an acceptable test device. Equation (1.76) provides the tool for checking this uncertainty as it pertains to the device shown in Figure 1-4. It should also be mentioned that the device in Figure 1-4 is a laboratory device that may or may not adequately represent an actual hydraulic control system. In situ measurements of the fluid bulk modulus have been considered in previous research [4]; however, these measurements have been shown to be expensive and highly uncertain thus rendering these methods somewhat doubtful. In summary, the reader should be impressed with the range of variability that the effective fluid bulk modulus may undergo and with the difficulty in obtaining confidence in its value even from an experimental point of view. Indeed, when simulating hydraulic control systems the specification for the fluid bulk modulus becomes one of the most arbitrary aspects of the model, which is highly unfortunate as this parameter is often the deciding factor between a stable and an unstable hydraulic component or system. This influence of the fluid bulk modulus will be shown in subsequent chapters of this text.

1.4 THERMAL FLUID PROPERTIES

1.4.1 Coefficient of Thermal Expansion

Definition. The isobar coefficient of thermal expansion for a hydraulic fluid was presented in Equation (1.3) and is rewritten here for convenience it terms of both density and fluid volume:

$$\alpha = -\frac{1}{\rho} \frac{d\rho}{dT} = \frac{1}{V} \frac{dV}{dT}, \tag{1.77}$$

where T is the fluid temperature. The coefficient of thermal expansion α describes the change in fluid density (or volume) as the temperature of the fluid increases or diminishes. Since the fluid is comprised of air and liquid, and since the fluid container also exhibits expansion and contraction with changes in temperature, it is of use to consider the effective coefficient of thermal expansion for the entire fluid system.

The Effective Coefficient of Thermal Expansion. The effective coefficient of thermal expansion can be evaluated by considering the device of Figure 1-2 and the effective volume of the fluid V_e as given in Equation (1.19). This volume is comprised of air and liquid and is adjusted for small deflections of the container itself. Using the effective volume, we may calculate the effective coefficient of thermal expansion as

$$\alpha_e = \frac{1}{V_e} \frac{dV_e}{dT}. \tag{1.78}$$

Substituting the differential volume dV_e of Equation (1.20) into this result yields the following expression for the effective coefficient of thermal expansion:

$$\alpha_e = \frac{V_l}{V_e} \left(\frac{1}{V_l} \frac{dV_l}{dT} \right) + \frac{V_a}{V_e} \left(\frac{1}{V_a} \frac{dV_a}{dT} \right) - \frac{1}{V_e} \frac{dV_\delta}{dT}, \tag{1.79}$$

where V_l is the volume of the liquid, V_a is the volume of the air, and V_δ is the deflected volume of the container shown in Figure 1-2. By definition, the coefficient of thermal expansion for liquid and air are given respectively as

$$\alpha_l = \frac{1}{V_l} \frac{dV_l}{dT}, \qquad \alpha_a = \frac{1}{V_a} \frac{dV_a}{dT}, \tag{1.80}$$

and a convenient definition for the coefficient of thermal expansion for the container is

$$\alpha_c = \frac{1}{V_e} \frac{dV_\delta}{dT}. \tag{1.81}$$

Again, the reader will observe that the definition for the coefficient of thermal expansion for the container α_c is somewhat different in form compared to that of α_l and α_a. Substituting Equations (1.80) and (1.81) into Equation (1.79) yields the following result for the effective coefficient of thermal expansion:

$$\alpha_e = \frac{V_l}{V_e} \alpha_l + \frac{V_a}{V_e} \alpha_a - \alpha_c. \tag{1.82}$$

The volumetric ratios in this expression describe the fractional content of liquid and air within the container. Using Equation (1.19) with Equation (1.82) it may be shown that

$$\alpha_e = \left(1 + \frac{V_\delta}{V_e} \right) \alpha_l + \left(1 - \frac{\alpha_l}{\alpha_a} \right) \frac{V_a}{V_e} \alpha_a - \alpha_c. \tag{1.83}$$

Since $V_e \gg V_\delta$ Equation (1.83) may be closely approximated as

$$\alpha_e = \alpha_l + \left(1 - \frac{\alpha_l}{\alpha_a} \right) \frac{V_a}{V_e} \alpha_a - \alpha_c. \tag{1.84}$$

Table 1-3 shows a partial list of data that is currently available for estimating the coefficient of thermal expansion of mineral oil and water as it varies with temperature. More extensive data related to the coefficient of thermal expansion for hydraulic fluids has been published by the US Army and may be consulted for more accurate information [8].

The isobar coefficient of thermal expansion for air may be determined using the ideal gas law of Equation (1.1). A convenient form of this equation is given by

$$\rho T = \frac{P}{R} = \text{constant}, \tag{1.85}$$

Table 1-3. Volumetric coefficient of thermal expansion for various liquids that are used within hydraulic systems reported in units of $10^{-4}/(°C$ or $°K)$

Temperature $°C$	0	20	40	60	80	100	120
Mineral oil	7.00	7.00	7.00	7.00	7.00	7.00	7.00
Water	−0.68	1.73	3.60	5.04	6.24	7.28	8.40

where the pressure P is fixed to satisfy the isobar requirement. Differentiating this equation yields the following result

$$\rho\, dT + T\, d\rho = 0. \tag{1.86}$$

Rearranging this expression yields the following result for the coefficient of thermal expansion for air:

$$\alpha_a = -\frac{1}{\rho}\frac{d\rho}{dT} = \frac{1}{T}. \tag{1.87}$$

The reader will recall that this temperature T is given in the absolute scale; therefore, the coefficient of thermal expansion for air is given more familiarly as

$$\alpha_a = \frac{1}{(T_C + 273.15)\ °C\ \text{or}\ °K} = \frac{1}{(T_F + 459.67)\ °F\ \text{or}\ °R}, \tag{1.88}$$

where T_C and T_F are the fluid temperatures measured in the Celsius and Fahrenheit scales, respectively.

An expression for the coefficient of thermal expansion for the container is given in Equation (1.81). If we use the cylindrical container geometry of Figure 1-2, the instantaneous volume of the container may be described using Equations (1.54) and (1.55). Differentiation of V_δ as shown in Equation (1.55) shows that the change in the deformed volume of the container with respect to temperature is given by

$$\frac{dV_\delta}{dT} = \pi d_o L \frac{d\delta}{dT}, \tag{1.89}$$

where d_o is the original inside diameter of the container, L is the length of the container, and δ is the inside radial growth of the container that occurs due to an increase of the temperature. If we use the approximation of Equation (1.59) for the effective volume of the container then it may be shown that

$$\alpha_c = \frac{1}{V_e}\frac{dV_\delta}{dT} = 4\frac{1}{d_o}\frac{d\delta}{dT}. \tag{1.90}$$

By definition the linear coefficient of thermal expansion is given by

$$\gamma = \frac{1}{l}\frac{dl}{dT}, \tag{1.91}$$

where l is the linear dimension that grows due to an increase in temperature. Recognizing that the linearly growing dimension in this problem is the diameter of the container, it may be shown that

$$l = d_o + 2\delta \approx d_o, \qquad \frac{dl}{dT} = 2\frac{d\delta}{dT}. \tag{1.92}$$

Using this result with Equations (1.90) and (1.91) the volumetric coefficient of thermal expansion for the container may be approximated by

$$\alpha_c = 2\gamma, \tag{1.93}$$

where γ is the linear coefficient of thermal expansion of the container material that may be found in material handbooks. Using material handbooks and Equation (1.93) the volumetric coefficients of thermal expansion for typical hydraulic containers are given in Table 1-4.

Case Study

To consider the possible variation that may be expected in the coefficient of thermal expansion let us once again consider a petroleum-based fluid (mineral oil) that is at a temperature of 70°C within in a cylindrical container. From Table 1-3 it may be shown that the coefficient of thermal expansion for this liquid is given by $7 \times 10^{-4}/°C$. Using Equation (1.88) the coefficient of thermal expansion for the air is given by $3 \times 10^{-3}/°C$. Calculate the effective coefficient of thermal expansion of the fluid for the following two cases:

1. When the container is made of steel and the liquid has no entrained air.
2. When the container is made of high-pressure hose material and the liquid has 1% entrained air by volume ($V_a/V_e = 0.01$).

For the first case, the coefficient of thermal expansion for the steel container may be determined using Table 1-4. This table shows that $\alpha_c = 22 \times 10^{-6}/°C$ for the steel container. Using this result with Equation (1.84), the effective coefficient of thermal expansion may be calculated as

$$\alpha_e = \alpha_l - \alpha_c = \frac{7 \times 10^{-4}}{°C} - \frac{22 \times 10^{-6}}{°C} = \frac{6.78 \times 10^{-4}}{°C}.$$

Similarly, for the second case, the coefficient of thermal expansion for the high-pressure hose may be determined using Table 1-4.

This table shows that $\alpha_c = 447 \times 10^{-6}/°C$ for the hose. Using this result with Equation (1.84), the effective coefficient of thermal expansion may be calculated as

$$\alpha_e = \alpha_l + \left(1 - \frac{\alpha_l}{\alpha_a}\right) \frac{V_a}{V_e} \alpha_a - \alpha_c = \frac{7 \times 10^{-4}}{°C} + \left(1 - \frac{7 \times 10^{-4}}{3 \times 10^{-3}}\right) \times 0.01 \times \frac{7 \times 10^{-4}}{°C} - \frac{447 \times 10^{-6}}{°C}$$

$$= \frac{2.58 \times 10^{-4}}{°C}.$$

Table 1-4. Volumetric coefficients of thermal expansion for materials commonly used to make hydraulic containers

α_c	Steel	Ductile cast iron	Copper	Brass	Aluminum	High-pressure hose*	TPE**
$\frac{10^{-6}}{°C \text{ or } °K}$	22.0	26.4	34.0	38.0	48.0	447	460
$\frac{10^{-6}}{°F \text{ or } °R}$	12.2	14.7	18.8	21.2	26.6	248	256

*Flexible and reinforced with stainless steel braids (consult manufacturers for more accurate numbers)
**Thermoplastic elastomer (melt-processible rubber)

A comparison of these two results reveals a 62% difference in the calculated effective coefficient of thermal expansion for fluids that are presumably very similar. Again, most of this difference is a result of simply putting the fluid in a different container.

In the preceding case study, it has been shown that the effective coefficient of thermal expansion can vary significantly for fluids that are used within hydraulic control systems. As shown by this example the coefficient of thermal expansion is not greatly influenced by reasonable percentages of entrained air, and since this is the only parameter within our model that varies with temperature we can say that the effective coefficient of thermal expansion does not vary much with temperature. On the other hand, this example has shown that the effective coefficient of thermal expansion for the fluid is significantly influenced by the container properties, which will vary for the fluid as it flows through various conduit materials within the same hydraulic circuit.

Measuring the Coefficient of Thermal Expansion. Since temperature variations within hydraulic control systems are not nearly as pronounced as, say, pressure variations, an accurate value for the coefficient of thermal expansion is not nearly as important as having an accurate value for the fluid bulk modulus. For this reason, we do not include an extensive discussion on measurement techniques (and the associated uncertainty) for the effective coefficient of thermal expansion. The reader should recognize, however, that these techniques do not change in principle from the techniques that were described for measuring the fluid bulk modulus. Similar types of experiments as those described in Subsection 1.3.3 may be used and the uncertainty in each measurement may be calculated as well.

1.4.2 Thermal Conductivity

The rate at which heat is transferred through a material (fluid or solid) is governed by Fourier's Law. This law is based on observed phenomena and states that the heat flux (heat per unit area) is directly proportional to the temperature gradient in a direction normal to an isothermal surface. Figure 1-5 illustrates this phenomenon.

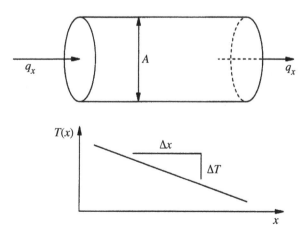

Figure 1-5. The relationship between the coordinate system, heat flow direction, and temperature gradient in one dimension.

By definition, the thermal conductivity of a material is given by

$$k = -\frac{q_x}{A} \frac{1}{(\partial T/\partial x)} \approx \frac{q_x}{A} \frac{\Delta x}{\Delta T}, \tag{1.94}$$

where q_x is the heat being transferred through the material, A is the cross-sectional area normal to the heat flow, x is the dimension in the direction of the heat flow, and T is the material temperature. Table 1-5 provides values of thermal conductivity for mineral oils and water as it varies with temperature.

1.4.3 Specific Heat

The specific heat is a thermodynamic property that is useful for determining the change of internal energy of a substance for small increases in temperature. As such, the specific heat for a substance that maintains a constant volume is given by

$$c_v \equiv \left.\frac{\partial u}{\partial T}\right|_v, \tag{1.95}$$

where u is the specific internal energy of the substance. If the internal energy is considered as a measure of added heat to the substance, we can see that the specific heat c_v describes the amount of additional heat that is required to raise the temperature of the substance by a unit of temperature (e.g., $1°C$ or $°K$). The thermodynamic property of enthalpy is given by the expression

$$h \equiv u + Pv, \tag{1.96}$$

where u is the specific internal energy, P is pressure, and v is the specific volume. By definition, the specific heat for a substance that maintains a constant pressure is given by

$$c_P \equiv \left.\frac{\partial h}{\partial T}\right|_P. \tag{1.97}$$

By differentiating Equation (1.96) with respect to temperature, it can be shown that

$$c_P = \left.\frac{\partial h}{\partial T}\right|_P = \left.\frac{\partial u}{\partial T}\right|_P + P\left.\frac{\partial v}{\partial T}\right|_P = \left.\frac{\partial u}{\partial T}\right|_P + Pv\alpha, \tag{1.98}$$

where α is the isobar coefficient of thermal expansion. Since α is very small for liquids (see Table 1-3) it can be shown that the change in internal energy is approximately equal to the change in enthalpy and; therefore,

$$c_v = c_P, \tag{1.99}$$

for liquids that are commonly used within hydraulic control systems. Table 1-6 shows the specific heat for mineral oil and water as it varies with fluid temperature. These results are used in the next chapter to calculate the temperature rise within a fluid due to a pressure drop across a flow passage.

Table 1-5. Values of thermal conductivity k for petroleum-based fluid and water reported in 10^{-3} W/(m °C)

	Temperature, °C						
	0	20	40	60	80	100	120
Mineral oil	147	145	145	141	138	137	135
Water	569	598	628	650	668	679	686

Table 1-6. Specific heat c_p for petroleum-based fluid and water reported in kJ/(kg°C)

Temperature °C	0	20	40	60	80	100	120
Mineral Oil	1.796	1.868	1.951	2.035	2.118	2.206	2.294
Water	4.217	4.184	4.178	4.184	4.195	4.214	4.239

1.5 FLUID VISCOSITY

1.5.1 Definitions

From solid mechanics one may recall that the shear stress within a solid is proportional to the shear strain within a region of linear deformation. In the case of solids this constant of proportionality relating the shear stress to the shear strain is called the shear modulus of elasticity and is often noted by the symbol G. When a shear stress is applied to a solid a fixed distortion of the material occurs and, until the shear stress is released, the material remains in a fixed distorted shape while storing potential energy much like a mechanical spring. Fluids do not behave this way. When a shear stress is applied to a fluid the fluid will not deform into a fixed shape while storing potential energy. Rather, for a constantly applied shear stress a fluid will deform continuously while dissipating energy in the form of heat. In this case it has been experimentally observed that the shear stress within the fluid is proportional to the rate of shear strain and the constant of proportionality in this relationship is called the absolute viscosity of the fluid noted by the symbol μ. Mathematically we describe this relationship as

$$\tau = \mu \dot{\gamma}, \tag{1.100}$$

where τ is the shear stress within the fluid and $\dot{\gamma}$ is the rate of shear strain. *Note*: Equation (1.100) has been shown to be valid for both liquids and gasses.

Figure 1-6 shows two parallel plates separated by a thin film of fluid. The bottom plate is stationary while the top plate is being pulled across the fluid medium at a velocity given by U. From experiments it has been observed that the fluid "sticks" to the surfaces of both plates so that the fluid velocity at the bottom plate is zero while the fluid velocity at the top plate is given by U. Between the two plates the fluid velocity u varies as a function of y. Two particles, p_1 and p_2, are shown to be moving within the fluid medium.

At one instant in time these particles are perfectly in line with each other while at the next instant in time one particle is ahead of the other. The second instant in time is shown in Figure 1-6 using the primed notation. The angle $\delta\gamma$ formed by the triangular relationship of these particles is called the shear strain of the fluid and this definition is common with the language of solid mechanics as well. From geometry it can be shown that

$$\delta\gamma \approx \tan(\delta\gamma) = \frac{\delta x}{\delta y}. \tag{1.101}$$

From the motion of the fluid particles it is clear that the position of each particle at the second instant in time is given by

$$x_1' = u_1 \, \delta t, \quad x_2' = u_2 \, \delta t, \tag{1.102}$$

where u_1 and u_2 are the linear velocities of particles 1 and 2, respectively, and δt is the separation of time between the two instances. Subtracting the second position of particle

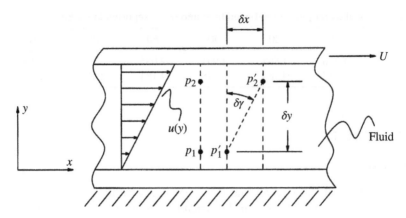

Figure 1-6. A schematic illustrating the shear strain within a fluid.

one from the second position of particle two yields the following result for their relative displacement at the second instant in time:

$$\delta x = \delta u \, \delta t, \tag{1.103}$$

where δu is given by $u_2 - u_1$. Substituting Equation (1.103) into Equation (1.101) yields the following relationship between the fluid shear strain, time, fluid velocity, and the dimensional variable y:

$$\frac{\delta \gamma}{\delta t} = \frac{\delta u}{\delta y}. \tag{1.104}$$

In the limit as δt and δy go to zero, it can be shown that

$$\dot{\gamma} = \frac{d\gamma}{dt} = \frac{du}{dy}, \tag{1.105}$$

where $\dot{\gamma}$ is the rate of shear strain and du/dy is the velocity gradient in the y-direction. Using Equations (1.100) and (1.105) it may now be shown that the shear stress within the fluid is given by

$$\tau = \mu \, \frac{du}{dy}. \tag{1.106}$$

From this result we define the absolute fluid viscosity as

$$\mu \equiv \frac{\tau}{\partial u / \partial y}. \tag{1.107}$$

In accordance with this definition, a plot of shear stress versus the rate of shear strain should be linear where the slope is equal to the viscosity. The actual value of the viscosity depends on the type of fluid and, for a particular fluid, it is also highly dependent on the temperature. Fluids for which the shear stress is linearly related to the rate of shear strain are called Newtonian fluids. Fluids for which the shear stress is not linearly related to the rate of shear strain are designated as non-Newtonian fluids. The kinematic viscosity is defined as $v = \mu / \rho$ where ρ is the fluid density. Physically speaking, the viscosity of the fluid is a type of friction coefficient associated with the motion of the fluid itself.

1.5.2 Viscous Drag Coefficient

The viscous drag coefficient is a lumped parameter that is commonly used in the modeling of dynamical systems to describe energy dissipation effects. This model is based on the principle of fluid shear and is therefore appropriately discussed in this section. If we consider the two plates shown in Figure 1-6, it is apparent that a certain force is required to drag the top plate across the fluid film that separates the two plates from each other. This drag force may be written as

$$F = \tau A, \tag{1.108}$$

where τ is the fluid shear stress at the contact surface of the top plate and A is the surface area of contact at this plate. Using Equation (1.106) it may be shown that the shear stress at the top plate is given by $\mu U / h$, where μ is the fluid viscosity, U is the sliding velocity of the top surface, and h is the fluid film thickness that exists between the two plates. Using this expression for the fluid shear, the following expression may be written for the viscous drag force on the top plate:

$$F = cU, \tag{1.109}$$

where the viscous drag coefficient is given by

$$c = \frac{\mu A}{h}. \tag{1.110}$$

From this expression, it may be seen that the viscous drag coefficient increases with fluid viscosity and the surface contact area A. Furthermore, this coefficient decreases as the fluid film thickness h increases. Within the machinery that is used to operate hydraulic control systems the fluid film thickness between moving parts is often on the order of microns; therefore, in a number of applications the viscous drag coefficient becomes an important parameter for describing viscous energy dissipation.

1.5.3 Viscosity Charts and Models

Viscosity is a bulk property of the fluid, which is generally measured rather than being analytically predicted. A chart showing this measured data for typical SAE oils is given in Figure 1-7. A very common hydraulic fluid is an SAE 10W fluid. Units of absolute viscosity are shown in Figure 1-7 in Pa-s and μReyn.

It is also common to report the absolute viscosity in the SI system as cP (centi-poise). These units are related to each other according to the following definitions:

$$1 \times 10^6 \; \mu\text{Reyn} = 1 \text{ psi-s}, \quad 1 \times 10^3 \text{ cP} = 1 \text{ Pa-s}. \tag{1.111}$$

As with all bulk properties of hydraulic fluids, the viscosity of the fluid varies significantly with exposed conditions. For instance, the absolute viscosity of a liquid tends to decrease markedly with temperature and increases mildly with pressure. Both of these variations occur exponentially and their combined effects may be mathematically written as

$$\mu = \mu_o \text{Exp}[\alpha(P - P_o) - \lambda(T - T_o)], \tag{1.112}$$

where μ_o is the fluid viscosity at a pressure and temperature given by P_o and T_o, and α and λ are constants for a particular fluid.

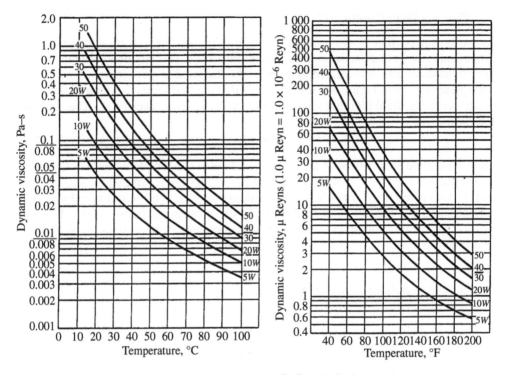

Figure 1-7. Fluid viscosity for SAE-grade fluids.

Note: α in this expression is not the coefficient of thermal expansion. For mineral oils it has been shown empirically that the pressure coefficient is related to the viscosity at zero gauge pressure by the dimensional relationship

$$\alpha = [0.6 + 0.965\log_{10}(\mu_o)] \times 10^{-8}, \tag{1.113}$$

where μ_o is measured in centi-poise and the units of α are in Pa^{-1} [5]. Figure 1-7 shows a graph of the fluid viscosities for SAE grade fluids as they vary with temperature at standard atmospheric pressure. The viscosity index is used to describe how quickly the viscosity drops with temperature. If the viscosity does not decrease rapidly with temperature the fluid is said to have a high-viscosity index. On the other hand, if the fluid viscosity drops off rapidly with temperature the fluid is said to have a low viscosity index.

This subsection has focused on the viscosity of mineral oils as they are predominantly used to operate hydraulic control systems. However, other fluids may be used as well and viscosity data for those fluids may be obtained from the fluid manufacturer. Regardless of the fluid type that is used it is recommended that fluid viscosities remain within the following range for the continuous operation of a hydraulic control system:

$$7.8\ cP < \mu < 95.7\ cP, \quad 1.1\ \mu Reyn < \mu < 13.8\ \mu Reyn. \tag{1.114}$$

The optimal operating viscosity is given by $\mu = 11.3$ cP $= 1.6$ μReyn. Maximum cold-start viscosities are recommended to be less than 1390 cP (201 μReyn) and minimum intermittent viscosities are recommended to be greater than 5.0 cP (0.7 μReyn).

1.6 VAPOR PRESSURE

There is a combination of pressure and temperature at which all liquids will tend to vaporize and change phase. The pressure at which this occurs is called the vapor pressure of the liquid and must be specified at a given temperature. For instance, water tends to vaporize (boil) at atmospheric pressure when its temperature reaches 100°C. That is to say, the vapor pressure of water at 100°C is 101 kPa absolute. If the pressure drops below 1 atmosphere (101 kPa) then water will vaporize at a lower temperature. Similarly, if the pressure increases above 1, atmosphere water will only vaporize at some temperature above 100°C. This discussion of vapor pressure is important because vapor bubbles do not tend to dissolve in solution like air; rather, vapor bubbles collapse. When a vapor bubble collapses near a surface it can cause erosion and extreme damage to mechanical parts of hydraulic machinery. This phenomenon is familiarly referred to as cavitation, a term coined by R. E. Froude (1810–1879). Due to the localized reduction in pressure that tends to occur in high-velocity jet streams, vapor bubbles are formed at moderate temperatures within hydraulic systems. Once these bubbles are exposed to a higher pressure, say, when the fluid velocity is reduced, the bubbles collapse with violent motions, which create large transient pressures. If the collapse occurs near a surface these transient pressures can be large enough to fail the surface by inducing high stresses in the solid material. By repeatedly exposing the surface to cavitating conditions the surface will eventually erode and before long the structural part may fail completely. The vapor pressure for petroleum-based fluids operating near 65°C is extremely low; that is, less than 350 Pa absolute [6]. Due to the high percentage of water contained in water glycol solutions and water-in-oil emulsions these fluids have a much higher vapor pressure than that of petroleum-based fluids.

1.7 CHEMICAL PROPERTIES

In this section various terms that are commonly used to describe fluid chemical properties are defined. These terms do not have quantitative definitions as given in the previous discussions; however, their qualitative meanings are of use and are therefore included here [1, 7, 9].

Emulsivity. This quality describes the fluid's ability to form emulsions. An emulsion is a fluid that is formed by suspending oily liquid in another liquid often by means of a gummy substance.

Lubricity. This quality of the fluid describes the "oiliness" of the fluid and refers to its adequacy when used as a lubricant. Many oils naturally contain some molecular species with boundary lubricating properties. Some vegetable oils, such as castor oil and rapeseed oil, contain more natural boundary lubricants than mineral oils. Additives are, therefore, usually incorporated into mineral oils for the purposes of improving the lubricity. Lack of adequate lubrication properties promotes wear and shortens the life of hydraulic components.

Thermal stability. This quality of the fluid describes the fluid's ability to resist chemical reactions and decomposition at high temperatures. Fluids react more vigorously as temperature is increased and may form solid reaction products.

Oxidative stability. This quality of the fluid describes an ability to resist reactions with oxygen-containing materials, especially air. Again, these reactions may form solid byproducts within the fluid.

Hydrolytic stability. This quality of the fluid describes an ability to resist reactions with water. Undesirable formation of solids may result or a stable water-in-oil emulsion may be formed which degrades lubricating ability and promotes rusting and corrosion. Demulsifier additives are often used to inhibit emulsion formations.

Compatibility. This quality of the fluid is a "catch-all" that describes the fluid's ability to resist chemical reactions with any material that may be used in the system to which it is exposed. For instance, some fluids tend to soften seals and gaskets, which may cause them to be incompatible. Water is incompatible with steel since it causes the steel to corrode.

Foaming. This term is used to describe the fluid's ability to combine with gases, principally air, and to form emulsions. Entrained air reduces the lubricating ability and bulk modulus of a liquid. A reduction in the bulk modulus can severely limit dynamic performance and for this reason fluids should have the ability to release air without forming emulsions. Antifoamant additives are used to encourage this ability.

Flash point. This is the lowest temperature at which the vapor of a volatile oil will ignite with a flash.

Pour point. This is the lowest temperature at which a fluid will flow.

Handling properties. These properties refer to the toxicity, odor, color, and storage characteristics of a fluid. These characteristics can be dangerous or annoying thereby making the handling or use of the fluid somewhat undesirable.

1.8   FLUID TYPES AND SELECTION

1.8.1   Petroleum-Based Fluids

Petroleum-based fluids are by far the most common fluids used in hydraulic systems. They are a complex mixture of hydrocarbons that must be refined to produce a fluid with the appropriate characteristics that are suitable for hydraulic control systems. Various additive packages are sold for petroleum-based fluids. These packages include inhibitors against oxidation, foaming, and corrosion. Additives can be used to increase the viscosity index and to improve the lubricity of the fluid as well.

1.8.2   Synthetic Fluids

Synthetic fluids are used to provide a fire-resistant alternative to petroleum based fluids and are named after their base stock which is the predominant material used to make them. Examples of these fluids are phosphate esters and silicate esters.

1.8.3   Biodegradable Fluids

Biodegradation is the ability of a substance to be broken down into innocuous products by the action of living things. Due to environmental concerns biodegradable fluids have become an important alternative for use in hydraulic control systems. This is especially true in the mobile hydraulic industry where the unintentional spill of mineral oil may result in long-term soil and water contamination in the vicinity of the spill itself. Various types of biodegradable fluids are available for use. The most basic forms of biodegradable fluids are the vegetable oils, especially those extracted from rapeseed. Other biodegradable fluids with higher performance are ester-based synthetic fluids. While the ester-based synthetic fluids provide a wider and more robust range of performance (especially at cold temperatures) they are very costly which prohibits their use in many applications.

1.8.4 Water

Water hydraulics is a growing topic of discussion among the engineering community. The reasons for this are somewhat obvious. Water is available in abundance and is therefore very inexpensive. Water is also very friendly to the environment and therefore alleviates the concerns for contamination in the event of a hydraulic failure or spill. On the other hand, water has several disadvantages that must be overcome before it is widely used as a medium for transmitting power hydraulically. Among these disadvantages water has the ability to sustain life in the form of bacterial growth. This growth causes inherent contamination within the hydraulic system, which may in turn cause a failure. Another disadvantage of water is that it has a very narrow temperature range between phase changes. Water freezes at 0°C and boils at 100°C (at standard atmospheric pressures). This narrow range of temperature over which water will remain liquid is not easily overcome by many applications. Water also exhibits a poor lubricity and a low viscosity. These characteristics make water a poor lubricating medium, which is often necessary for maintaining a longer life for the hydraulic system. Another disadvantage of water is that it is corrosive to ferrous materials that are commonly used to build hydraulic machinery. In order to use water as the working medium machine parts must be coated with polymer-type materials, which means that they must be subject to low surface stresses to keep the lower strength polymers from failing. This drives the hydraulic system toward low-pressure applications and causes the power density of the system to be sacrificed.

1.8.5 Fluid Selection

When selecting a hydraulic fluid one must keep in mind the following nine characteristics:

1. The fluid must exhibit good lubricity with compatible materials that are used for bearings and sealing surfaces.
2. The fluid must exhibit a high-viscosity index over a wide range of operating temperatures.
3. The fluid must provide a long service life (at least 5000 hours). This means that it must be stable against heat, water, oxidation, and shear.
4. The fluid must be compatible with environmental requirements. This may or may not require the use of biodegradable fluids.
5. The fluid must have a high bulk modulus for a satisfactory dynamic response of the hydraulic system. Generally, this means that the fluid should resist absorption of air and exhibit a low tendency to foam.
6. The fluid must generally be low cost and highly available.
7. The fluid must exhibit a resistance to flammability.
8. The fluid should exhibit a high thermal conductivity for the purposes of transferring heat away from the system.
9. The fluid should exhibit a low vapor pressure and a high boiling temperature to avoid cavitation.

1.9 CONCLUSION

In conclusion, this chapter has been written to consider the various physical properties that are commonly discussed for hydraulic fluids. These properties have been related to the mass density of the fluid, its thermal characteristics, the fluid viscosity, and other important topics of cavitation potential and chemical compatibilities. It must be mentioned that this brief overview is far from exhaustive and the reader is referred to sundry texts that have fluid property data scattered throughout. For the remainder of this text the fluid properties mentioned here are referred to and used for describing the overall performance of hydraulic control systems. The following chapter on fluid mechanics assumes a working knowledge of the terms and definitions that have been presented here.

1.10 REFERENCES

[1] Merritt, H. E. 1967. *Hydraulic Control Systems*. John Wiley & Sons, New York.

[2] Boresi, A. P., R. J. Schmidt, and O. M. Sidebottom. 1993. *Advanced Mechanics of Materials*, 5th ed. John Wiley & Sons, New York.

[3] Doebelin, E. O. 1990. *Measurement Systems: Application and Design*, 4th ed. McGraw-Hill, New York.

[4] Manring, N. D. 1997. The effective fluid bulk-modulus within a hydrostatic transmission. *ASME Journal of Dynamic Systems, Measurement, and Control* 119:462–66.

[5] Hutchings, I. M. 1992. *Tribology: Friction and Wear of Engineering Materials*. CRC Press, London.

[6] Esposito, A. 2000. *Fluid Power with Applications*, 5th ed. Prentice Hall, Upper Saddle River, NJ.

[7] Yeaple, F. 1996. *Fluid Power Design Handbook*, 3rd ed. Marcel Dekker, New York.

[8] United States Army Material Command. 2000. *Engineering Design Handbook: Hydraulic Fluids*. University Press of Hawaii, Honolulu.

[9] Radhakrishnan, M. 2003. *Hydraulic Fluids: A Guide to Selection, Test Methods and Use*. American Society of Mechanical Engineers, New York.

1.11 HOMEWORK PROBLEMS

1.11.1 Fluid Mass Density

1.1   The density of water at the standard reference state (pressure $= 1$ atm., temperature $= 25°C$) is $1000 \, \text{kg/m}^3$. Calculate the water's fluid density when the pressure is increased to 20 MPa and the temperature is increased to 80°C. *Note*: At the reference state, the bulk modulus of water is given by 22 kbar and the coefficient of thermal expansion is given by $200 \times 10^{-6}/°C$.

1.2   The density of hydraulic fluid at atmospheric pressure and a temperature of 100°C is $842 \, \text{kg/m}^3$. Calculate the percent change in the fluid's density when the pressure is increased to 40 MPa and the temperature is reduced to 0°C. *Note*: At the original state, the bulk modulus of the hydraulic fluid is given by 12 kbar and the coefficient of thermal expansion is given by $7 \times 10^{-4}/°C$.

1.3 An open container contains 1 gallon of hydraulic fluid weighing 7 lbf. The temperature of the hydraulic fluid is increased by 100°F and the volume of fluid increases to 1.01 gallons. What are the original and final densities of the hydraulic fluid? What is the coefficient of thermal expansion for this fluid?

1.11.2 Fluid Bulk Modulus

1.4 It is shown from experiments that for a fixed mass of water glycol the fluid pressure varies according to the following equation:

$$P = \frac{K_o(V_o - V)}{V_o - m\,(V_o - V)},$$

where K_o and m are experimental constants, V_o is the fluid volume when the pressure is zero, and V is the fluid volume associated with the pressure P. Using the definition for the secant bulk modulus K and the tangent bulk modulus β, develop an expression for each. What is the percent difference between these two results for $m = 4.5$ and $V = 0.98\ V_o$?

1.5 Calculate the effective fluid bulk modulus for mineral oil with 1% of entrained air (by volume) inside a brass tube with an outside diameter of 20 mm and a wall thickness of 2.5 mm. Assume that the fluid pressure is 2100 kPa and that the operating temperature is 90°C. All processes are assumed to be isothermal.

1.6 A measurement is taken in the laboratory to determine the effective fluid bulk modulus of a fluid within a container similar to what is shown in Figure 1-4. A 6000 psi pressure transducer (accurate within ± 1.5% of its full-scale reading) is used with a ½-inch LVDT (accurate within ± 1.5% of its full-scale reading) to make this measurement. The original depth of the container is 1 foot ± 0.1 in. The experimental records are given in the following table:

Fluid pressure P[psi]	Displacement x[in.]	Bulk modulus β [psi]	Maximum error ε_{max}
1500	0.203		
3000	0.274		
4500	0.336		
6000	0.389		

Using this data, fill in the remainder of the table for the measured fluid bulk modulus and the uncertainty associated with the measurement. At what point would you begin to consider the measurements to be useful?

1.11.3 Thermal Fluid Properties

1.7 Calculate the effective coefficient of thermal expansion for a hydraulic fluid with 3% entrained air within a high-pressure hose. The operating temperature is 40°C.

1.8   A heat flux of 550 W/m² is measured across a stagnant film of hydraulic fluid 2.54 mm thick. The temperature drop across the fluid film is given by 10°C. Calculate the coefficient of thermal conductivity for the hydraulic fluid.

1.11.4   Fluid Viscosity

1.9   Using Figure 1-7, determine the fluid viscosity of a 10W petroleum based fluid at an operating temperature of 50°C. Using Equation (1.111) convert this result to the μReyn and the centi-poise scale.

1.10   A 10W hydraulic fluid is nominally used at 70°F and atmospheric pressure. Using Equation (1.112) and Figure 1-7 calculate the fluid property λ if the temperature is increased to 200°F. If the pressure is also increased to 6000 psi what is the new viscosity? Does the pressure have much effect in your opinion?

2

FLUID MECHANICS

2.1 INTRODUCTION

Fluid flow is a fundamental concept for hydraulic control systems. Unfortunately, the physics of fluid flow is not a trivial matter and therefore an entire chapter of this book must be devoted to the groundwork associated with practical fluid flow problems. In this chapter we present the Navier-Stokes equations that govern the dynamics of fluid motion. We also discuss the flow regimes in which the equations may be approximated and solved depending on the Reynolds number that characterizes the flow. From these approximations we generally solve the equations for typical boundaries and geometries that are found within hydraulic control systems. Transient pressure phenomenon also is considered, as is the derivation of hydraulic power. Finally, this chapter concludes with fundamental concepts that are germane to lubrication theory.

2.2 GOVERNING EQUATIONS

2.2.1 Navier-Stokes Equations

The Navier-Stokes equations are given in vector form:

$$\rho\left(\frac{\partial \mathbf{u}}{\partial t} + \mathbf{u}\ \nabla\mathbf{u} - \mathbf{f}\right) = -\nabla p + \mu\ \nabla^2\mathbf{u}, \tag{2.1}$$

where ρ is the fluid density, $\mathbf{u}$ is the velocity vector of the fluid, t is time, $\mathbf{f}$ is a vector containing all the body forces that are acting on the fluid, p is the fluid pressure, and μ is the fluid viscosity. In Equation (2.1), ∇ is the gradient operator and ∇^2 is the Laplace operator. Since there are potentially several different scales of phenomenon occurring within the flow field it is useful to write the Navier-Stokes equations in nondimensional form. To do this we introduce definitions:

$$\mathbf{u} = \hat{\mathbf{u}}U, \qquad t = \hat{t}\tau, \qquad \mathbf{f} = \hat{\mathbf{f}}F, \qquad p = \hat{p}P,$$

$$\nabla = \hat{\nabla}\frac{1}{L}, \qquad \nabla^2 = \hat{\nabla}^2\frac{1}{L^2}, \tag{2.2}$$

where the carets denote dimensionless quantities, U is a characteristic velocity of the flow field, τ is a quantity of time that is typical of some transient behavior within the flow, F is a characteristic body force acting on the fluid, P is a characteristic pressure of the fluid, and L is a characteristic length dimension within the flow field. Substituting Equation (2.2) into Equation (2.1) yields the following dimensionless result for the Navier Stokes equations:

$$\left(\frac{\rho L^2}{\mu \tau}\right)\frac{\partial \hat{\mathbf{u}}}{\partial \hat{t}} + \text{Re } \hat{\mathbf{u}} \hat{\nabla}\hat{\mathbf{u}} - \hat{\nabla}^2\hat{\mathbf{u}} = \left(\frac{\rho L^2 F}{\mu U}\right)\hat{\mathbf{f}} - \left(\frac{PL}{\mu U}\right)\hat{\nabla}\hat{p}, \tag{2.3}$$

where Re is the Reynolds number given by

$$\text{Re} = \frac{\rho U L}{\mu}. \tag{2.4}$$

For steady flow in the absence of body forces the Navier-Stokes equations reduce to the following form:

$$\text{Re } \hat{\mathbf{u}}\hat{\nabla}\hat{\mathbf{u}} - \hat{\nabla}^2\hat{\mathbf{u}} = -\left(\frac{PL}{\mu U}\right)\hat{\nabla}\hat{p}. \tag{2.5}$$

This equation is complicated to solve and generally requires the intensive use of a computer to generate numerical results; therefore, in the following subsections we consider regimes of fluid flow for which Equation (2.5) may be solved analytically.

2.2.2   High Reynolds Number Flow

From the N-S Equations.   When we speak of a Reynolds number being either high or low, we are speaking of a relative quantity. The Reynolds number must be high or low compared to something. From the dimensionless form of the Navier-Stokes equations we can see that proper scaling of terms will cause the quantities with carets to be approximately equal to unity. In this case, the Reynolds number is used to amplify or attenuate the convective acceleration term $\hat{\mathbf{u}} \hat{\nabla}\hat{\mathbf{u}}$ and therefore weights that term in relationship to all other terms in the equation. If, in fact, the Reynolds number is much, much greater than unity, it can be shown from Equation (2.5) that

$$\text{Re } \hat{\mathbf{u}} \hat{\nabla}\hat{\mathbf{u}} = -\left(\frac{PL}{\mu U}\right)\hat{\nabla}\hat{p}. \tag{2.6}$$

In dimensional form, this equation may be expressed as

$$\rho\mathbf{u} \nabla\mathbf{u} = -\nabla p. \tag{2.7}$$

If we consider a one-dimensional flow field in, say, the x-direction, Equation (2.7) may be expressed as

$$\rho u \frac{du}{dx} = -\frac{dp}{dx}, \tag{2.8}$$

where u is the fluid velocity in the x-direction. Direct integration of this equation yields the following result:

$$p + \rho\frac{1}{2}u^2 = \text{constant}, \tag{2.9}$$

which is commonly known as the Bernoulli equation (without body forces). Thus, the Bernoulli equation is an equation that describes fluid flow for Reynolds numbers that are much greater than unity and this equation must be satisfied along streamlines within a high Reynolds number steady flow field.

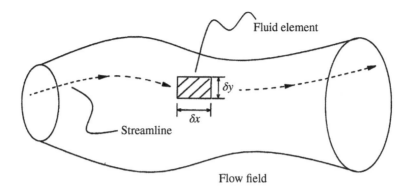

Figure 2-1. Free-body-diagram of a fluid element undergoing high Reynolds number flow.

From the Free-Body-Diagram. To lend physical insight into the Bernoulli equation it is helpful to consider the forces acting on a fluid element within the flow field. The free-body diagram of Figure 2-1 shows a fluid element with pressure acting in the x-direction.

Because the Reynolds number is high, we neglect the shear forces that are acting along the sides of the element and set the pressure forces equal to the time-rate-of-change of linear momentum in accordance with Newton's second law. This result is given by

$$\delta m \frac{du}{dt} = p \, \delta y \, \delta z - \left(p + \frac{dp}{dx} \delta x \right) \delta y \, \delta z, \tag{2.10}$$

where δm is the mass of the fluid element, u is the fluid velocity in the x-direction, t is time, p is the fluid pressure, and $x, y,$ and z are the Cartesian coordinates describing the fluid flow field. By definition the mass of the fluid element is given by

$$\delta m = \rho \, \delta x \, \delta y \, \delta z, \tag{2.11}$$

where ρ is the fluid density. For a steady flow problem we recognize that the fluid velocity u does not depend explicitly on time t. However, the velocity certainly may vary with the coordinate location given by x and this location changes over time. Therefore, using the chain rule of calculus, the time rate of change of linear momentum for the fluid element may be written as

$$\frac{du}{dt} = \frac{du}{dx} \frac{dx}{dt} = u \frac{du}{dx}. \tag{2.12}$$

Substituting Equations (2.11) and (2.12) into Equation (2.10) yields the following result for describing the motion of the fluid element:

$$\rho u \frac{du}{dx} = -\frac{dp}{dx}, \tag{2.13}$$

which is the same governing equation presented in Equation (2.8). The corresponding solution to this equation is given as the Bernoulli equation and is presented in Equation (2.9).

2.2.3 Low Reynolds Number Flow

From the N-S Equations.   For low Reynolds number flow, the Reynolds number must be much, much less than unity. Under such conditions it can be seen from Equation (2.5) that the governing equation of motion for the fluid is given by

$$\hat{\nabla}^2 \hat{\mathbf{u}} = \left(\frac{PL}{\mu U} \right) \hat{\nabla} \hat{p}. \tag{2.14}$$

In dimensional form this result is written as

$$\mu \, \nabla^2 \mathbf{u} = \nabla p. \tag{2.15}$$

If we consider a one-dimensional flow field where the pressure varies only in the x-direction (the direction of flow) and the velocity varies only in the y-direction (normal to the direction of flow) Equation (2.15) may be written more explicitly as

$$\mu \, \frac{d^2 u}{dy^2} = \frac{dp}{dx}. \tag{2.16}$$

Integrating this result twice with respect to y yields the following result for the fluid velocity

$$u = \frac{1}{2\mu} \frac{dp}{dx} \, y^2 + C_1 \, y + C_2, \tag{2.17}$$

where C_1 and C_2 are "constants" of integration that generally depend on x. If the velocity is specified in two places within the flow field, say, $u(y_1) = u_1$ and $u(y_2) = u_2$, the constants of integration may be solved as

$$C_1 = \frac{(u_1 - u_2)}{(y_1 - y_2)} - \frac{1}{2\mu} \frac{dp}{dx} (y_1 + y_2), \quad C_2 = \frac{y_1 u_2 - y_2 u_1}{(y_1 - y_2)} + \frac{1}{2\mu} \frac{dp}{dx} y_1 y_2. \tag{2.18}$$

Substituting Equation (2.18) into Equation (2.17) produces the following result for the fluid velocity

$$u = \frac{1}{2\mu} \frac{dp}{dx} (y - y_1)(y - y_2) + u_1 \frac{(y - y_2)}{(y_1 - y_2)} - u_2 \frac{(y - y_1)}{(y_1 - y_2)}. \tag{2.19}$$

The volumetric flow rate per unit width through a gap defined by the coordinates y_1 and y_2 is given by

$$Q = \int_{y_1}^{y_2} u \, dy = \frac{1}{12\mu} \frac{dp}{dx} (y_1 - y_2)^3 - \frac{(u_1 + u_2)}{2} (y_1 - y_2). \tag{2.20}$$

From the conservation of mass for an incompressible fluid we know that $dQ/dx = 0$. Using the conservation of mass, it can be shown from Equation (2.20) that

$$\frac{d}{dx} \left(\frac{h^3}{\mu} \frac{dp}{dx} \right) = 12 \, \overline{U} \, \frac{dh}{dx}, \tag{2.21}$$

where the fluid gap is given by $h = y_2 - y_1$ and the average fluid velocity is given by $\overline{U} = \frac{u_2 + u_1}{2}$. *Note*: In this equation the average fluid velocity is assumed to be independent of the coordinate x. Equation (2.21) is known as the classical Reynolds equation for one-dimensional flow. If we assume that h is independent of x, Equation (2.21) may be used to show that

$$\frac{d^2p}{dx^2} = 0. \tag{2.22}$$

The solution to this equation is given by

$$p = C_3 x + C_4, \tag{2.23}$$

where C_3 and C_4 are constants of integration. This result describes a linear change in pressure with respect to x. By specifying the fluid pressure in two places within the flow field, say, $p(x_1) = p_1$ and $p(x_2) = p_2$, the constants of integration may be solved for as

$$C_3 = \frac{(p_1 - p_2)}{(x_1 - x_2)}, \quad C_4 = p_1 - \frac{(p_1 - p_2)}{(x_1 - x_2)}x_1. \tag{2.24}$$

Substituting these results back into Equation (2.23) yields the following expression for the pressure:

$$p = p_1 + \frac{(p_1 - p_2)}{(x_1 - x_2)}(x - x_1). \tag{2.25}$$

This result describes the pressure field for a fluid that is contained between, say, two flat plates. More interesting results are obtained by relaxing the requirement that the fluid gap h must be independent of x. Some of these results are shown later in Section 2.7 of this chapter.

From the Free-Body-Diagram. To lend more physical insight into the low Reynolds number problem, a free-body-diagram of a fluid element is shown in Figure 2-2. This fluid element is taken from a one-dimensional flow field with a pressure gradient in the x-direction and a velocity gradient in the y-direction. The forces acting on the fluid element come from the pressure gradient in the x-direction and the fluid shear gradient in the y-direction.

Since the Reynolds number is low the time-rate-of-change of linear momentum will be neglected and the forces will be summed and set equal to zero. This result is given by

$$0 = p \, \delta y \, \delta z - \left(p + \frac{dp}{dx}\delta x \right) \delta y \, \delta z + \left(\tau + \frac{d\tau}{dy}\delta y \right) \delta x \, \delta z - \tau \, \delta x \, \delta z, \tag{2.26}$$

where p is the fluid pressure and τ is the fluid shear. Simplifying Equation (2.26) yields the following result:

$$\frac{d\tau}{dy} = \frac{dp}{dx}. \tag{2.27}$$

For Newtonian fluids the fluid shear is given by

$$\tau = \mu\frac{du}{dy}, \tag{2.28}$$

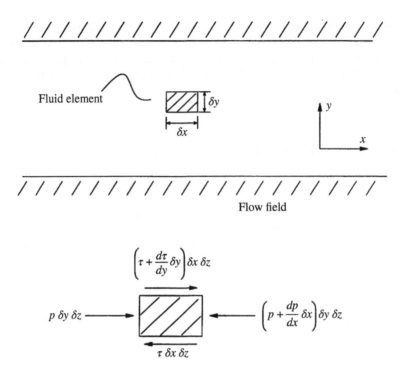

Figure 2-2. Free-body-diagram of a fluid element undergoing low Reynolds number flow as before.

where μ is the absolute fluid viscosity and u is the fluid velocity in the x-direction. Substituting Equation (2.28) into Equation (2.27) yields the following result:

$$\mu \frac{d^2u}{dy^2} = \frac{dp}{dx}, \qquad (2.29)$$

which is the same governing equation derived from the Navier-Stokes equations. See Equation (2.16). The solution for this equation follows exactly.

Case Study

The figure shows a cylindrical flow passage with a drastic change in the cross-sectional area at position 2. The density of the hydraulic fluid is 850 kg/m^3 and the viscosity is 0.02 Pa-s. The constant flow rate through the passage is shown by the symbol Q and is given 0.10 liters/min. The two diameters shown in the figure are given by $d_1 = 1$ m and $d_2 = 1$ mm.

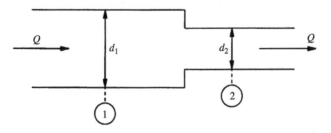

Calculate the Reynolds number as sections 1 and 2 in the figure. What are the equations that may be used to describe the flow through each section? What are the physical meanings behind each equation?

For section 1 the velocity of the fluid flow is given by $4\,Q/(\pi\,d_1^2)$. Using the data given in the problem statement this velocity is calculated to be 2.1×10^{-6} m/s. Using this result, the Reynolds number in section 1 of the flow passage may be calculated as

$$\text{Re} = \frac{\rho U_1 d_1}{\mu} = \frac{850\,\text{kg/m}^3 \times 2.1 \times 10^{-6}\text{m/s} \times 1\,\text{m}}{0.02\,\text{Pa-s}} = 0.09.$$

Since this number is much less than unity, the flow through section 1 of the passage is characterized by a low Reynolds number and therefore the Reynolds equation must be used to model the fluid motion in this section of the pipe. In this case the fluid motion is heavily influenced by the fluid viscosity and the inertial effects of the fluid motion are insignificant.

For section 2 the velocity of the fluid flow is given by $4\,Q/(\pi\,d_2^2)$. Using the data given in the problem statement, this velocity is calculated to be 2.1 m/s. Using this result the Reynolds number in section 2 of the flow passage may be calculated as

$$\text{Re} = \frac{\rho U_2 d_2}{\mu} = \frac{850\,\text{kg/m}^3 \times 2.1\,\text{m/s} \times 0.001\,\text{m}}{0.02\,\text{Pa s}} = 90.$$

Since this number is much greater than unity the flow through section 2 of the passage is characterized by a high Reynolds number and therefore the Bernoulli equation must be used to model the fluid motion in this section of the pipe. In this case the fluid motion is heavily influenced by the fluid inertia and the viscosity effects of the fluid motion are insignificant.

2.2.4 Turbulent versus Laminar Flow

Turbulent flow is defined as a fluid flow in which the velocity at a given point varies erratically in magnitude and direction. Laminar flow is defined as streamline flow near a solid boundary. *Note*: A streamline is a line that is everywhere tangent to the velocity field. In experiments conducted by Osborne Reynolds (1842–1912) it was observed that flow in a round pipe with a Reynolds number less than 2100 was consistently laminar. It was also noted that the same flow was consistently turbulent for Reynolds numbers greater than 4000. If the flow was characterized by a Reynolds number between 2100 and 4000, the flow was observed to be in a transitional stage since it randomly changed between laminar and turbulent conditions. This classical experiment is often used to predict the onset of turbulence as a function of Reynolds number.

In the previous discussion of low and high Reynolds number flow we distinguished between Reynolds numbers that were much, much less than one and Reynolds numbers that were much, much greater than one. In the jargon of fluid flow there is a tendency to refer to low Reynolds number flow as being laminar and high Reynolds number flow as being turbulent. In some flow regimes this language may be accurate but it certainly does not apply everywhere. For instance, we expect all low Reynolds number flow to be laminar since Reynolds numbers much, much less than one are certainly less than a Reynolds number of 2100, which was the Reynolds number of turbulence onset observed by Osborne Reynolds in round pipes. On the other hand, all high Reynolds number flow is not necessarily turbulent. A Reynolds number much, much greater than one may still be less than 2100, which suggests that turbulence and random velocity patterns may not be present in the flow field even for high Reynolds number flow. Still, for a high Reynolds number flow that exceeds a Reynolds number of 4000 it is likely that turbulence exists; therefore, this high Reynolds number flow may be classified as turbulent. In summary, it may be said that all low Reynolds number flow is laminar; however, high Reynolds number flow may be either laminar or turbulent depending on the Reynolds number itself.

2.2.5   Control Volume Analysis

Reynolds Transport Theorem.   The Navier-Stokes equations govern the motion of a particular fluid element as it flows through a given flow field. These equations are necessary for understanding a detailed level of information regarding the flow field itself; however, this level of detail is often more than we need to know. Sometimes we only care about the effect of the flow field on surrounding objects that define a given space or control volume. To study this overall effect of the fluid as it acts on the control volume we must use an analytical tool known as the Reynolds Transport Theorem. The following development of this theorem may be found in standard fluid mechanics textbooks [1].

Figure 2-3 shows a fixed control volume (defined by the dashed lines) at two instances in time. At time t the control volume is shown to contain a system of fluid particles that are characterized by some fluid property given by B.

In general, the fluid property of interest could be velocity, mass, temperature, and so on. At this point we consider a generalized fluid property and specify its physical definition as the analysis dictates. Figure 2-3 also shows a snapshot of the original system of particles at a second instant in time $t + \delta t$. At this second instant in time the original system of particles has started to move out of the control volume and a new group of fluid particles has started to enter into the control volume. At the first instant in time we recognize that the fluid property

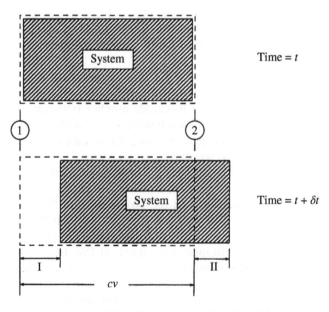

Figure 2-3. A control volume shown by dashed lines with a system of particles moving across boundary 2.

of the original system of particles is identical to the fluid property within the control volume since they occupy the exact same space. We write this mathematically as

$$B_{sys}(t) = B_{cv}(t). \tag{2.30}$$

At the second instant in time it can be shown that the fluid property of the original system is given by the following expression:

$$B_{sys}(t + \delta t) = B_{cv}(t + \delta t) + B_{II}(t + \delta t) - B_{I}(t + \delta t), \tag{2.31}$$

where B_{I} is the fluid property associated with the new particles that have just entered the control volume and B_{II} is the fluid property associated with the particles of the original system that have just exited the control volume. The change in the fluid property of the original system with respect to time is given by

$$\frac{\delta B_{sys}}{\delta t} = \frac{B_{sys}(t + \delta t) - B_{sys}(t)}{\delta t}. \tag{2.32}$$

Substituting Equations (2.30) and (2.31) into Equation (2.32) yields the following result:

$$\frac{\delta B_{sys}}{\delta t} = \frac{B_{cv}(t + \delta t) - B_{cv}(t)}{\delta t} + \frac{B_{II}(t + \delta t)}{\delta t} - \frac{B_{I}(t + \delta t)}{\delta t}. \tag{2.33}$$

As δt goes to zero we can see that

$$\lim_{\delta t \to 0} \frac{\delta B_{sys}}{\delta t} = \frac{D B_{sys}}{Dt}, \tag{2.34}$$

where large D has been used to denote the total or material derivative of the fluid property B. This notation acknowledges the Lagrangian nature of the derivative representing the time-rate-of-change of the fluid property B as the system of fluid particles move through the space of the control volume. The control volume property B_{cv} may be written as

$$B_{cv} = b_{cv} \, m_{cv} = \int_{V_{cv}} b_{cv} \, \rho_{cv} \, dV, \tag{2.35}$$

where b_{cv} is the specific fluid property within the control volume, m_{cv} is the mass of fluid within the control volume, ρ_{cv} is the fluid mass density within the control volume, and V_{cv} is the volume of the control volume itself. Using this result, the first term on the right-hand side of Equation (2.33) may be evaluated in the limit as δt goes to zero. This result is

$$\lim_{\delta t \to 0} \frac{B_{cv}(t + \delta t) - B_{cv}(t)}{\delta t} = \frac{\partial B_{cv}}{\partial t} = \frac{\partial}{\partial t} \int_{V_{cv}} b_{cv} \, \rho_{cv} \, dV. \tag{2.36}$$

To consider the second term on the right-hand side of Equation (2.33) we must consider the amount of the system fluid property that is leaving the control volume. This amount of the fluid property may be written as

$$B_{II}(t + \delta t) = b_2 \, \delta m_2 = b_2 \, \rho_2 \, \delta V_2, \tag{2.37}$$

where b_2 is the specific fluid property at boundary 2, ρ_2 is the fluid mass density at boundary 2, and δm_2 and δV_2 are the mass and volume that have crossed boundary 2 in the period of time δt. Recognizing that

$$\delta V_2 = \delta t \int_{A_2} u_2 \, dA, \tag{2.38}$$

where u_2 is the fluid velocity crossing boundary 2, it can be shown that

$$B_{II}(t + \delta t) = b_2 \, \delta m_2 = \delta t \int_{A_2} b_2 \, \rho_2 \, u_2 \, dA. \tag{2.39}$$

Similarly, the third term on the right-hand side of Equation (2.33) may be written as

$$B_{I}(t + \delta t) = b_1 \, \delta m_1 = \delta t \int_{A_1} b_1 \, \rho_1 \, u_1 \, dA, \tag{2.40}$$

where the subscript 1 denotes properties that relate to boundary 1 of the control volume. Using the results of Equations (2.34), (2.36), (2.39), and (2.40) with Equation (2.33) yields the following result for the total time rate of change of the fluid property B as a system of fluid passes through the control volume:

$$\frac{DB_{sys}}{Dt} = \frac{\partial}{\partial t} \int_{V_{cv}} b_{cv} \, \rho_{cv} \, dV_{cv} + \int_{A_2} b_2 \, \rho_2 \, u_2 \, dA_2 - \int_{A_1} b_1 \, \rho_1 \, u_1 \, dA_1. \tag{2.41}$$

This is the Reynolds Transport Theorem for the simple control volume shown in Figure 2-3. For the more general case where flow may be exiting and entering the control volume from many different locations, and where the flow and the fluid property of interest may be three dimensional in nature, the following expression may be written for the Reynolds Transport Theorem:

$$\frac{D\mathbf{B}}{Dt} = \frac{\partial}{\partial t} \int_{c.v.} \rho \, \mathbf{b} \, dV + \int_{c.s.} \rho \, \mathbf{b} \, (\mathbf{u} \cdot \hat{\mathbf{n}}) \, dA, \tag{2.42}$$

where $\mathbf{B}$ is the fluid property of the system, $\mathbf{b}$ is the specific fluid property of the control volume, $\mathbf{u}$ is the velocity vector of flow entering and exiting the control volume, and $\hat{\mathbf{n}}$ is a unit vector pointing normally outward from the control volume at the location where the fluid is crossing the boundary of the control surface. The bold notation in this equation indicates a vector quantity. This result is valid for a fixed control volume. If $\mathbf{u}$ is considered to be an expression of relative velocity, Equation (2.42) also applies for a control volume that moves with a constant velocity. The first term on the right-hand side of Equation (2.42) describes the time-rate-of-change of the fluid property within the control volume itself. The second term on the right-hand side of Equation (2.42) describes the inflow and outflow of the fluid property across the control volume boundary with respect to time. As it turns out, the Reynolds Transport Theorem is an extremely powerful tool for designing and evaluating hydraulic control valves. As such, this tool is used extensively in Chapter 4 to analyze the fluid forces that are exerted on the mechanical parts of the valve. The following paragraphs serve to illustrate practical uses of the Reynolds Transport Theorem.

Conservation of Mass.   To consider the conservation of mass for a control volume we use the Reynolds Transport Theorem and the fluid property of mass itself. In other words,

$$\mathbf{B} = M, \quad \text{and} \quad \mathbf{b} = 1, \tag{2.43}$$

where M is the mass of the fluid system within the control volume. Using these expressions with Equation (2.42) we see that the total time rate of change of mass within the system is given by

$$\frac{D}{Dt}M = \underbrace{\frac{\partial}{\partial t} \int_{c.v.} \rho \, dV}_{\substack{\text{time-rate-of-change of} \\ \text{mass within the } c.v.}} + \underbrace{\int_{c.s.} \rho(\mathbf{u} \cdot \hat{\mathbf{n}}) dA.}_{\substack{\text{inflow of mass across} \\ \text{the } c.v. \text{ boundary}}} \qquad (2.44)$$

Unless there is a mass source within the system, the left-hand side of Equation (2.44) must always be zero. Therefore, the conservation of mass for the system is written as

$$0 = \frac{\partial}{\partial t} \int_{c.v.} \rho \, dV + \int_{c.s.} \rho(\mathbf{u} \cdot \hat{\mathbf{n}}) dA. \qquad (2.45)$$

This result states that the net change in the fluid mass within the control volume must be equal to the sum of all mass flow that occurs across the boundary of the control volume. *Note*: These flow rates are both positive and negative depending on the direction of flow in and out of the control volume.

Conservation of Fluid Momentum. To consider the conservation of fluid momentum for a control volume we use the Reynolds Transport Theorem and the fluid property of momentum. In this case,

$$\mathbf{B} = \mathbf{p}, \quad \text{and} \quad \mathbf{b} = \mathbf{u}, \qquad (2.46)$$

where $\mathbf{p}$ is the fluid momentum vector and $\mathbf{u}$ is the fluid velocity vector. Using these expressions with Equation (2.42), the total time-rate-of-change of fluid momentum for the system may be written as

$$\frac{D}{Dt}\mathbf{p} = \underbrace{\frac{\partial}{\partial t} \int_{c.v.} \rho \mathbf{u} \, dV}_{\substack{\text{time-rate-of-change of} \\ \text{momentum within the } c.v.}} + \underbrace{\int_{c.s.} \rho \mathbf{u}(\mathbf{u} \cdot \hat{\mathbf{n}}) dA.}_{\substack{\text{inflow of momentum} \\ \text{across the } c.v. \text{ boundary}}} \qquad (2.47)$$

From Newton's second law we know that the system's time rate of change of fluid momentum must equal the net force acting on the control volume by its surrounding environment. Therefore, the conservation of fluid momentum for the control volume may be written as

$$\mathbf{F} = \frac{\partial}{\partial t} \int_{c.v.} \rho \mathbf{u} \, dV + \int_{c.s.} \rho \mathbf{u}(\mathbf{u} \cdot \hat{\mathbf{n}}) dA, \qquad (2.48)$$

where $\mathbf{F}$ is the vector force acting on the control volume itself. As previously stated this result is of great practical importance to us in this text for considering the forces that are exerted on hydraulic control valves. The general use of Equation (2.48) is much more apparent in Chapter 4.

Case Study

To illustrate the use of the Reynolds Transport Theorem, consider the flow case shown in the figure. This figure shows an arc-shaped conduit in which the volumetric flow rate Q_1 enters the conduit at surface 1 and the volumetric flow rate Q_2 exits the conduit at surface 2. The direction of the fluid exiting the conduit is determined by the angle θ. Unit vectors pointing normally outward from the dashed control volume surface are shown in the figure by $\hat{\mathbf{n}}_1$ and $\hat{\mathbf{n}}_2$.

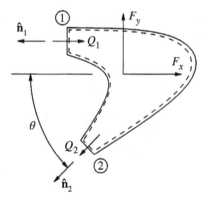

Since the fluid flow imparts momentum to the conduit, equal and opposite applied forces acting on the dashed control volume by the conduit are shown in the figure using the scalar quantities F_x and F_y. Using the Reynolds Transport Theorem you are to develop an analytical expression for the reaction forces F_x and F_y.

As shown by the general expression for the Reynolds Transport Theorem given in Equation (2.42), it is necessary to establish vector expressions for the fluid velocities crossing the control volume boundary and for the unit vectors that point normally outward from these control volume surfaces. Using the figure these expressions are given by

$$\mathbf{u}_1 = \frac{Q_1}{A_1}\,\hat{\mathbf{i}} + 0\,\hat{\mathbf{j}}, \qquad \mathbf{u}_2 = -\frac{Q_2}{A_2}\,\cos(\theta)\,\hat{\mathbf{i}} - \frac{Q_2}{A_2}\,\sin(\theta)\,\hat{\mathbf{j}},$$

$$\hat{\mathbf{n}}_1 = -1\,\hat{\mathbf{i}} + 0\,\hat{\mathbf{j}}, \qquad \hat{\mathbf{n}}_2 = -\cos(\theta)\,\hat{\mathbf{i}} - \sin(\theta)\,\hat{\mathbf{j}},$$

where A_1 and A_2 are cross-sectional areas of the flow passages at surfaces 1 and 2 and unit vectors are shown by the symbols with carets. For steady flow conditions the time derivative term in the Reynolds Transport Theorem vanishes. Using Equation (2.45) for the conservation of mass it may be shown for the figure that

$$\int_{c.s.} \rho(\mathbf{u} \cdot \hat{\mathbf{n}})\,dA = \int_{A_1} \rho(\mathbf{u}_1 \cdot \hat{\mathbf{n}}_1)\,dA_1 + \int_{A_2} \rho(\mathbf{u}_2 \cdot \hat{\mathbf{n}}_2)\,dA_2 = 0$$

$$= -\int_{A_1} \rho\,\frac{Q_1}{A_1}\,dA_1 + \int_{A_2} \rho\,\frac{Q_2}{A_2}\,dA_2 = -\rho\,Q_1 + \rho\,Q_2.$$

From this expression it can be seen that for steady flow through the control volume, the conservation of mass requires

$$Q_1 = Q_2 = Q.$$

The steady form of Equation (2.48) may now be used to calculate the net force exerted on the control volume by the conduit. This result is given by

$$\mathbf{F} = \int_{c.s.} \rho\mathbf{u}(\mathbf{u}\cdot\hat{\mathbf{n}})\,dA = \int_{A_1} \rho\mathbf{u}_1(\mathbf{u}_1\cdot\hat{\mathbf{n}}_1)\,dA_1 + \int_{A_2} \rho\mathbf{u}_2(\mathbf{u}_2\cdot\hat{\mathbf{n}}_2)\,dA_2$$

$$= -\int_{A_1} \rho\left(\frac{Q^2}{A_1^2}\hat{\mathbf{i}} + 0\hat{\mathbf{j}}\right)dA_1 - \int_{A_2} \rho\left(\frac{Q^2}{A_2^2}\cos(\theta)\,\hat{\mathbf{i}} + \frac{Q^2}{A_2^2}\sin(\theta)\,\hat{\mathbf{j}}\right)dA_2.$$

Recognizing that $\mathbf{F} = F_x\,\hat{\mathbf{i}} + F_y\,\hat{\mathbf{j}}$ this result may be used to show that the scalar forces exerted on the control volume are

$$F_x = -\rho\,Q^2\left(\frac{1}{A_1} + \frac{\cos(\theta)}{A_2}\right) \quad\text{and}\quad F_y = -\rho\,Q^2\,\frac{\sin(\theta)}{A_2}.$$

The negative sign on these results shows that the actual reaction force exerted on the control volume by the arc-shaped conduit is in the opposite direction of the arrows shown in the figure. Furthermore, the steady momentum forces may be altered by either changing the volumetric flow rate through the conduit or by changing the conduit geometry itself.

2.3 FLUID FLOW

2.3.1 The Reynolds Number

Flow Characterization. As previously discussed, the Reynolds number is used for characterizing the type of fluid flow under consideration and for identifying the appropriate governing equations for analysis. The Reynolds number was generally presented in Equation (2.4) and is rewritten here for convenience:

$$\text{Re} = \frac{\rho U L}{\mu}, \tag{2.49}$$

where ρ is the fluid density, U is a representative fluid velocity within the flow field, L is a representative length dimension within the flow field, and μ is the absolute fluid viscosity. To be able to calculate the Reynolds number for flow within hydraulic control systems, we must recognize that all fluid flow within these systems is contained within a passage that is completely filled with fluid; that is, no open channel flow exists. These fluid passages are used to transport a certain volume of fluid per unit time, which relates to the average fluid velocity of the flow. Furthermore, the size and geometry of the passage may be used to characterize a length dimension within the flow field. From the passage geometry we specify both the representative fluid velocity and the representative length dimension within the flow field as

$$U = \frac{Q}{A} \quad\text{and}\quad L = D_h, \tag{2.50}$$

where Q is the volumetric flow rate of the fluid, A is the cross-sectional area of the flow passage, and D_h is the hydraulic diameter of the flow passage to be discussed shortly. Substituting these expressions into Equation (2.49) yields the following expression for the Reynolds number that is to be used for characterizing flow within hydraulic control systems:

$$\text{Re} = \frac{\rho Q D_h}{\mu A}. \tag{2.51}$$

For characterizing the type of flow within the fluid a passage in question, the following classifications will be reemphasized:

$$\text{Re} = \frac{\rho Q D_h}{\mu A} \gg 1 \qquad \text{high Reynolds number flow,}$$

$$\text{Re} = \frac{\rho Q D_h}{\mu A} \ll 1 \qquad \text{low Reynolds number flow.} \qquad (2.52)$$

For flow that is characterized by high Reynolds number flow the governing equation of fluid motion is given by the Bernoulli equation shown in Equation (2.7). For flow that is characterized by low Reynolds number flow the governing equation of fluid motion is given by the Reynolds equation shown in Equation (2.15). These classifications of fluid flow are used in the following subsections to develop important equations that may be used to model the fluid flow within a hydraulic control system; but first, we must consider the geometry of the flow passage itself.

Hydraulic Diameter.   The hydraulic diameter is a concept that is useful for handling flow passages that are of a noncircular cross-section. If we consider the equation for the area of a circular flow passage we can see that

$$A = \frac{\pi}{4} D_h^2, \qquad (2.53)$$

where the hydraulic diameter D_h is used for the circular diameter of the flow passage. It is useful at this point to recall the definition of the number π given from the geometry of a perfect circle. This definition is

$$\pi \equiv \frac{\text{circle circumference}}{\text{circle diameter}}. \qquad (2.54)$$

By making the following replacement in Equation (2.53):

$$\pi \rightarrow \frac{\text{flow passage perimeter}}{\text{hydraulic diameter}} = \frac{S}{D_h}, \qquad (2.55)$$

the hydraulic diameter of a generally shaped flow passage may be written as

$$D_h = \frac{4A}{S}, \qquad (2.56)$$

where A is the cross-sectional area and S is the flow passage perimeter. *Note*: For a circular shaped flow passage, the hydraulic diameter simply equals the diameter of a circle. The hydraulic diameter shown in Equation (2.56) should be used for calculating the Reynolds number in Equation (2.51).

2.3.2 Bernoulli Flow and the Orifice Equation

Developing the Orifice Equation.   One of the most practical equations to be developed for the modeling of hydraulic control systems is the classical orifice equation. This equation is based on the Bernoulli equation and is therefore applicable for steady, incompressible, high Reynolds number flow. Figure 2-4 shows a flow passage with a sharp edged orifice separating two sides of the flow passage. The cross-sectional area of the orifice is given by A_o; however, a reduced area A_2 gives the cross-sectional area of the fluid streamlines that

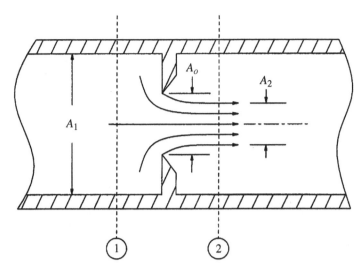

Figure 2-4. High Reynolds number flow through a sharp edged orifice, depicts the area of the vena contracta.

are passing through the orifice. This reduction in the cross-sectional flow area is known as the vena contracta and is a result of the fluid's inability to turn a sharp 90° corner as it moves around the walls of the flow chamber toward the opening of the orifice. The curved streamlines in Figure 2-4 indicate that a pressure gradient exists in a direction normal to the centerline of the flow passage with the highest pressure at the center of the flow passage and the lowest pressures existing at the edges of the orifice.

The ratio of the area of the vena contracta to the area of the orifice is called the contraction coefficient. Mathematically the definition of this coefficient is given by

$$C_c \equiv \frac{A_2}{A_o}. \tag{2.57}$$

The contraction coefficient is determined experimentally for various geometries of the flow passage. Some of these experimental results are illustrated in Figure 2-5.

From the flow field of Figure 2-4 and the Bernoulli equation presented in Equation (2.9) it may be shown that the relationship between the fluid pressure p and the fluid velocity u measured at cross sections 1 and 2 is given by

$$u_2^2 - u_1^2 = \frac{2}{\rho}(p_1 - p_2). \tag{2.58}$$

For an incompressible fluid the conservation of mass requires that

$$Q = u_1 A_1 = u_2 A_2, \tag{2.59}$$

where Q is the volumetric flow rate through the orifice. Substituting Equations (2.57) and (2.59) into Equation (2.58) yields the following result known as the classical orifice equation:

$$Q = A_o C_d \sqrt{\frac{2}{\rho}(p_1 - p_2)}, \tag{2.60}$$

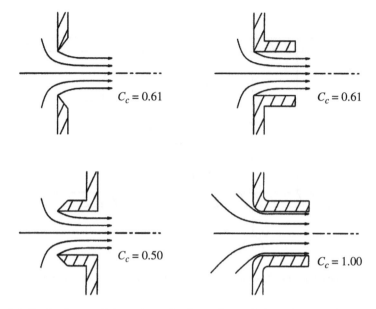

Figure 2-5. Contraction coefficients determined experimentally for various circular geometries.

where C_d is called the discharge coefficient and is given by

$$C_d = \frac{C_c}{\sqrt{1 - C_c^2 \left(\frac{A_o}{A_2}\right)^2}}. \tag{2.61}$$

Since A_o is usually much, much less than A_2 the discharge coefficient is usually very close to the contraction coefficient C_c. In general, however, the discharge coefficient must be determined experimentally for specific geometry cases. When the geometry is not exactly known or when experimental results are not available a sharp-edged orifice result of 0.62 is typically used for the discharge coefficient.

Uses of the Orifice Equation.   As previously mentioned Equation (2.60) is only valid for steady, incompressible, high Reynolds number flow (i.e., Re $\gg$ 1). If these conditions are met Equation (2.60) may be used with confidence. For the purposes of modeling, however, the assumptions behind the orifice equation are often ignored and Equation (2.60) is used to calculate the volumetric flow rate for unsteady flow, or slightly compressible flow, or even low Reynolds number flow. The reason for violating the fundamental assumptions behind the orifice equation is often a result of emphasizing mathematical expediency rather than physical accuracy. The orifice equation provides a concise calculation that is easily carried out either by hand or by computer and is therefore the dominant equation used for modeling fluid flow within hydraulic control systems.

2.3.3   Poiseuille Flow and the Annular Leakage Equation

General.   If the Reynolds number is much, much less than one the resistance to flow will be dominated by viscous shear effects as opposed to fluid momentum. In this case it is necessary to adopt the low Reynolds number flow equation presented in Equation (2.15) for our analysis. In the following discussion we present Poiseuille's law for a circular flow passage and the annular leakage equation, which is useful for modeling leakage within hydraulic systems. To conclude the discussion of this subsection noncircular duct geometry is presented as well.

Poiseuille's Law. Figure 2-6 shows a circular cross-section of a tube through which low Reynolds number flow occurs.

For one-dimensional flow through cylindrical geometry Equation (2.15) may be explicitly written as

$$\mu \frac{1}{r} \frac{\partial}{\partial r} \left(r \frac{\partial u}{\partial r} \right) = \frac{\partial p}{\partial z}, \tag{2.62}$$

where r and z are the radial and axial dimensions of the passage. The axial dimension is normal to the paper. In this equation the fluid pressure p is independent of the radial dimension and the fluid velocity u is independent of the axial dimension. Integrating Equation (2.62) once with respect to r yields the following result:

$$\frac{du}{dr} = \frac{r}{2\mu} \frac{dp}{dz} + C_1 \frac{1}{r}, \tag{2.63}$$

where C_1 is a constant of integration. Integrating this result once again with respect to r yields the following result for the fluid velocity:

$$u = \frac{r^2}{4\mu} \frac{dp}{dz} + C_1 \ln(r) + C_2, \tag{2.64}$$

where C_2 is another constant of integration. By imposing the following boundary conditions on the velocity field: $u'(r = 0) = 0$ and $u(R) = 0$, where $u' = du/dr$ and R is the outside radius of the cylindrical passage, it can be shown that

$$C_1 = 0, \quad C_2 = -\frac{R^2}{4\mu} \frac{dp}{dz}. \tag{2.65}$$

Substituting these results into Equation (2.64) yields the following velocity profile for the low Reynolds number flow as it passes through a circular-shaped passage:

$$u = -\frac{1}{4\mu} \frac{dp}{dz} (R^2 - r^2). \tag{2.66}$$

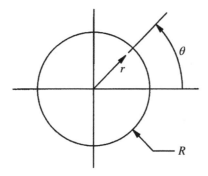

Figure 2-6. Circular flow passage through which low Reynolds number flow occurs.

Using this result the volumetric flow rate through the passage is given by

$$Q = \int_A u \, dA = 2\pi \int_0^R u \, r \, dr = -\frac{\pi R^4}{8\mu} \frac{dp}{dz}. \tag{2.67}$$

To determine the pressure gradient with respect to z we note that the conservation of mass for an incompressible fluid requires that

$$\frac{dQ}{dz} = 0. \tag{2.68}$$

From this requirement and Equation (2.67) we learn that the pressure varies linearly with respect to z and that the slope of variation is given by

$$\frac{dp}{dz} = -\frac{p_1 - p_2}{z_2 - z_1} = -\frac{\Delta p}{l}, \tag{2.69}$$

where p_1 and p_2 are fluid pressures at locations z_1 and z_2 within the flow passage. Substituting this result into Equation (2.67) yields the following expression for the fluid flow through the cylindrically shaped passage:

$$Q = \frac{\pi R^4}{8\mu} \frac{\Delta p}{l}. \tag{2.70}$$

This result is commonly known as Poiseuille's law or the Hagen-Poiseuille law and is valid for circular flow passages undergoing low Reynolds number flow.

Annular Leakage Equation. As it turns out, one of the most useful applications of the low Reynolds number flow equation for hydraulic control systems is the annular leakage equation. The left-hand side of Figure 2-7 shows the end view of a circular tube with a round insert. In this figure Poiseuille flow is occurring between the circular tube and the round insert in a direction that is normal to the paper. The insert is located off-center from the tube by the eccentricity dimension e. The dimension r_o locates the outside surface of the insert while the dimension ρ locates the inside surface of the tube.

The equation for the inside surface of the tube may be expressed as follows using the Cartesian coordinate system of Figure 2-7:

$$x^2 + (y - e)^2 = R^2, \tag{2.71}$$

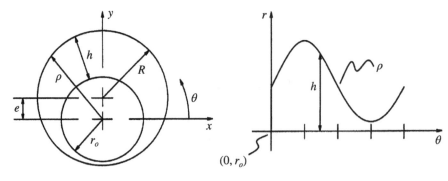

Figure 2-7. Schematic of an annular leak path.

where R is the inside radius of the tube. From the polar coordinate system shown in Figure 2-7 it is clear that

$$x = \rho \, \cos(\theta), \quad y = \rho \, \sin(\theta). \tag{2.72}$$

Substituting Equation (2.72) into Equation (2.71) yields the following quadratic expression that describes the inside surface of the tube as a function of ρ and θ:

$$\rho^2 - (2 \, \sin(\theta) \, e)\rho - (R^2 - e^2) = 0. \tag{2.73}$$

Solving Equation (2.73) for ρ and selecting the only reasonable root of the quadratic equation yields the following expression for the inside radius of the tube with respect to the coordinate system shown in Figure 2-7:

$$\rho = e \, \sin(\theta) + \sqrt{R^2 - (e \, \cos(\theta))^2}. \tag{2.74}$$

For small values of e Equation (2.74) may be approximated as

$$\rho = R + \sin(\theta) \, e. \tag{2.75}$$

The right-hand side of Figure 2-7 shows a plot of the fluid film thickness h between the insert and the tube where $h = \rho - r_o$. This side of the figure may be thought of as a Cartesian version of the radial geometry that is shown on the left. Using the "Cartesian" variables given by θ and $(r - r_o)$ Equation (2.15) may be used to show that the fluid velocity within the gap is

$$u = -\frac{1}{2\mu} \frac{dp}{dz} \, (\rho - r) \, (r - r_o), \tag{2.76}$$

where the no-slip boundary conditions at $r = r_o$ and $r = \rho$ have been applied and where z is the Cartesian coordinate extending out of the paper. Equation (2.76) assumes that the pressure p varies only in the z-direction and that ρ may vary as a function of θ. Once the fluid velocity profile is known the volumetric flow rate through the gap of the annular geometry shown in Figure 2-7 may be generally written as

$$Q = \int_A u \, dA = \int_0^{2\pi} \int_{r_o}^{\rho} u \, r \, dr \, d\theta. \tag{2.77}$$

By substituting the result of Equation (2.75) into Equation (2.76) the solution to Equation (2.77) is given by

$$Q = -\frac{\pi(R - r_o)^3 (R + r_o)}{12\mu} \frac{dp}{dz} \left\{ 1 + 3 \left[\frac{e}{(R - r_o)} \right]^2 \frac{R}{(R + r_o)} + \frac{3}{8} \left[\frac{e}{(R - r_o)} \right]^4 \frac{(R - r_o)}{(R + r_o)} \right\}. \tag{2.78}$$

By inspection it can be that the last term in this equation is negligible. By making the following definition for the eccentricity ratio:

$$\varepsilon = \frac{e}{R - r_o}, \tag{2.79}$$

and realizing that for a thin gap $R + r_o \approx 2R$, the volumetric flow rate through the annular gap of Figure 2-7 may be more concisely expressed as

$$Q = -\frac{\pi R (R - r_o)^3}{6\mu} \frac{dp}{dz} \left(1 + \frac{3}{2} \, \varepsilon^2\right). \tag{2.80}$$

Since the conservation of mass requires that the volumetric flow rate in the z-direction be a constant, it is also clear that the pressure gradient in this direction must also be constant. Therefore, the volumetric flow rate may be further reduced to the following form:

$$Q = \frac{\pi R(R - r_o)^3}{6\mu} \frac{\Delta p}{l} \left(1 + \frac{3}{2}\varepsilon^2\right), \tag{2.81}$$

where Δp is the pressure drop in the z-direction (out of the paper) and l is the length of the flow passage in the z-direction.

Equation (2.81) is the classical annular leakage equation that finds tremendous application in the modeling of leakage flow within hydraulic control systems. The reason for this is because round flow passages are among the easiest geometries to manufacture when building hydraulic equipment. When flow through one of these passages is not desired, as may be the intermittent case of a flow control valve, the easiest way to stop or attenuate the flow is by inserting a solid round device into the circular flow passage for the purposes of blocking the flow. To get the device into the flow passage a clearance must necessarily exist between the inside wall of the flow passage and the round insert.

Furthermore, the round insert and the flow passage are seldom concentric with respect to one another and therefore the eccentricity ratio between these two parts must be considered. From Equation (2.79) it can be shown that the eccentricity ratio must be between 0 and 1 and from Equation (2.81) it may be shown that the eccentricity ratio has a significant impact on the amount of flow that occurs within the gap. For an eccentricity ratio of 1 it can be seen that the flow is 1.5 times as much as would be expected from a perfectly centered case where the eccentricity ration is zero. The user of Equation (2.81) must recall that this result is only valid (strictly speaking) for fully developed, low Reynolds number flow. The Reynolds number itself should be checked using the discussion previously presented in Subsection 2.3.1. For computing the Reynolds number for the annular geometry of Figure 2-7 it may be shown that the hydraulic diameter is closely approximated by $D_h = 2(R - r_o)$.

Case Study

A 25 mm radius piston is designed to fit inside a circular bore with a length of 75 mm. The fluid pressure on one end of the piston is 16 Mpa, while the fluid pressure on the opposite end of the piston is zero gauge pressure. The radial clearance between the piston and the bore is 10 microns, the fluid viscosity is 0.03 Pa-s, and the fluid density is 850 kg/m^3. Assuming that the eccentricity ratio is equal to 1.0 calculate the volumetric flow rate through the gap using the annular leak equation presented in Equation (2.81). Check the Reynolds number to verify that this is indeed a low Reynolds number flow. What is the volumetric flow rate and new Reynolds number when the eccentricity ratio is zero?

Using Equation (2.81) the volumetric flow rate through the annular gap with an eccentricity ratio of 1 may be calculated as

$$Q = \frac{\pi R(R - r_o)^3}{6\mu} \frac{\Delta p}{l} \left(1 + \frac{3}{2}\varepsilon^2\right)$$

$$= \frac{\pi \times 25 \text{ mm} \times (10 \text{ }\mu\text{m})^3}{6 \times 0.03 \text{ Pa s}} \times \frac{16 \text{ MPa}}{75 \text{ mm}} \times \left(1 + \frac{3}{2} \times 1.0^2\right)$$

$$= 0.014 \frac{\text{liters}}{\text{min}}.$$

To verify that this is indeed a low Reynolds number flow Equation (2.51) may be used to show that

$$\text{Re} = \frac{\rho Q D_h}{\mu A} = \frac{850 \text{ kg/m}^3 \times 0.014 \text{ lpm} \times (2 \times 10 \ \mu m)}{0.03 \text{ Pa s} \times (2 \pi \times 25 \text{ mm} \times 10 \ \mu m)} = 0.084.$$

Since the Reynolds number is much less than unity the flow is a low Reynolds number flow and the annular leakage equation used to calculate the volumetric flow rate is valid. If the eccentricity ratio goes to zero for a perfectly centered piston the volumetric flow rate and Reynolds number will decrease by a factor of 2/5. This means that the new volumetric flow rate is 0.0056 lpm and the new Reynolds number is 0.034.

Other Duct Geometry. Figure 2-8 shows a generalized area through which Poiseuille flow may occur. A Cartesian coordinate system is shown in this figure and the flow is assumed to occur in the z-direction (out of the paper).

Using the general form of Equation (2.15) it may be shown that the governing equation for the fluid flow is given by

$$\frac{\partial^2 u}{\partial x^2} + \frac{\partial^2 u}{\partial y^2} = \frac{1}{\mu}\frac{\partial p}{\partial z}. \tag{2.82}$$

As implied in this formulation it is assumed that the fluid velocity does not vary in the z-direction and that the fluid pressure does not vary in the x and y-direction. The conservation of mass will require that the pressure gradient in the z-direction be a constant; therefore, the right-hand side of Equation (2.82) must be a constant as well. In this case Equation (2.82) is in the standard form of Poisson's equation and by applying the no-slip boundary condition to the surface of the arbitrarily shaped duct; that is, u at the surface equals zero, this equation represents the classical Dirichlet problem with many known solutions. Some of these solutions are presented in Figure 2-9 for some of the more common flow passage geometries that exist within hydraulic machinery. In this figure Δp is the pressure drop in the z-direction (out of the paper) and l is the length of the flow passage in the z-direction. A special case for the elliptical geometry shown in Figure 2-9 occurs when a and b are equal. In this case the geometry of the flow passage becomes a perfect circle and the volumetric flow equation reduces to Poiseuille's law given in Equation (2.70).

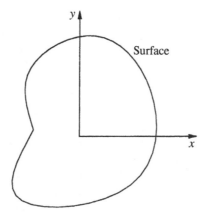

Figure 2-8. An arbitrarily shaped flow passage.

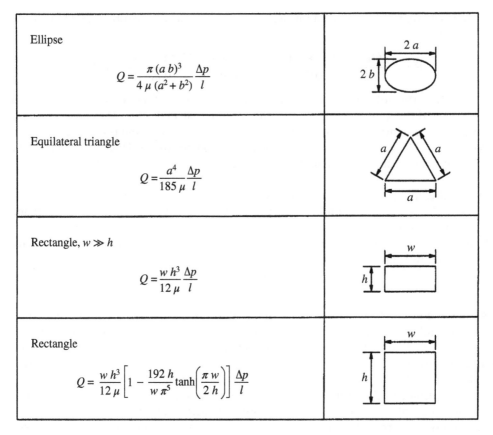

Ellipse $$Q = \frac{\pi (a\,b)^3}{4\,\mu\,(a^2 + b^2)}\,\frac{\Delta p}{l}$$	
Equilateral triangle $$Q = \frac{a^4}{185\,\mu}\,\frac{\Delta p}{l}$$	
Rectangle, $w \gg h$ $$Q = \frac{w\,h^3}{12\,\mu}\,\frac{\Delta p}{l}$$	
Rectangle $$Q = \frac{w\,h^3}{12\,\mu}\left[1 - \frac{192\,h}{w\,\pi^5}\tanh\left(\frac{\pi\,w}{2\,h}\right)\right]\frac{\Delta p}{l}$$	

Figure 2-9. Poiseuille flow through variously shaped flow passages.

2.3.4 Pipe Flow

Fully Developed Flow. The previous subsection has assumed that fluid flow occurs through a pipe with a constant cross-sectional area and of a given length l. The pressure drop across the pipe has been given by the quantity Δp and the fluid flow occurs in the opposite direction of the pressure gradient. Also, in the previous subsection the flow has been assumed to be of the low Reynolds number type. Another assumption that has been used without being explicitly stated is that the fluid flow is fully developed. In this subsection we define what is meant by a fully developed flow for both turbulent and laminar flow regimes.

Figure 2-10 shows the entrance region of a pipe. As the fluid enters the pipe the velocity of the fluid is essentially uniform across the entire cross-section of the flow passage.

As the flow progresses through the pipe the fluid begins to adhere to the internal surfaces thus creating a frictional resistance to the flow that increases for particles that are closer to the internal boundaries. Within the entrance length of the pipe l_e there exists a diminishing core of inviscid flow in which the fluid velocity remains uniform. Once the inviscid core vanishes the flow is said to be fully developed. The entrance length of the pipe l_e that is required to achieve fully developed flow tends to correspond well with the Reynolds number

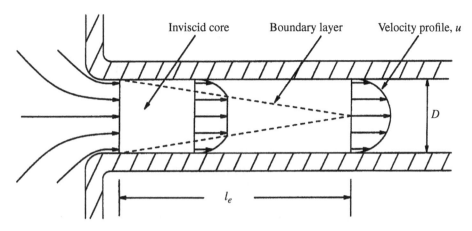

Inviscid core Boundary layer Velocity profile, u

D

l_e

Figure 2-10. Schematic of pipe flow near the entrance region.

for both turbulent and laminar flow. These entrance lengths have been determined through experimentation, which has revealed the following relationships:

$$\frac{l_e}{D} = 0.06 \, \text{Re} \qquad \text{laminar flow}$$
$$\frac{l_e}{D} = 4.40 \, \text{Re}^{1/6} \qquad \text{turbulent flow.} \tag{2.83}$$

The reader will recall that laminar flow is typically characterized by a Reynolds number that is less than 2100, whereas consistent turbulent flow has been observed for Reynolds numbers that are in excess of 4000. For Reynolds numbers between 2100 and 4000, the flow is said to be transitional and may be characterized by either laminar or turbulent conditions.

Dimensional Analysis. Dimensional analysis is the process by which the dimensional variables that describe a physical problem may be combined into dimensionless groups to simplify (and gain insight into) a given problem. This analysis is a powerful tool for creating physical equations when their exact forms are not known from first principles. It is also a valuable tool for reducing the number of parameters that must be measured for experimental validation of a theory. Here we present the concepts behind dimensional analysis and apply these concepts to the subject of pipe flow.

In the early part of the twentieth century, Edgar Buckingham (1867–1940) presented his theorem that demonstrates the key features of dimensional analysis. This theorem is commonly known as the Buckingham Pi Theorem and simply states the following: If there are m dimensional variables describing a physical phenomenon (e.g., density, mass, length, velocity) given by $A_1, A_2, \ldots, A_m$, there exists a functional relationship between these variables such that

$$\psi(A_1, A_2, \ldots, A_m) = 0. \tag{2.84}$$

In other words, this theorem states that if our list of variables $A_1, A_2, \ldots, A_m$ is complete and relevant there exists a solution to the problem in nature. Obviously, this means that we must have some a priori knowledge of the contributing variables to the problem and without this complete set of relevant parameters we cannot successfully conduct the dimensional analysis. Assuming that we do have a complete and relevant list of variables for our problem it may be shown that these variables contain and share basic physical dimensions such as length, time, and mass. If there are n dimensions contained in the list of variables, then the Buckingham Pi Theorem states that there exists a set of $m - n$ dimensionless groups of variables that may be used to describe the same physical problem. These dimensionless groups are given by $\Pi_1, \Pi_2, \ldots, \Pi_{(m-n)}$ and the new functional relationship may be expressed

$$\Psi(\Pi_1, \Pi_2, \ldots, \Pi_{(m-n)}) = 0. \tag{2.85}$$

It should be pointed out that the dimensionless groups that satisfy Equation (2.85) are not unique but certain groups may be preferred for practical and perhaps historical reasons. For instance, in Fluid Mechanics, the Reynolds number is a classical number that has been used for many investigations. Therefore, if at all possible, the Reynolds number should represent a preferred dimensionless group for the investigation of a problem in Fluid Mechanics. Other dimensionless groups may also be identified and used in conjunction with the Reynolds number.

If we consider the flow of fluid through a pipe similar to that shown in Figure 2-10, we may intelligently guess that this flow depends on the following parameters:

Parameter	Symbol	Dimensions
Pressure drop	Δp	$\dfrac{M}{T^2 L}$
Fluid velocity	V	$\dfrac{L}{T}$
Pipe diameter	D	L
Pipe length	l	L
Fluid density	ρ	$\dfrac{M}{L^3}$
Fluid viscosity	μ	$\dfrac{M}{TL}$
Pipe surface roughness	ε	L

where the dimensions M, T, and L are mass, time, and length, respectively. According to this list we have assumed that there are seven parameters that are relevant for describing the problem of fluid flow through the pipe and that there are only three dimensions that are used to describe those parameters; namely, mass, time, and length. Therefore, in this problem $m = 7$ and $n = 3$ and thus the problem can be described using four dimensionless groups. In other words, the physical problem may be described as

$$\psi(\Delta p, V, D, l, \rho, \mu, \varepsilon) = 0, \tag{2.86}$$

or an equivalent description would be

$$\Psi(\Pi_1, \Pi_2, \Pi_3, \Pi_4) = 0. \tag{2.87}$$

By comparing Equations (2.86) and (2.87), one of the strengths of dimensional analysis should be quite evident; namely, that the complexity of the physical problem has been reduced from seven independent variables down to four. From an experimental point of view this is tremendously useful since an investigation of this problem only requires the measurement of four dimensionless groups rather than seven different parameters. To continue our consideration of the fluid flowing through the pipe of Figure 2-10 the following nonunique dimensionless groups are selected:

$$\Pi_1 = \text{Re} = \frac{\rho V D}{\mu}, \qquad \Pi_2 = \frac{\Delta p}{\frac{1}{2}\rho V^2}, \qquad \Pi_3 = \frac{l}{D}, \qquad \Pi_4 = \frac{\varepsilon}{D}. \qquad (2.88)$$

Again, the motivation for selecting the first dimensionless group as the Reynolds number is due to the historical usage of this number in Fluid Mechanics. The selection of the second dimensionless group is based on the ratio of the two basic terms in the Bernoulli equation (Equation (2.9)) and the selection of the last two dimensionless groups is straightforward. From physical intuition we would surmise that the pressure drop through the pipe is perhaps proportional to the length of the pipe itself, which suggests that a wise arrangement of the dimensionless groups is given by

$$\frac{\Delta p}{\frac{1}{2}\rho V^2} = \frac{l}{D} \, \varphi\left(\text{Re}, \frac{\varepsilon}{D}\right). \qquad (2.89)$$

In Equation (2.89) the function φ is unknown and must generally be determined from experiments. This function is normally defined as the friction factor, which may be determined from Equation (2.89) as

$$f = \varphi\left(\text{Re}, \frac{\varepsilon}{D}\right) = \frac{\Delta p}{\frac{1}{2}\rho V^2}\frac{D}{l}. \qquad (2.90)$$

By rearranging Equation (2.90), the pressure drop across the pipe may be expressed as

$$\Delta p = f\frac{1}{2}\rho V^2 \frac{l}{D}. \qquad (2.91)$$

For low Reynolds number flow, this pressure drop may also be determined from Poiseuille's law, as given in Equation (2.70). Rearrangement of Poiseuille's law yields

$$\Delta p = \frac{8\mu Q l}{\pi R^4} = \frac{32\mu V l}{D^2}, \qquad (2.92)$$

where it has been recognized that $Q = \pi R^2 V$ and $D = 2R$. Setting Equation (2.91) and (2.92) equal to each other, the friction factor for low Reynolds number flow may be determined in closed form as

$$f = \frac{64\mu}{DV\rho} = \frac{64}{\text{Re}}. \qquad (2.93)$$

Notice that this result states that the friction factor for low Reynolds number flow is independent of the surface roughness ε. Experimentally, this has proven to be true for low Reynolds number flow; however, surface roughness has shown itself to be important for high Reynolds number flow. These experimental results have been classically presented in the well-known Moody diagram shown in Figure 2-11. Notice in this diagram the friction factor becomes independent of the Reynolds number and strongly dependent on the surface roughness for pipe flow that is characterized by turbulent flow.

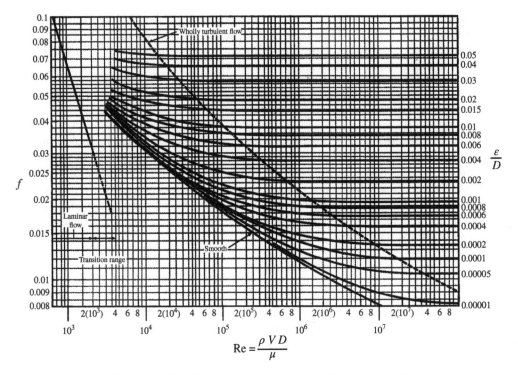

Figure 2-11. The Moody diagram adapted from Munson et al. [1].

2.4 PRESSURE LOSSES

2.4.1 Major Losses

In the previous section the pressure drop in a straight pipe was considered. This pressure drop in a straight section of pipe is normally referred to as a major loss within the hydraulic system and is most generally expressed according to Equation (2.91). Notice that this pressure loss is generally a function of both geometry and the Reynolds number. Though most hydraulic systems do contain longer sections of straight pipe they also usually consist of several shorter and more tortuous sections through which fluid is required to flow. The pressure drop that is experienced in the shorter and more tortuous sections is often referred to as a minor loss and is addressed in the following subsection.

2.4.2 Minor Losses

As previously mentioned, minor pressure losses are those losses that occur through shorter fluid passages that are generally quite tortuous. These losses result from the existence of sharp geometry changes that exist within the system and are typically introduced by the presence of a valve, a hydraulic fitting, or a sharp bend in the flow path. The classification of these losses as being minor is often a misnomer since they are frequently larger than the major losses and must usually be considered when analyzing the power loss and heat generation of the hydraulic control system.

Equation (2.88) shows the dimensionless groups that were used for determining the major pressure losses that occur within a hydraulic system. In this formulation the length of the pipe was used to describe a major geometry feature of the flow path; however, for more twisted geometries the significant geometry characteristics are more difficult to determine. For the consideration of minor losses it is convenient to model the pressure loss using a more general and experimentally dependent expression given by

$$\frac{\Delta p}{\frac{1}{2}\rho V^2} = \varphi(\text{Re}, \text{geometry}). \tag{2.94}$$

This expression should be compared with Equation (2.89), which has been written for the consideration of straight pipe sections within the system. Again, the evaluation of φ must be carried out experimentally for the specific flow geometry being considered. The experimental result of Equation (2.94) is typically referred to as the loss coefficient given by the symbol K_L and the minor pressure loss is given by

$$\Delta p = K_L \frac{1}{2}\rho V^2. \tag{2.95}$$

As shown in Equation (2.94) the loss coefficient is assumed to be dependent on both the Reynolds number and the characteristic geometry features of the flow path. For most of the minor loss flow geometry within hydraulic systems the fluid velocities are quite high due to the restricted flow area associated with a normally fixed volumetric flow rate. This means that the Reynolds number is also high and in such cases an inference may be drawn from the Moody chart shown in Figure 2-11, which states that the loss coefficient itself is more dependent on the geometry of the flow passage as opposed to the Reynolds number. This is shown in Figure 2-11 by the relatively flat lines for the friction factor (loss coefficient) in the high Reynolds number regime. In other words

$$K_L = \varphi(\text{geometry}). \tag{2.96}$$

Figure 2-12 shows four geometry conditions and their associated loss coefficients for fluid that enters a pipe from a large reservoir. In this situation the pressure drop results mainly from the shear stresses along the wall of the geometry change. For more abrupt geometries the loss coefficient is larger than it is for the more gradually changing geometries.

Indeed, simply rounding the edges of the entrance geometry can significantly reduce the loss coefficient. This effect is illustrated in more detail in Figure 2-13 where experimental results for the loss coefficient are reported as a function of the radial geometry at the entrance region of the pipe.

Figure 2-14 shows four geometry conditions and their associated loss coefficients for fluid that exits a pipe and enters into a large reservoir of the same fluid.

In this case the fluid flow is not resisted so much by the shear stress along the surface of an abrupt geometry change as it is by the turbulent mixing of the fluid in the reservoir. This resistance is still a viscous-type loss and since the velocity of the fluid comes to a near rest within the reservoir all of the kinetic energy associated with the fluid is dissipated in the form of frictional heat. In the case of fluid entering a large reservoir the loss coefficient is given by unity.

Minor losses are also experienced in pipe flow where there is a sudden expansion or contraction of the pipe itself. These loss coefficients are shown in Figure 2-15 for both types of geometry change.

Note: The geometry associated with the sudden expansion of the pipe has a closed-form approximation that agrees nicely with experimental data. The geometry associated with

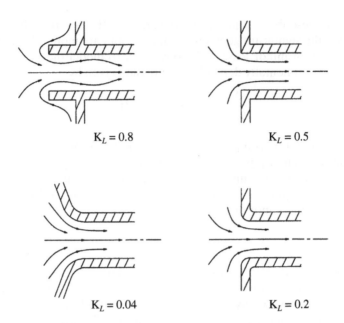

Figure 2-12. Loss coefficients for entrance flow conditions.

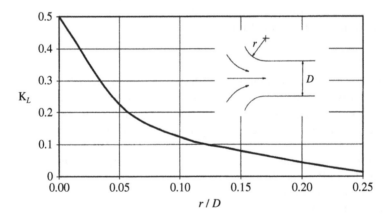

Figure 2-13. Loss coefficients for rounded entrance flow conditions.

the sudden contraction requires experimental determination [1]. In Figure 2-15 it can be seen when $A_1 \ll A_2$, the case is approaching an infinite reservoir condition much like that shown in Figures 2-12 and 2-14. When $A_1 = A_2$ there is no geometry change and so the loss coefficient is zero.

Figure 2-16 shows the loss coefficient associated with a pipe bend as reported by a previous author [2]. In this data it can be shown that the pressure loss increases for larger bends but that a minimum loss for a given bend is achieved for a bend-radius to pipe-diameter ratio of about 4.5. Other loss coefficients that are typically observed within hydraulic systems are given in Table 2-1.

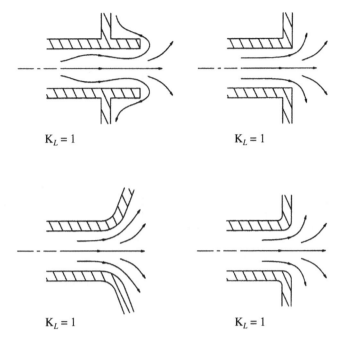

Figure 2-14. Loss coefficients for exit flow conditions.

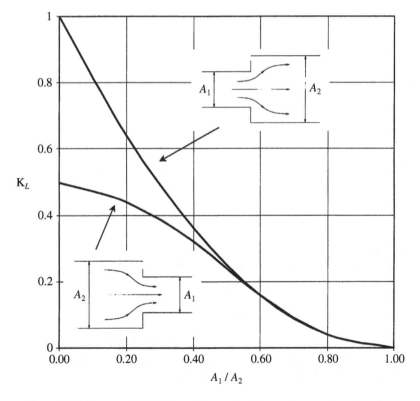

Figure 2-15. Loss coefficients for sudden contractions and expansions in a pipe.

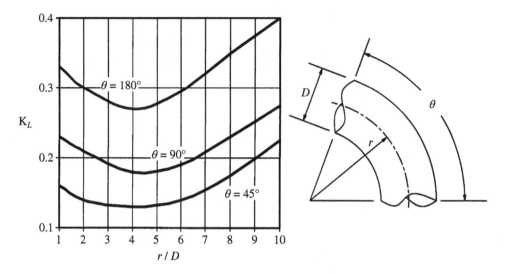

Figure 2-16. Loss coefficients for a pipe bends.

Table 2-1. Typical loss coefficients within hydraulic systems.

Device	K_L	
	Typical value	Typical range
Tee	0.9	0.2–2.0
Diffuser	0.8	0.2–1.2
Check Valve	0.8	0.5–1.5
Gate Valve (fully open)	0.15	0.1–0.3
Gate Valve (1/4 closed)	0.26	–
Gate Valve (1/2 closed)	2.1	–
Gate Valve (3/4 closed)	17.0	–
Ball Valve (fully open)	0.05	–
Ball Valve (1/3 closed)	5.5	–
Ball Valve (1/3 closed)	210.0	–
Butterfly Valve (fully open)	0.2	0.2–0.6
Globe Valve (fully open)	4.0	3.0–10.0

Case Study

The figure shows two identical segments of cylindrical conduit connected using a 180°
elbow. The parameters in this figure are given by

$$Q = 45 \text{ lpm}, \qquad D = 20 \text{ mm}, \qquad L = 1 \text{ m},$$

$$d = 8 \text{ mm}, \qquad r = 32 \text{ mm}.$$

The fluid density is given by 850 kg/m³ and the fluid viscosity is 0.02 Pa-s. The pipe
roughness is 0.5 mm everywhere. Calculate the total pressure drop to be expected
between the inlet and the discharge of the conduit.

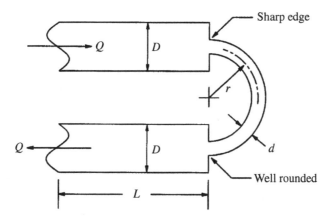

The pressure drop through the straight pipe sections may be calculated using the major pressure drop equation given in Equation (2.91). This equation depends on a friction factor, which generally depends on the pipe roughness and the Reynolds number. Using Equation (2.51) the Reynolds number may be calculated as 2029. This is laminar flow, which means that the friction factor can be calculated as 0.0315 using Equation (2.93). The fluid velocity through the pipe is the volumetric flow rate Q divided by the cross-sectional area of the pipe. This velocity is given by 2.387 m/s. Using these results with Equation (2.91) the total pressure drop through the two straight pipe segments is given by\

$$\Delta p = 2 \times f \frac{1}{2} \rho V^2 \frac{l}{D} = 2 \times 0.0315 \times \frac{1}{2} \times 850 \frac{\text{kg}}{\text{m}^3} \times \left(2.387 \frac{\text{m}}{\text{s}}\right)^2$$
$$\times \frac{1 \text{ m}}{0.02 \text{ m}} = 7.6 \text{ kPa}.$$

In this calculation, the factor of 2 is used to account for both straight segments of the pipe.

Three minor pressure drops exist within the system shown in the figure: a pressure drop at the sharp edge, a pressure drop in the elbow, and a pressure drop at the well rounded edge. From Figure 2-13 the loss coefficient at the sharp edge is given by 0.5. Using Figure 2-16 for $r/d = 4$ the loss coefficient through the 180° elbow is given by 0.275. Using Figure 2-14 the loss coefficient at the well-rounded edge is given by unity. All of these loss coefficients may be added together for a net loss coefficient of 1.775. Using Equation (2.95) and calculating the fluid velocity through the elbow as 14.9 m/s the total pressure drop through the elbow may determined as

$$\Delta p = K_L \frac{1}{2} \rho V^2 = 1.775 \times \frac{1}{2} \times 850 \frac{\text{kg}}{\text{m}^3} \times \left(14.9 \frac{\text{m}}{\text{s}}\right)^2 = 167.5 \text{ kPa}.$$

A comparison of this calculation with the previous one shows that the minor loss is greater than the major loss. The total pressure loss through the system is 175 kPa. At the volumetric flow rate of 45 lpm, it can be shown that this pressure drop results in 132 Watts of energy being dissipated in the form of heat.

2.5 PRESSURE TRANSIENTS

2.5.1 Hydraulic Conduits

General Equations. The transient phenomenon within hydraulic conduits is analyzed by using the unsteady momentum and conservation of mass equations with the fluid equation of state which describes the compressibility of the fluid with respect to pressure. Figure 2-17 shows an infinitesimal control volume within a hydraulic conduit. The total length of the conduit is given by the dimension L and the cross-sectional area of the conduit is given by A.

The infinitesimal length of the control volume is δx. The control volume is fixed in space and hydraulic fluid is shown to pass through the control volume in the x-direction. Due to the short length of the control volume the pressure, density, and fluid velocity are shown to vary linearly across the length of the control volume itself. By applying Equation (2.44) to the control volume of Figure 2-17 the conservation of mass for the hydraulic conduit may be expressed as

$$\frac{\partial}{\partial t} \int_0^{\delta x} \left(\rho + \frac{\partial \rho}{\partial x} x \right) A \, dx - u \rho A + \left(\rho + \frac{\partial \rho}{\partial x} \delta x \right) \left(u + \frac{\partial u}{\partial x} \delta x \right) A = 0. \tag{2.97}$$

Recognizing that the cross-sectional area of the control volume is a constant, and neglecting terms that are quadratically small, the conservation of mass for the control volume shown in Figure 2-17 is given by

$$\frac{\partial \rho}{\partial t} + \frac{\partial (u \rho)}{\partial x} = 0. \tag{2.98}$$

Similarly, by applying Equation (2.48) to the control volume of Figure 2-17 the conservation of fluid momentum for the hydraulic conduit may be expressed as

$$PA - \left(P + \frac{\partial P}{\partial x} \delta x \right) A = \frac{\partial}{\partial t} \int_0^{\delta x} \left(\rho + \frac{\partial \rho}{\partial x} x \right) \left(u + \frac{\partial u}{\partial x} x \right) A \, dx$$

$$- \rho u^2 A + \left(\rho + \frac{\partial \rho}{\partial x} \delta x \right) \left(u + \frac{\partial u}{\partial x} \delta x \right)^2 A. \tag{2.99}$$

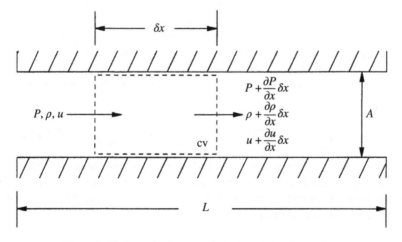

Figure 2-17. Control volume analysis for a hydraulic conduit.

By neglecting quadratically small terms the equation may be simplified as follows:

$$\frac{\partial P}{\partial x} + \frac{\partial(u\rho)}{\partial t} + \frac{\partial(u^2\rho)}{\partial x} = 0. \tag{2.100}$$

From the equation of state for a liquid, changes in density and pressure with respect to distance and time may be written as

$$\frac{1}{\rho}\frac{\partial\rho}{\partial x} = \frac{1}{\beta}\frac{\partial P}{\partial x} \quad\text{and}\quad \frac{1}{\rho}\frac{\partial\rho}{\partial t} = \frac{1}{\beta}\frac{\partial P}{\partial t}, \tag{2.101}$$

where β is the fluid bulk modulus of elasticity.

Nondimensionalized Equations. To gain insight into the dominant physical contributors of the pressure transients with the hydraulic conduit it will be convenient to nondimensionalize the general equations by scaling each physical quantity by a value that characterizes that parameter within the problem. This scaling of parameters is given by

$$P = \hat{P}P_o, \quad u = \hat{u}u_o, \quad \rho = \hat{\rho}\rho_o, \quad x = \hat{x}L, \quad t = \hat{t}\tau, \tag{2.102}$$

where the parameters with carets are dimensionless and of the order one while P_o, u_o, and ρ_o are characteristic pressures, velocities, and densities of the fluid within the problem, L is the length of the conduit shown in Figure 2-17, and τ is a characteristic time scale of the phenomenon occurring within the conduit. By substituting Equation (2.102) into Equation (2.98) the dimensionless mass equation may be written as

$$\frac{\partial\hat{\rho}}{\partial\hat{t}} + \left(\frac{u_o\tau}{L}\right)\frac{\partial(\hat{u}\hat{\rho})}{\partial\hat{x}} = 0. \tag{2.103}$$

Similarly, by substituting Equation (2.102) into Equation (2.100) the dimensionless momentum equation may be written as

$$\frac{\partial\hat{P}}{\partial\hat{x}} + \left(\frac{\rho_o u_o L}{\tau P_o}\right)\frac{\partial(\hat{u}\hat{\rho})}{\partial\hat{t}} + \left(\frac{\rho_o u_o^2}{P_o}\right)\frac{\partial(\hat{u}^2\hat{\rho})}{\partial\hat{x}} = 0. \tag{2.104}$$

Similarly, the nondimensional equation of state may be expressed as

$$\frac{1}{\hat{\rho}}\frac{\partial\hat{\rho}}{\partial\hat{x}} = \frac{P_o}{\beta}\frac{\partial\hat{P}}{\partial\hat{x}} \quad\text{and}\quad \frac{1}{\hat{\rho}}\frac{\partial\hat{\rho}}{\partial\hat{t}} = \frac{P_o}{\beta}\frac{\partial\hat{P}}{\partial\hat{t}}. \tag{2.105}$$

By combining the equation of state with the mass equation the following dimensionless expression may be written for the conservation of mass:

$$\frac{\partial\hat{P}}{\partial\hat{t}} + \left(\frac{u_o\tau}{L}\right)\hat{u}\frac{\partial\hat{P}}{\partial\hat{x}} + \left(\frac{\beta u_o\tau}{P_o L}\right)\frac{\partial\hat{u}}{\partial\hat{x}} = 0. \tag{2.106}$$

Similarly, by combining the equation of state with the momentum equation the following dimensionless expression may be written for the conservation of fluid momentum:

$$\left(\frac{1}{\hat{\rho}} + \frac{\rho_o u_o^2}{\beta}\hat{u}^2\right)\frac{\partial\hat{P}}{\partial\hat{x}} + \left(\frac{\rho_o u_o L}{\tau P_o}\right)\frac{\partial\hat{u}}{\partial\hat{t}} + \left(\frac{\rho_o u_o L}{\tau\beta}\right)\hat{u}\,\frac{\partial P}{\partial\hat{t}} + \left(\frac{\rho_o u_o^2}{P_o}\right)\frac{\partial(\hat{u}^2)}{\partial\hat{x}} = 0. \tag{2.107}$$

If the following ratios of the characteristic dimensions are satisfied

$$\frac{u_o\tau}{L} \ll 1 \ll \frac{\beta}{P_o}, \quad \frac{\rho_o u_o^2}{P_o} \ll 1 \approx \frac{\rho_o u_o L}{\tau P_o}, \tag{2.108}$$

then it can be shown from Equations (2.106) and (2.107) that the dimensionless mass and momentum equations reduce to the following forms:

$$\frac{\partial \hat{P}}{\partial \hat{t}} + \left(\frac{\beta u_o \tau}{P_o L} \right) \frac{\partial \hat{u}}{\partial \hat{x}} = 0 \qquad \text{mass,}$$

$$\frac{\partial \hat{P}}{\partial \hat{x}} + \left(\frac{\rho_o u_o L}{\tau P_o} \right) \frac{\partial \hat{u}}{\partial \hat{t}} = 0 \qquad \text{momentum,} \tag{2.109}$$

where $\hat{\rho}$ has been taken to equal unity. *Note*: Equation (2.109) is valid for situations where Equation (2.108) is satisfied. Under these conditions the fluid velocities are generally small and the transient phenomenon within the hydraulic conduit occurs over a short period of time; that is, u_o and τ are small. This result illustrates the power of nondimensional analysis. Without this type of analysis we would not be able to eliminate unimportant terms in the equation and the complexity of the mass and momentum equations would remain as shown in Equations (2.106) and (2.107).

Equation Summary.   By substituting Equation (2.102) back into Equation (2.109), the mass and momentum equations may be redimensionalized as:

$$\frac{\partial P}{\partial t} \frac{1}{\beta} + \frac{\partial u}{\partial x} = 0 \quad \text{mass,} \qquad \frac{\partial P}{\partial x} \frac{1}{\rho} + \frac{\partial u}{\partial t} = 0 \quad \text{momentum.} \tag{2.110}$$

These equations are more recognizably known as the wave equations where ρ is now treated as a constant within the solution process. Solutions methods for these equations with nonzero boundary conditions have been attempted in previous literature [3].

2.5.2 Water Hammer

Water hammer is the phenomenon that occurs within a hydraulic pipeline when the velocity of the fluid suddenly changes. This sudden change in velocity causes a pressure wave to propagate within the pipe and often creates a noise similar to that of a hammer beating on the pipe itself. Therefore, this phenomenon has been aptly referred to as "water hammer." The previous analysis that was done for generating the wave equations within a hydraulic conduit is generally applicable to the water hammer problem; however, in this section we derive a more "hands on" result for the maximum pressure within the pressure wave using a simple energy balance. This maximum pressure quantity is typically the desired result for water hammer analysis since a reduction in the maximum water hammer pressure corresponds directly to a reduction in the difficulties that pertain to the water hammer phenomenon; namely, loud noises and bursting pipelines. Figure 2-18 shows a column of fluid that is approaching a solid boundary at an initial velocity u_o.

The kinetic energy associated with this motion is given by

$$KE = \frac{1}{2} M u_o^2 = \frac{1}{2} \rho_o V_o u_o^2, \tag{2.111}$$

where M is the mass of the fluid column, ρ_o is the initial fluid density of the column, and V_o is the initial fluid volume of the column. As the fluid column hits the solid boundary the velocity of the fluid is suddenly reduced to zero and the fluid column deforms to absorb the energy associated with the initial motion of the fluid. Since the fluid is nearly incompressible the deformations are small and the energy stored within the fluid during the deformation

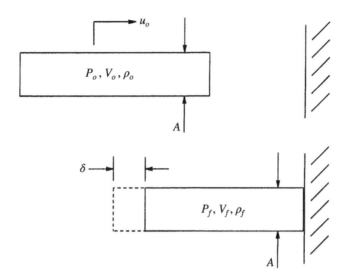

Figure 2-18. A sudden change in the velocity of a fluid column.

process may be treated similar to that of a mechanical spring. The classical expression for the potential energy stored in a mechanical spring is given by

$$PE = \frac{1}{2} k \delta^2, \tag{2.112}$$

where k is the spring constant and δ is the deformation of the spring or fluid column as shown in Figure 2-18. Using Hooke's law the spring rate itself may be written as

$$k = \frac{F}{\delta} = \frac{(P_f - P_o)A}{\delta}, \tag{2.113}$$

where F is the "spring force" acting on the fluid column, P_f is the final pressure of the fluid column at the wall, P_o is the initial pressure of the fluid column, and A is the unchanging cross-sectional area of the fluid column. From the equation of state for a liquid we learn that the change in volume of the fluid column is given by

$$A\delta = (V_o - V_f) \approx \frac{V_o}{\beta}(P_f - P_o), \tag{2.114}$$

where β is the fluid bulk modulus. Using Equations (2.113) and (2.114) with Equation (2.112) the following result may be written for the potential energy stored in the column of compressed fluid:

$$PE = \frac{1}{2} \frac{V_o}{\beta} (P_f - P_o)^2. \tag{2.115}$$

Setting Equations (2.111) and (2.115) equal to each other and rearranging terms yields the following result for the maximum fluid pressure resulting from a sudden stoppage of flow within a traveling column of fluid:

$$P_f = P_o + u_o \rho_o c, \tag{2.116}$$

where c is the speed of sound within the liquid medium given by

$$c = \sqrt{\beta/\rho_o}. \tag{2.117}$$

From Equation (2.116) it may be seen that the maximum pressure due to a water hammer phenomenon is the result of fluid properties and the initial fluid velocity only. This result does not depend on the length of the fluid column and therefore the only real adjustment available to the engineer for reducing water hammer effects is to reduce the initial velocity of the fluid itself. For a given volumetric flow rate this can be done by using a pipe with a larger cross-sectional area A.

2.5.3   Pressure Rise Rates Within a Varying Control Volume

In this section we derive one of the most useful equations to be used in the analysis and modeling of hydraulic control systems. This equation is called the pressure-rise-rate equation. Figure 2-19 shows a control volume filled with a homogeneous fluid undergoing hydrostatic pressurization. In other words, the pressure and fluid density within the control volume is uniform throughout. The piston on the right-hand side of Figure 2-19 is used to adjust the size of the control volume and a volumetric flow rate of fluid given by Q is shown to be entering the control volume from the left.

The instantaneous mass of fluid within the control volume is

$$M = \rho V, \tag{2.118}$$

where ρ is the fluid density and V is the instantaneous volume of fluid within the control volume. By differentiating Equation (2.118) once with respect to time the time-rate-of-change of mass within the control volume may be expressed as

$$\frac{dM}{dt} = \frac{d\rho}{dt}V + \rho\frac{dV}{dt}. \tag{2.119}$$

From the conservation of mass it can be shown that the left-hand side of Equation (2.119) is given by

$$\frac{dM}{dt} = \rho Q, \tag{2.120}$$

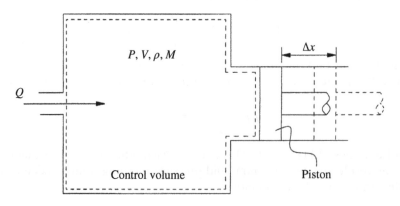

Figure 2-19. Schematic of a varying control volume.

where Q is the total volumetric flow rate of fluid into the control volume. From the equation of state for a liquid we know that

$$\frac{d\rho}{dt} = \frac{\rho}{\beta}\frac{dP}{dt},$$ (2.121)

where β is the fluid bulk modulus. Substituting Equations (2.120) and (2.121) into Equation (2.119) and rearranging terms yields the following equation for the pressure-rise-rate within the varying control volume of Figure 2-19:

$$\frac{dP}{dt} = \frac{\beta}{V}\left(Q - \frac{dV}{dt}\right).$$ (2.122)

This is the pressure-rise-rate equation. This equation shows that the pressure change within the varying control volume is due to the tradeoff between fluid entering and exiting the chamber and the rate of change of the control volume itself. This principle of pressure change is simple and may be illustrated using the crudest examples of a control volume. For instance, consider a balloon. As air is blown into the balloon the pressure in the balloon rises. In response to this pressure rise the balloon expands and changes its volume, thereby tending to reduce the pressure. When the volume can no longer change at a rate proportional to the volumetric flow rate into the balloon the pressure increases dramatically and causes the balloon to break. This basic phenomenon occurs throughout hydraulic circuitry and must be considered everywhere a volume of fluid is being pressurized. The basic form of Equation (2.122) is used frequently in subsequent sections of this book.

Case Study

A new tube of toothpaste is rolled up from the bottom end until the original volume of 30 ml is decreased by 2%. Suddenly, the toothpaste tube breaks and develops a low Reynolds number leak such that toothpaste squirts out of the tube at a volumetric flow rate that is proportional to the toothpaste pressure. The constant of proportionality for the leakage is given by 1.2 ml/(MPa-s) and the bulk modulus for the toothpaste is 90 MPa. Calculate the maximum toothpaste pressure that is achieved. After the toothpaste tube breaks, how long does it take for the toothpaste pressure to reach 98% of its equilibrium value?

Both questions in this case study may be answered using the basic form of Equation (2.122). To determine the maximum pressure that is achieved within the tube of toothpaste as it is being rolled up from the bottom end, the volumetric flow rate in Equation (2.122) is set to zero and the pressure-rise-rate equation is solved in closed-form. This result is

$$P = P_o + \beta \ln\left(\frac{V_o}{V}\right),$$

where P_o is the original toothpaste pressure, β is the toothpaste bulk modulus, and V_o is the original volume of the toothpaste. From the problem statement we know that when the toothpaste tube breaks $V_o/V = 1.02$. Setting the original toothpaste pressure equal to zero we calculate the burst pressure of the toothpaste tube to be 1.8 MPa.

To determine how long it takes for the toothpaste pressure to reach 98% of its equilibrium value after the tube has broken, the volume time-rate-of-change in Equation (2.122) is ignored and the volumetric flow rate is modeled as $Q = -KP$, where K is the proportional coefficient given in the problem statement and P is the instantaneous toothpaste pressure. Using this model with Equation (2.122) the governing differential equation for the transient pressure problem is

$$\frac{V}{K\beta}\dot{P} + P = 0,$$

where the dot notation denotes a derivative with respect to time. The solution to this equation shows that 98% of the transient response dies away when

$$t = 4\frac{V}{K\beta} = 4 \times \frac{0.98 \times 30 \text{ ml}}{1.2 \text{ ml/(MPa-s)} \times 90 \text{ MPa}} = 1.09 \text{ s.}$$

The solution to a first-order differential equation like the one presented here is discussed more extensively in the next chapter.

2.6   HYDRAULIC ENERGY AND POWER

2.6.1   Fluid Power

Power is a concept that involves quantities of both effort and the rate at which effort is applied. For instance, mechanical systems power is defined as force times velocity or torque times angular velocity. For electrical systems power is defined as voltage times current, and for hydraulic systems fluid power is defined as pressure times volumetric flow rate. In this subsection we briefly show this power relationship using the first principles of work.

Figure 2-20 shows a column of fluid that is being forced in the x-direction by the applied force $F = PA$. The cross-sectional area of the fluid column is given by A and the pressure of the fluid at x is given by P. By definition an infinitesimal amount of work being done on the fluid is given by the applied force times an infinitesimal displacement of the fluid.

This definition is expressed mathematically as

$$dW = F \, dx, \tag{2.123}$$

where W is work and F and x are shown in Figure 2-20. Recognizing that $F = PA$ Equation (2.123) may be used to show that the work done on the fluid by the fluid pressure P is given by

$$dW = PA \, dx = P \, dV, \tag{2.124}$$

where dV is an infinitesimal volume of displaced fluid in the x-direction. The definition of power is given by the time-rate-of-change of work. If we divide each side of this equation by

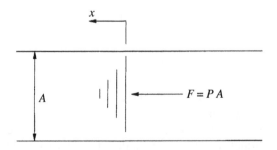

Figure 2-20. Schematic of a fluid column with an applied force.

an infinitesimal amount of time dt the power transmitted hydraulically through the channel shown in Figure 2-20 may be written as

$$\frac{dW}{dt} = P\frac{dV}{dt} = PQ, \tag{2.125}$$

where Q is defined as the volumetric flow rate of the fluid. This result is used extensively in the following chapters to discuss the efficiency of hydraulic components and control systems.

2.6.2   Heat Generation in Hydraulic Systems

Like all systems that perform useful work, hydraulic systems perform their task with an efficiency that is less than 100%. This means that some of the power within a hydraulic control system is diverted away from the primary stream of power and ultimately ends up creating heat that must be dissipated. Within real hydraulic systems this need for heat dissipation results in the presence of coolers and heat transfer devices that accompany the overall control system.

To consider this issue of heat generation and temperature rise, consider Figure 2-21, which shows fluid passing through a restricted orifice.

The restricted orifice creates a pressure drop across the flow passage and the power required to push the fluid through the orifice is given by

$$\dot{W} = Q \ (P_1 - P_2), \tag{2.126}$$

where Q is the volumetric flow rate through the orifice and $P_1 - P_2$ is the pressure drop across the orifice. If it is assumed that all of this energy goes into the fluid and that none of it is dissipated into the surrounding atmosphere of the orifice it may be shown from thermodynamics that the temperature increase across the valve is given by

$$T_2 - T_1 = \frac{\dot{W}}{c_p \dot{m}}, \tag{2.127}$$

where c_p is the specific heat of the fluid and $\dot{m}$ is the mass flow rate of fluid through the orifice. Recognizing that the mass flow rate is given by $\dot{m} = \rho Q$, where ρ is the mass density of the fluid, Equations (2.126) and (2.127) may be used to show that the temperature rise in the fluid passing through the orifice in Figure 2-21 is given by

$$T_2 - T_1 = \frac{P_1 - P_2}{c_p \ \rho}. \tag{2.128}$$

From Table 1-6 of Chapter 1 it may be seen that the specific heat of mineral oil at 80°C is given by 2.118 kJ/(kg °C). The density of mineral oil is approximately 850 kg/m³. Using these two numbers with Equation (2.128) it may be shown that the temperature of hydraulic fluid

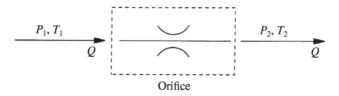

Figure 2-21. Temperature change across a flow restrictor.

will increase by 0.56°C for every MPa of pressure drop that exists across a flow passage, which is the equivalent of 2.55°F for every pressure drop of 1000 psi. Obviously, pressure drops that occur in series throughout a hydraulic system will step the temperature up in the fluid dramatically. This generated heat must eventually be rejected using conductive or convective methods, which is why hydraulic systems are generally equipped with cooling systems.

2.7 LUBRICATION THEORY

General.   In tribological systems fluid lubricants are used to introduce a layer of material between solid sliding surfaces that has a lower shear strength than the surfaces themselves. This allows for ease of sliding without damaging the solid surfaces and often provides an effective bearing characterized by extreme power density; that is, the thin film of fluid can be used to carry a tremendous load. In this section we present the general equations that are used to describe the lubrication phenomenon between two sliding surfaces. These equations are of great practical importance to the design engineer who is trying to make sure that mechanical parts within hydraulic control systems are well lubricated.

Tilted Bearing Pad.   Figure 2-22 shows a tilted bearing pad that is sliding relative to a flat and stationary surface. The velocity of the bearing pad in the x-direction is given by U and the velocity in the negative y-direction is given by V. The Cartesian coordinate system moves with the bearing pad in the x-direction only and the x-axis remains along the flat surface. A fluid with a pressure that varies in magnitude in the x-direction (but not the y-direction) exists between the two surfaces. The fluid film thickness varies in the x-direction and is shown in Figure 2-22 by the dimension h.

To analyze the lubrication conditions of Figure 2-22 we use the conservation of momentum for a low Reynolds number flow and the conservation of mass. Equation (2.19) provides the fluid velocity profile in the x-direction that results from the momentum equation

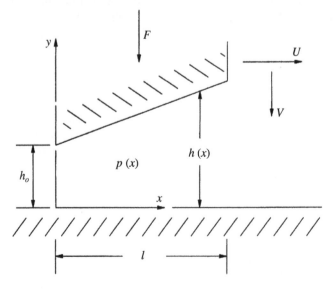

Figure 2-22. Nonflat lubricated surfaces of a bearing pad.

for a low Reynolds number flow. Using the geometry of Figure 2-22 this result may be expressed as

$$u = -\frac{1}{2\mu}\frac{dp}{dx}(h-y)\,y + U\frac{y}{h},\tag{2.129}$$

where μ is the fluid viscosity and both the pressure and fluid film thickness vary in the x-direction. Using this result the volumetric flow rate in the x-direction (per unit width) may be written as

$$\int_0^h u\,dy = -\frac{1}{12}\frac{dp}{\mu\,dx}h^3 + \frac{U}{2}h.\tag{2.130}$$

For incompressible flow the conservation of mass for two dimensional flow is given by

$$\frac{\partial u}{\partial x} + \frac{\partial v}{\partial y} = 0,\tag{2.131}$$

where v is the fluid velocity in the y-direction [4]. For the analysis of this subsection it is more convenient to write the conservation of mass in the following form:

$$\int_0^h \left(\frac{\partial u}{\partial x} + \frac{\partial v}{\partial y}\right)dy = \int_0^h \frac{\partial u}{\partial x}dy + \int_0^h \frac{\partial v}{\partial y}dy = 0.\tag{2.132}$$

From calculus we know that

$$\int_0^h \frac{\partial}{\partial x}g(x,y)\,dy = \frac{\partial}{\partial x}\int_0^h g(x,y)\,dy - (g|_{y=h})\frac{\partial h}{\partial x}.\tag{2.133}$$

Using this general form it can be shown that

$$\int_0^h \frac{\partial u}{\partial x}\,dy = \frac{\partial}{\partial x}\int_0^h u\,dy - U\frac{\partial h}{\partial x} \quad\text{and}\quad \int_0^h \frac{\partial v}{\partial y}\,dy = -V.\tag{2.134}$$

Substituting these results into Equation (2.132) produces the following form for the conservation of mass:

$$\frac{\partial}{\partial x}\int_0^h u\,dy = U\frac{\partial h}{\partial x} + V.\tag{2.135}$$

Combining this result with Equation (2.130) yields the following governing equation for the lubrication problem shown in Figure 2-22:

$$\frac{\partial}{\partial x}\left(-\frac{1}{12\mu}\frac{dp}{dx}h^3\right) = \frac{U}{2}\frac{\partial h}{\partial x} + V.\tag{2.136}$$

By integrating this result once with respect to x it may be shown that

$$\frac{dp}{dx} = -\frac{6U\mu}{h^2} - \frac{12\mu V(x+C_1)}{h^3},\tag{2.137}$$

where C_1 is a constant of integration. Since h can generally vary as a function of x, its form must be known before proceeding further. For simplicity a first approximation of the nonconforming surfaces that are shown in Figure 2-22 may be given by a linear variation of the fluid film thickness. An expression for this would be

$$h = h_o + mx, \tag{2.138}$$

where h_o is shown in Figure 2-22 and m is the linear slope of variation. For a perfectly flat condition $m = 0$. Substituting Equation (2.138) into Equation (2.137) and integrating yields the following result for the fluid pressure between the two surfaces:

$$p(x) = \frac{6\,U\mu}{m(h_o + mx)} + \frac{6\,V\mu[h_o + m(C_1 + 2x)]}{[m(h_o + mx)]^2} + C_2. \tag{2.139}$$

By applying the following boundary conditions:

$$p(0) = P_o \qquad \text{and} \qquad p(l) = P_l, \tag{2.140}$$

The constants of integration in Equation (2.139) may be determined as

$$C_1 = \frac{h_o^2(h_o + lm)^2\ (P_o - P_l)}{6l\ (2h_o + lm)\,V\mu} - \frac{h_o\,l}{2h_o + lm} - \frac{h_o(h_o + lm)\ U}{(2h_o + lm)\ V},$$

$$C_2 = -\frac{h_o^2\ (P_o - P_l)}{lm\ (2h_o + lm)} - \frac{6\ (mU + 2V)\mu}{m^2\ (2h_o + lm)} + P_l. \tag{2.141}$$

Substituting these results back into Equation (2.139) yields the following result for the fluid pressure:

$$p(x) = \frac{x\,[h_o^2(P_l - P_o)(2h_o + mx) + 6l(mU + 2V)\ (l - x)\mu]}{l\ (2h_o + ml)\ (h_o + mx)^2}$$

$$+ \frac{h_o^2 P_o + mx(2h_o + mx)P_l}{(h_o + mx)^2}. \tag{2.142}$$

The load carrying capacity of the fluid between the two surfaces shown in Figure 2-22 is given by

$$F = \int_0^l p(x)\ dx = \frac{6(m\,U + 2\,V)\mu}{m^3}\,\ln\left(1 + \frac{lm}{h_o}\right)$$

$$+ \frac{l[lm^3 P_l + h_o m^2(P_l + P_o) - 12\ (m\,U + 2\,V)\mu]}{m^2(2h_o + lm)}. \tag{2.143}$$

Equation Summary.   The previous equations describe the pressure and load characteristics for the tilted bearing pad shown in Figure 2-22. For most lubrication problems, the tilt of the bearing pad is very small and this effect can therefore be considered by examining the first and second terms of the Taylor series written for small values of the slope of variation m. Using this technique the pressure and load carrying capacity for small values of m may be expressed as shown in Equations (2.144) and (2.145).

From these equations it can be seen that three types of lubrication are evident in the lubrication process. The first type is called hydrostatic lubrication and is shown to be independent of any velocity effects. This type of lubrication is therefore a result of the boundary pressures that tend to force fluid through the gap as shown in Figure 2-22. The flow that results from the hydrostatic pressure gradient is called Poiseuille flow and in the case of flat surfaces this lubrication term becomes independent of the fluid film thickness h_o between the two plates. Even so, it must be kept in mind that a certain fluid film thickness must be present for the results to be valid.

$$p(x) = P_o + \underbrace{\frac{[2\,h_o + 3\,m\,(l-x)]\,x}{2\,h_o\,l}\,(P_l - P_o)}_{\text{Hydrostatic}} + \underbrace{\frac{3\,m\,(l-x)\,x\,\mu}{h_o^3}\,U}_{\text{Hydrodynamic}}$$

$$+ \underbrace{\frac{3(l-x)[2h_o - m(l+4x)]x\mu}{h_o^4}\,V,}_{\text{Squeeze Film}} \tag{2.144}$$

$$F = \underbrace{\frac{l}{2}(P_l + P_o) + \frac{l^2 m}{4\,h_o}(P_l - P_o)}_{\text{Hydrostatic}} + \underbrace{\frac{l^3 m\,\mu}{2\,h_o^3}\,U}_{\text{Hydrodynamic}} + \underbrace{\frac{l^3(2\,h_o - 3\,lm)\mu}{2\,h_o^4}V.}_{\text{Squeeze Film}} \tag{2.145}$$

The second type of lubrication shown in Equations (2.144) and (2.145) is called hydrodynamic lubrication and is shown to be directly proportional to the relative sliding velocity between the two surfaces U. Flow generated by the relative sliding velocity is called Couette flow and results from dragging fluid into the gap that is attached to the surfaces according to the no-slip boundary conditions that exist there. For this type of lubrication to be effective it may be seen from the governing equations that the surfaces must be nonflat; that is, $m \neq 0$. It should also be noted that this lubrication type depends heavily on the fluid film thickness h_o and that for small values of the parameter the hydrodynamic lubrication term can become quite large.

Finally, the third type of lubrication shown in Equations (2.144) and (2.145) is known as squeeze film lubrication. This lubrication term results from the relative downward velocity that exists between the surfaces, which creates a squeezing effect on the fluid film itself. Notice: this type of lubrication does not depend on a nonflat surface but always remains heavily dependent on the fluid film thickness h_o. Since downward motions of the tilted bearing pad shown in Figure 2-22 are only transient occurrences the squeeze film lubrication effect is only a temporary contribution to the lubrication process and therefore cannot be depended on for steady lubrication of the bearing interface.

2.8 CONCLUSION

This chapter attempts to provide the reader with a basic overview of the most fundamental concepts in Fluid Mechanics that are germane to hydraulic control systems. As such, this chapter falls far short of an exhaustive treatment of this subject, and for a more in-depth treatment of Fluid Mechanics the reader is referred to full texts on the subject [1, 4]. In the chapters that follow many of the concepts presented here will be heavily relied on. For instance, the Reynolds Transport Theorem are relied on to conduct repetitive analysis on various valve types and to determine the fluid forces that must be overcome in order to actuate the valve. The classical orifice equation is used extensively to model both large and small scale flow and the low Reynolds number flow equations are the basis for modeling leakage phenomenon within the hydraulic control system. Again, the pressure-rise-rate equation is used on multiple occasions to evaluate transient pressure phenomenon within hydraulic control systems that have a tendency to create instability. In short, the subject of Fluid Mechanics plays a basic role in understanding the behavior of a hydraulic control system and the engineer finds the topics covered in this chapter to be a valuable resource for improving the performance of these systems.

2.9 REFERENCES

[1] Munson, B. R., D. F. Young, and T. H. Okiishi. 1998. *Fundamentals of Fluid Mechanics*, 3rd ed. John Wiley & Sons, New York.

[2] Merritt, H. E. 1967. *Hydraulic Control Systems*. John Wiley & Sons, New York.

[3] Blackburn, J. F., G. Reethof, and J. L. Shearer. 1960. *Fluid Power Control*. The Technology Press of the Massachusetts Institute of Technology and John Wiley & Sons, New York.

[4] Currie, I. G. 1993. *Fundamental Mechanics of Fluids*, 2nd ed. McGraw-Hill, St. Louis.

2.10 HOMEWORK PROBLEMS

2.10.1 Governing Equations

2.1   You are interested in modeling the fluid flow that occurs through a long pipe that is 40 mm in diameter. The average velocity of the fluid through the pipe is expected to be 1 m/s. The fluid is hydraulic oil with an absolute viscosity of 0.02 Pa-s and a density of 850 kg/m^3. Calculate the Reynolds number and decide whether this is a high or low Reynolds number flow. Will this flow be turbulent or laminar? What is the equation that will govern the fluid motion in this case?

2.2   For the preceding problem, say that you are interested in modeling the fluid flow over a 20 ms time period. In this case, can the transient effects of the fluid flow be neglected? What is the equation of motion for the fluid in this case? Can this problem be solved by hand? Explain your answer.

2.3   A fire hose nozzle has an inlet area given by A_i and discharge area given by A_d. A 25 psi pressure drop exists across the nozzle. Assuming that the flow rate is of the high Reynolds number type, what is the area ratio A_d/A_i that will produce a discharge velocity of 100 ft/s? What is the inlet velocity of the nozzle?

2.4   Using the information of Problem 2.3 and the Reynolds Transport Theorem, determine the force required to hold the nozzle in place. Assume that the flow is steady and that the inlet area to the nozzle is 1.5 in^2.

2.10.2 Fluid Flow

2.5   A 1 mm × 0.5 mm rectangular-shaped orifice is used to throttle hydraulic fluid that is flowing from a 20 MPa pressurized hose to a reservoir at atmospheric pressure. Using the classical orifice (equation with a discharge coefficient of 0.62 and a fluid density of 850 kg/m^3), calculate the volumetric flow rate through this passage. What is the Reynolds number that characterizes this flow rate assuming that the fluid viscosity is 0.01 Pa-s? Is this flow laminar or turbulent? Does the orifice equation apply? Why or why not?

2.6   Using Problem 2.5 calculate the volumetric flow rate that would be characteristic of a Reynolds number equal to 0.05 (a low Reynolds number). Assuming that the channel of flow is 10 mm long, what is the pressure drop across the channel?

2.7   Using dimensional analysis, you are to develop an expression for the volumetric flow rate of fluid flowing through an orifice at a high Reynolds number. The relevant list of physical parameters for this exercise are: the volumetric flow rate Q, the pressure drop across the orifice Δp, the cross-sectional area of the orifice A_o, the cross-sectional area of the vena contracta A_v, and the fluid density ρ. How many dimensionless groups may be written to describe this problem? How does your result compare to the classical orifice equation?

2.10.3 Fluid Pressure

2.8   The figure shows a section of piping that is used to convey fluid at a volumetric flow rate of 45 gpm. The pipe diameter is 1 inch and the following length dimensions apply: $L_1 = 36$ in, $L_2 = 24$ in, and $L_3 = 18$ in. The mean bend radius of the pipe is 2 inches while the gate valve is half closed. The surface roughness of the pipe is 0.05 inches, the fluid density is 53 lbm/ft^3, and the fluid viscosity is 3 micro-Reyn. Calculate the pressure drop through the pipe.

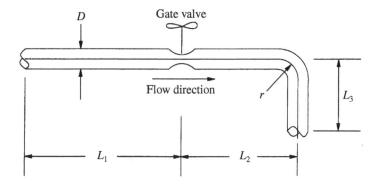

2.9   A 20 liter container is completely filled with hydraulic fluid and is initially pressurized at 2100 kPa. An accident smashes one side of the container creating a permanent deformation that reduces the volume by 4%. Assuming that the fluid bulk modulus is 1 GPa what is the new pressure in the container?

2.10   Using the results of Problem 2.9 calculate how much time it takes for the container to reach atmospheric pressure if the container develops a slow leak after being smashed. Assume that the leakage coefficient is 2×10^{-13} m^3/(Pa-s).

2.10.4 Fluid Power

2.11   For Problem 2.8, calculate the amount of power that is lost due to the pressure drop through the pipe assembly. Assuming that all of this power goes into heating the hydraulic fluid, estimate the temperature rise in the hydraulic fluid.

2.12   A hydraulic pump produces a volumetric displacement of 55 cm^3/rev and is driven at 2100 rpm. If the intake pressure is zero gauge pressure and the discharge pressure is 30 MPa calculate the hydraulic power that is delivered by this pump. If the pump is 100% efficient what is the input torque on the pump shaft?

3

DYNAMIC SYSTEMS AND CONTROL

3.1 INTRODUCTION

The subject matter of this text is hydraulic control systems. As the name implies, hydraulic control systems are fluid-powered systems that are designed to perform a specified control objective. As previously stated in the introduction of this book, hydraulic control systems have gained their place in the market primarily because of their unique dynamic characteristics. In particular, the favorable dynamic response of a hydraulic control system has aptly been referred to as "stiff" because of the tremendous effort that these systems are capable of producing with only minor inertial resistance and overshoot. In this chapter the basics of dynamic analysis and classical control theory are reviewed and summarized with an eye toward characterizing the dynamic response of hydraulic control systems in subsequent chapters. The following sections of this chapter include a brief overview of modeling of dynamical systems, linearization of these systems, and a discussion of first- and second-order dynamical system behavior. These sections are followed by three methods of analysis that may be employed for considering the characteristics of dynamic systems: state space analysis, block diagrams and the Laplace transform, and frequency response analysis. The concluding portion of this chapter is used to present the topic of basic feedback control. In this text the methods of feedback control include forms of controllers designed using continuous time domain, frequency domain, discrete time domain, and state space methods. This chapter serves as a reference throughout the remainder of the text as conventional response characteristics are defined and as basic tools of dynamical analysis are employed in subsequent chapters.

3.2 MODELING

3.2.1 General

To discuss the subject of control there is the presupposed existence of something that needs to be controlled. That something is generally referred to as the "plant" or the dynamic system through which the control objective is achieved. Without a rudimentary understanding of the nature and expected behavior of this dynamic system the subject of control becomes haphazard to say the least. As engineers we gain the understanding of dynamic systems through either experimental or analytical techniques of modeling. In this section we discuss the basics of modeling as it pertains to some relatively simple systems. These modeling examples are presented to illustrate analytical techniques and the basic structures of time varying systems that may be encountered within hydraulic control systems.

3.2.2 Mechanical Systems

One of the most classical systems to be discussed in a dynamics textbook is the swinging pendulum shown in Figure 3-1. Using the basic principles of rigid body mechanics the equation of motion for this pendulum may be determined by summing moments about the pivot point O and setting them equal to the time-rate-of-change of angular momentum for the moving body about this point. This result is written as

$$I_o \ddot{\theta} = T - C\dot{\theta} - Mgl\sin(\theta), \tag{3.1}$$

where I_o is the mass moment of inertia of the pendulum about point O, T is an external couple that is applied to the pendulum perhaps for the purposes of controlling its angular position, C is a viscous drag coefficient, M is the mass of the pendulum, g is the gravitational constant, and the dimensions l and θ are shown in Figure 3-1. This equation describes the physical behavior of the pendulum as it responds to the input effort T. This equation is also called the model of the system or plant as shown in Figure 3-1. Furthermore, although Equation (3.1) appears to be fairly benign at first glance, this equation is nonlinear in θ and generally requires a numerical technique for its solution. We discuss ways of simplifying this model in the next section of this chapter for determining an approximate closed-form

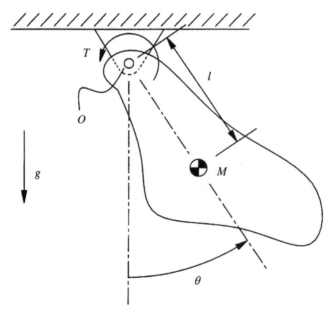

Figure 3-1. The swinging pendulum.

solution that remains valid within restricted bounds of pendulum operation. Putting the details of this model aside for a moment the reader should appreciate the fact that hydraulic control systems are made of moving mechanical parts and that they ultimately interface with a mechanical output. From this perspective the modeling of mechanical systems like the one shown in Figure 3-1 becomes very important for the subject of this text.

3.2.3 Hydromechanical Systems

Figure 3-2 shows another dynamical subsystem that is typically found within hydraulic control systems. This system is a simple hydraulic actuator that utilizes pressurized fluid to displace a piston. The pressurized fluid is resident within the chamber on the left-hand side of the piston while a spring and an external force F are applied to the right-hand side of the piston. As the force is applied fluid is controllably injected into the chamber at a volumetric flow rate given by Q and may be withdrawn from the chamber by making Q negative.

The instantaneous volume of this chamber is given by

$$V = V_o + Ax, \tag{3.2}$$

where V_o is the volume of the chamber when x equals zero and A is the cross-sectional area of the piston. Using this result the pressure-rise-rate equation from Chapter 2 may be written for the chamber as

$$\dot{P} = \frac{\beta}{V_o + Ax}\,(Q - C_l P - A\dot{x}), \tag{3.3}$$

where P is the chamber pressure, β is the fluid bulk modulus of elasticity, and C_l is a coefficient of leakage. Next, summing forces on the piston and setting them equal to the piston's time rate of change of linear momentum yields the following equation of motion for the piston:

$$M\ddot{x} = PA - C\dot{x} - kx - F, \tag{3.4}$$

where M is the mass of the piston, C is the viscous drag coefficient, k is the spring rate, and F is an applied force by a load which may be acting on the piston. Equations (3.3) and (3.4) are used to describe the transient behavior of the physical system shown in Figure 3-2. From these equations it can be seen that the system is a third-order system with a nonlinear coupling that occurs between the pressure, velocity, and displacement of the piston. Again, this

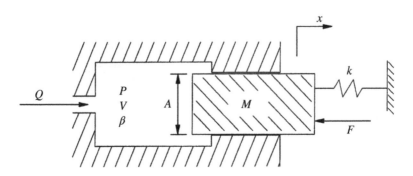

Figure 3-2. Hydraulic actuator.

system generally requires numerical methods to obtain a solution to the coupled differential equations. In the following section of this chapter these equations are linearized about a nominal operating point for the purposes of obtaining a solution within a restricted range of system operation.

3.2.4 Electromechanical Systems

Within modern hydraulic control systems an electronic interface is invariably used to provide the input signal for adjusting the controlled system output. Solenoids and torque motors are frequently used for such functions. Figure 3-3 shows an electromagnetic actuator that is used to displace an armature, which then acts against a mechanical system. The wire coil applies an electromagnetic force to the armature, which is proportional to the electric current flowing through the wire.

The constant of proportionality for the electromagnetic force is given by the magnetic coupling coefficient

$$\alpha = 2\pi r N B, \tag{3.5}$$

where r is the mean coil radius, N is the number of wire wraps that are used in the coil, and B is the magnetic flux-density of the armature. Summing forces on the armature and setting them equal to the armature's time rate of change of linear momentum yields the following governing equation for the motion of the armature:

$$M\ddot{x} = \alpha i - C\dot{x} - kx - F, \tag{3.6}$$

where M is the mass of the armature, i is the current in the coil, C is a viscous damping coefficient, k is the spring rate of the spring that resists the motion of the armature, and F is the force of the load that is being moved by the actuator. The electrical circuit for the coil is shown in Figure 3-4. The circuit shows the applied voltage across the coil v, the coil inductance and resistance L and R, and the induced electromagnetic force that results from the motion of the armature $e = \alpha\dot{x}$.

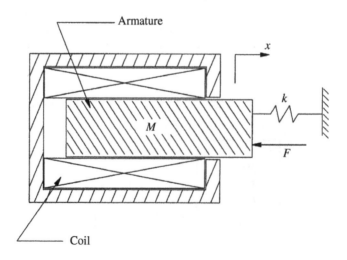

Figure 3-3. Electromagnetic actuator.

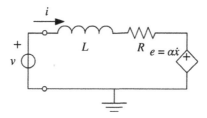

Figure 3-4. Electrical circuit for the wire coil shown in Figure 3-3.

Applying Kirchhoff's voltage law the following differential equation may be written to describe the time variation of the electrical current through the wire:

$$L\dot{i} + Ri = v - \alpha\dot{x}. \tag{3.7}$$

Equations (3.6) and (3.7) are used to describe the transient behavior of the physical system shown in Figure 3-3. From these equations it can be seen that this system is a third-order system coupled through the magnetic coupling coefficient α. Unlike the previous two examples, this system of equations is linear and may be solved without the use of numerical methods.

3.2.5 Summary

As previously stated, modeling of a physical system is a prerequisite for controlling the system intelligently. From the previous examples it is clear that the physical models depend on the type of system being controlled and that these models are generally of different orders with potential nonlinear characteristics. As the order of the model increases, or as the nonlinear components propagate, the complexity of the physical system grows, which makes the control problem more formidable. In the following section we describe the method of linearization, which may be used to simplify the model within a restricted domain of operation. In a subsequent section we also show that our understanding of the first- and second-order model is far superior to that of higher-order systems and therefore a legitimate reduced-order model of the first- or second-order type can be used to provide more physical insight to the control problem and the transient behavior of the system. In short, we may say that the first- and second-order linear models are among the class of greatest use to us and therefore throughout this book we attempt to represent our physical systems in this way whenever possible. Of course, oversimplified systems are of limited use as well and therefore we only recommend this approach for situations in which the impact of linearization and order reduction is well understood.

3.3 LINEARIZATION

3.3.1 General

Linearization is the technique for making nonlinear governing equations of a system linear and reasonably representative of the original set of equations that have been generated to describe the system. This technique is particularly useful for hydraulic control systems as these systems contain a higher degree of nonlinearity than most dynamical systems. Before discussing techniques of linearization, it will be helpful to present and illustrate what is meant by a linear versus a nonlinear equation.

A linear equation with n unknowns is an equation of the following form:

$$f(x_1, x_2, \ldots, x_n) = a_1 x_1 + a_2 x_2 + \ldots + a_n x_n - b_1, \tag{3.8}$$

where a_1, a_2, $\ldots$, a_n, and b_1 are real numbers and x_1, x_2, $\ldots$, x_n are variables. For instance, the equation $f(x_1) = \sin(x_1)$ is a nonlinear equation since it doesn't appear in the form of Equation (3.8). As is illustrated shortly, for small values of x_1 the function $\sin(x_1) \approx x_1$, which is a linear equation of the form shown in Equation (3.8). Again, mathematical expediency is the motivation for using a linear equation as opposed to a nonlinear equation. This expediency arises from our comprehensive knowledge of linear equations whereas nonlinear equations are much more difficult for us to solve and handle. In this section we demonstrate the method used for linearizing nonlinear equations.

3.3.2 The Taylor Series Expansion

The Taylor series is a power series that is used as the mathematical tool for linearizing a nonlinear equation having at least one derivative with respect to the independent variable x. In general, the Taylor series may be written as

$$f(x) = \sum_{k=0}^{\infty} \frac{f^k(x_o)}{k!} (x - x_o)^k$$

$$= f(x_o) + f'(x_o)(x - x_o) + \frac{f''(x_o)}{2!}(x - x_o)^2 + \ldots + \frac{f^n(x_o)}{n!}(x - x_o)^n + \ldots, \tag{3.9}$$

where x_o is the nominal value of the independent variable, ! is the factorial operator given by $n! = 1 \cdot 2 \cdot 3 \cdot \ldots \cdot n$, and the prime notation denotes differentiation with respect to the independent variable x. When x_o is zero, the Taylor series becomes known as the Maclaurin series. Clearly, when the independent variable x remains in the vicinity of the nominal value x_o the higher-order terms of the Taylor series become diminishingly small. In other words, $(x - x_o)^2 > (x - x_o)^3 > \ldots > (x - x_o)^n$. This means that if the second-order term $(x - x_o)^2$ is sufficiently small then the higher-order terms of the Taylor series can been reasonably neglected and the Taylor series expansion of Equation (3.9) becomes

$$f(x) \approx f(x_o) + f'(x_o)(x - x_o). \tag{3.10}$$

As one can see by comparison with Equation (3.8) this form of the Taylor series is linear by definition; however, it only remains as a reasonable representation of the original function $f(x)$ as long as x does not deviate too far from x_o. When the function f depends on more than one independent variable, the Taylor series for this function is given by

$$f(x_1, x_2, \ldots, x_n) \approx f(x_{1_o}, x_{2_o}, \ldots, x_{n_o}) + \left.\frac{\partial f}{\partial x_1}\right|_o (x_1 - x_{1_o}) + \left.\frac{\partial f}{\partial x_2}\right|_o (x_2 - x_{2_o}) + \ldots$$

$$+ \left.\frac{\partial f}{\partial x_n}\right|_o (x_n - x_{n_o}), \tag{3.11}$$

where the partials are shown to be evaluated at the nominal conditions of the independent variables $x_{1_o}, x_{2_o}, \ldots, x_{n_o}$ and deviations from these values are considered to be small.

3.3.3 Examples of Linearization

Mathematical Example. To illustrate the use of the Taylor series expansion for the linearization of a function, consider the nonlinear tangent function given by

$$f(x) = \tan(x). \tag{3.12}$$

To linearize this function in the vicinity of x_o, we can use the general form of Equation (3.10) to show that

$$f(x) \approx \tan(x_o) + \sec^2(x_o)\,(x - x_o). \tag{3.13}$$

Within dynamical systems it is very common to linearize trigonometric functions for "small values" of the argument. This language means that the function is evaluated as a Maclaurin series or in other words x_o is so small that it is considered to be zero. For the case shown here one may see that "for small values of x"

$$f(x) = \tan(x) \approx x. \tag{3.14}$$

Similarly, the student should gain confidence in this technique by showing that the following Maclaurin series can be written for both the sine and cosine function:

$$\sin(x) \approx x, \quad \cos(x) \approx 1. \tag{3.15}$$

Mechanical System Example. The swinging pendulum shown in Figure 3-1 was mathematically modeled using the following equation:

$$I_o\ddot{\theta} = T - C\dot{\theta} - Mgl\sin(\theta), \tag{3.16}$$

where I_o is the mass moment of inertia of the pendulum, T is a couple that is applied to the pendulum, C is a viscous drag coefficient, M is the mass of the pendulum, g is the gravitational constant, and the dimensions l and θ are shown in Figure 3-1.

As previously mentioned, this equation does not have a straightforward solution in its present form due to the nonlinearity of the sine function. For this reason numerical solutions are often sought for this simple system; however, a closed-form solution utilizing the Jacobian elliptic function of the first kind may be determined for this system assuming that energy is conserved; that is, $C = 0$. For the derivation of this result see Greenwood [1]. While the nonlinear form of Equation (3.16) is somewhat problematic a linear form may be determined by using the linearization technique that we have previously described. If we assume, for instance, that the pendulum does not deviate far from rest then θ will be small and the Maclaurin series for the sine function may be written as

$$\sin(\theta) \approx \theta. \tag{3.17}$$

Using this result with Equation (3.16), the following linear equation, which describes the motion of the pendulum in the vicinity of $\theta = 0$, may be written:

$$I_o\ddot{\theta} = T - C\dot{\theta} - Mgl\theta. \tag{3.18}$$

Due to our strong understanding of linear systems, this equation poses no formidable problem for finding a closed-form solution and is therefore highly preferred over Equation (3.16). The solution to this second-order linearized system is discussed in subsequent sections.

Hydromechanical System Example.   The governing equations for the hydromechanical system shown in Figure 3-2 were presented in Equations (3.3) and (3.4). The nonlinearity of these equations results from the time varying volume as shown in the pressure-rise-rate equation. For convenience this equation is rewritten here as

$$\dot{P} = \frac{\beta}{V_o + Ax} (Q - C_l P - A\dot{x}), \tag{3.19}$$

where V_o is the volume of the chamber when x equals zero, A is the cross-sectional area of the hydraulic piston, P is the chamber pressure, β is the fluid bulk modulus of elasticity, and C_l is a coefficient of leakage. If we linearize the right-hand side of this equation with respect to x, Q, P, and $\dot{x}$, it may be shown that

$$\dot{P} = \frac{\beta}{V_o} (Q - C_l P - A\dot{x}), \tag{3.20}$$

where the nominal values for the linearization are given by $x = \dot{x} = 0$, $Q = C_l P_o$, and $P = P_o$. The pressure-rise-rate equation is now in a linear form and may be solved using standard linear solution techniques. Since the momentum equation for this system (Equation 3.4) is already in linear form the total system may be linearly described as:

$$\dot{P} = \frac{\beta}{V_o} (Q - C_l P - A\dot{x}), \quad M\ddot{x} = PA - C\dot{x} - kx - F, \tag{3.21}$$

where M is the mass of the piston, C is the viscous drag coefficient, k is the spring rate, and F is an applied force by a load which may be acting on the piston. Again, it must be emphasized that these equations are only reasonably valid for operating conditions that have been specified in the linearization of the equations. In other words, they are only valid for small and slow motions of the piston and fluid injection rates that are near leakage rates.

Electromechanical System Example.   The electromechanical system shown in Figure 3-3 is an example of a system that is linear in its original form. Inspection of Equations (3.6) and (3.7) shows this to be the case. Therefore, this system does not need to be linearized and can be studied using linear techniques from the start. Limits of validity for these equations are determined only by the limits imposed on the original model.

3.4   DYNAMIC BEHAVIOR

3.4.1   First-Order Response

General.   In this subsection we consider the first-order spring-damper system shown in Figure 3-5. Since a first-order linear model is the simplest and most reduced form that we may possibly use to describe a system's time response, this model proves to be instructive for a basic understanding of the time and frequency response for many systems. In particular the steady-state response and time constant for this system is of interest to us. In Figure 3-5 the spring rate of the system is shown by the symbol k and the viscous damping coefficient is shown by the symbol c. The displacement of the system $x(t)$ is shown to vary with time and the force applied to the system is also time varying in general. In the following paragraphs we consider the response of this system to an unforced and harmonic excitation.

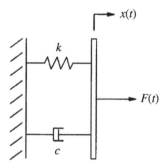

Figure 3-5. First-order spring-damper system.

Unforced Time Response. By neglecting the external force applied to the system in Figure 3-5, the equation of motion for the free response of this system is given by

$$c\dot{x} + kx = 0, \tag{3.22}$$

which is a linear first-order differential equation of the homogenous type. Using a standard solution technique for this system, we seek a solution by letting

$$X = Ae^{st} \quad \text{and} \quad \dot{x} = Ase^{st}, \tag{3.23}$$

where s is known as the eigenvalue of the system. By substituting Equation (3.23) into Equation (3.22) the characteristic equation for this system may be written as

$$cs + k = 0, \tag{3.24}$$

which as a single root given by

$$s = -\frac{k}{c} = -\frac{1}{\tau}. \tag{3.25}$$

In this result τ is called the time constant of the system. Substituting Equation (3.25) into Equation (3.23) yields the following time varying result for the system displacement:

$$x = x_o e^{-t/\tau}, \tag{3.26}$$

where x_o is the initial displacement of the system from its position of equilibrium. The unforced time response of the first-order system is shown in Figure 3-6. This response is

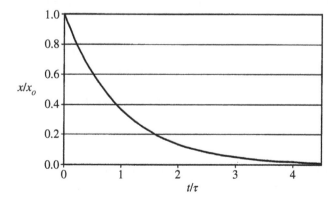

Figure 3-6. First-order time response of the unforced spring-damper system.

shown to decay in time with approximately 98% of the response completed within four time constants of the system.

Time Response to Harmonic Excitation.   The equation of motion for the first-order system shown in Figure 3-5 with applied force, $F(t)$, is given by,

$$c\dot{x} + kx = F(t). \tag{3.27}$$

To consider the time response of the first-order system to a harmonic excitation it is convenient to let the applied force of the system shown in Figure 3-5 be given by

$$F(t) = kA \cos(\omega t), \tag{3.28}$$

where k is the spring constant of the system, A is the displacement amplitude for the system, and ω is the forcing frequency. Using this expression with Equation (3.27), the equation of motion for the system may be written as

$$c\dot{x} + kx = kA \cos(\omega t). \tag{3.29}$$

In general, the solution to this equation is determined by adding the "transient solution" to the "steady solution." The transient solution is given in Equation (3.26) as the topic of our previous discussion. The steady solution for a harmonic excitation may be determined by using the method of undetermined coefficients and a solution of the following form:

$$x = C_1 \sin(\omega t) + C_2 \cos(\omega t), \quad \dot{x} = C_1\omega \cos(\omega t) - C_2\omega \sin(\omega t), \tag{3.30}$$

where C_1 and C_2 are the coefficients that have yet to be determined. Substituting Equation (3.30) into Equation (3.29) yields the following expression:

$$(c\omega C_1 + kC_2) \cos(\omega t) - (c\omega C_2 - kC_1) \sin(\omega t) = kA \cos(\omega t). \tag{3.31}$$

For this equation to be satisfied it can be seen that

$$c\omega C_1 + kC_2 = kA \quad \text{and} \quad c\omega C_2 - kC_1 = 0. \tag{3.32}$$

A simultaneous solution of these two equations yields the following results for the coefficients C_1 and C_2:

$$C_1 = (\tau\omega) \frac{A}{(\tau\omega)^2 + 1}, \quad C_2 = \frac{A}{(\tau\omega)^2 + 1}, \tag{3.33}$$

where, as in Equation (3.25), $\tau = c/k$. Substituting Equation (3.33) into Equation (3.30) yields the following solution for the displacement of the harmonically excited system:

$$x = \frac{A}{(\tau\omega)^2 + 1} [(\tau\omega) \sin(\omega t) + \cos(\omega t)]. \tag{3.34}$$

This is known as the "steady solution" to Equation (3.29) since it exhibits repeated sinusoidal behavior as time increases. Figure 3-7 shows an example of a time response of the first-order system to harmonic excitation.

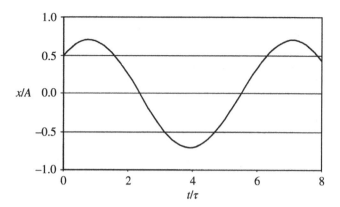

Figure 3-7. Example of a first-order time response of the harmonically forced spring-damper system where $\tau\omega = 1$ oscillation has been shifted slightly to the right. As shown, the example is for a system characterized by $\tau\omega = 1$.

Frequency Response. By using the appropriate trigonometric identities it can be shown that another representation for Equation (3.34) is given by

$$x = X\,\cos(\omega t - \varphi),\qquad\qquad(3.35)$$

where

$$X = \frac{A}{\sqrt{(\tau\omega)^2 + 1}}\qquad \text{and}\qquad \varphi = \tan^{-1}(\tau\omega).\qquad\qquad(3.36)$$

In this result X is the amplitude of the harmonic response and φ is the phase angle. As shown in Equation (3.36) both of these quantities are functions of the excitation frequency ω. Figure 3-8 shows a plot of the normalized amplitude and phase angle as a function of $\tau\omega$. The physical insight gained from this plot is that as the forcing frequency grows considerably, the system stops responding to the input as shown by the amplitude plot. Under these same conditions the phase angle approaches $\pi/2$. Figure 3-8 is known as the frequency response plot.

One of the performance specifications for a first-order system is given in terms of frequency response. This specification is known as the bandwidth frequency, which is defined

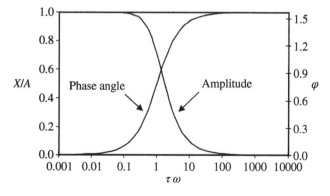

Figure 3-8. First-order frequency response.

as the maximum input frequency at which the output of a system will track the input sinusoid in a satisfactory manner [2]. By convention the bandwidth frequency is taken to be the input frequency ω at which the output amplitude X is attenuated by a factor of $\sqrt{2}/2$ times the input amplitude A. Solving Equation (3.36) for the case when $X/A = \sqrt{2}/2$ it may be shown that the bandwidth frequency for the first-order system is given by

$$\omega_{bw} = \frac{1}{\tau},\tag{3.37}$$

where the time constant τ is given in Equation (3.25). Generally, large bandwidth frequencies and small time constants are desirable for systems to respond quickly.

3.4.2 Second-Order Response

General. In the previous section we considered a first-order system, which is the simplest dynamical system to be discussed. In this section we consider the second-order spring-mass-damper system shown in Figure 3-9. This system has increased complexity due to the added consideration of inertia. Whereas the time constant was of importance in the first-order system, the natural frequency and damping ratio are of importance when considering a second-order system. In Figure 3-9 the spring rate of the system is given by the symbol k, the viscous damping coefficient is given by the symbol c, and the mass of the system is shown by the symbol m. The displacement of the system $x(t)$ is shown to vary with time and the force applied to the system $F(t)$ is also time varying in general.

Unforced Time Response. By neglecting the external force applied to the system in Figure 3-9 the equation of motion for the free response of this system is given by

$$m\ddot{x} + c\dot{x} + kx = 0,\tag{3.38}$$

which is a linear second-order differential equation of the homogenous type. A more convenient and standard form of this equation is given by

$$\ddot{x} + 2\zeta\omega_n\dot{x} + \omega_n^2 x = 0,\tag{3.39}$$

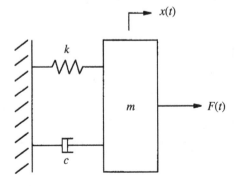

Figure 3-9. Second-order spring-mass-damper system.

where the undamped natural frequency and dimensionless damping ratio are given respectively as

$$\omega_n = \sqrt{\frac{k}{m}} \quad \text{and} \quad \zeta = \frac{c}{2m\omega_n}. \tag{3.40}$$

Following a similar solution technique that was used for the first-order system, we seek a solution to Equation (3.39) of the form

$$x = Ae^{st}, \quad \dot{x} = Ase^{st}, \quad \ddot{x} = As^2e^{st}, \tag{3.41}$$

where s is called an eigenvalue of the system. Substituting Equation (3.41) into Equation (3.39) yields the following characteristic equation for the system:

$$s^2 + 2\zeta\omega_n s + \omega_n^2 = 0. \tag{3.42}$$

The two solutions to this equation are given by

$$s_1 = (-\zeta + \sqrt{\zeta^2 - 1})\,\omega_n \quad \text{and} \quad s_2 = (-\zeta - \sqrt{\zeta^2 - 1})\,\omega_n. \tag{3.43}$$

Since our system is linear and since there are two solutions to the characteristic equation the general solution to the unforced equation of motion is

$$x = A_1 e^{s_1 t} + A_2 e^{s_2 t}, \tag{3.44}$$

where s_1 and s_2 are given in Equation (3.43) and A_1 and A_2 may generally be complex numbers. Clearly, the nature of the second-order response depends on the value of the damping ratio ζ. Figure 3-10 shows a plot of the eigenvalues s_1 and s_2 in the real-imaginary plane, which is often called the s-plane. This figure plots the eigenvalue locus as a function of the

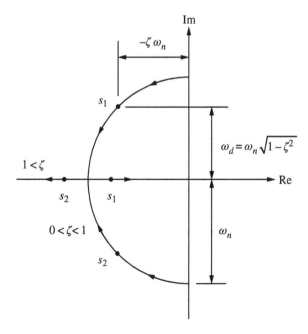

Figure 3-10. Eigenvalue plot for the second-order system.

damping ratio and thereby gives an indication of what the basic response characteristics will be as ζ varies. For instance, for $\zeta = 0$ it can be seen that the eigenvalues are strictly imaginary and given by $\pm\omega_n i$. This condition is known as marginal stability. For $0 < \zeta < 1$, the eigenvalues are characterized by complex conjugates that are located symmetrically with respect to the real axis on a circle with a radius given by ω_n. This system is often referred to as being underdamped. A critically damped condition occurs when ζ approaches unity and both eigenvalues are given by the value $-\omega_n$ on the real axis. The overdamped condition of the system occurs when ζ approaches infinity. Under these conditions, one eigenvalue tends toward zero while the other tends toward minus infinity. Each of the damped conditions is discussed in the following paragraphs.

Underdamped Response.   To consider conditions of marginal stability and the underdamped response we recall that these conditions are characterized by a damping ratio given by $0 \le \zeta < 1$. Under such conditions it is convenient to express the general solution of Equation (3.44) in the following form:

$$x = e^{-\zeta\omega_n t}(A_1 e^{\omega_d it} + A_2 e^{-\omega_d it}), \tag{3.45}$$

where the damped natural frequency shown in Figure 3-10 is given by

$$\omega_d = \omega_n \sqrt{1 - \zeta^2}. \tag{3.46}$$

At this point it is useful to employ Euler's identity, which is given by

$$e^{\theta i} = \cos(\theta) + i \, \sin(\theta). \tag{3.47}$$

From this identity it can be shown that

$$e^{\omega_d it} = \cos(\omega_d t) + i \, \sin(\omega_d t), \quad e^{-\omega_d it} = \cos(\omega_d t) - i \, \sin(\omega_d t). \tag{3.48}$$

Substituting these results into Equation (3.45) yields

$$x = e^{-\zeta\omega_n t}\{A_1[\cos(\omega_d t) + i \, \sin(\omega_d t)] + A_2[\cos(\omega_d t) - i \, \sin(\omega_d t)]\}. \tag{3.49}$$

As previously noted, the coefficients A_1 and A_2 are generally complex. If we express these coefficients in the following form:

$$A_1 = a_1 + b_1 i, \quad A_2 = a_2 + b_2 i, \tag{3.50}$$

then Equation (3.49) may be rewritten as

$$x = e^{-\zeta\omega_n t} [(a_1 + a_2)\cos(\omega_d t) - (b_1 - b_2) \, \sin(\omega_d t)]$$
$$+ e^{-\zeta\omega_n t} [(a_1 - a_2)\sin(\omega_d t) + (b_1 + b_2) \, \cos(\omega_d t)] \, i. \tag{3.51}$$

For the imaginary part of this solution to vanish it must be a requirement that

$$a_1 = a_2 = \frac{\alpha}{2} \quad \text{where} \quad b_1 = -b_2 = -\frac{\beta}{2}, \tag{3.52}$$

where α and β are strictly real. Substituting Equation (3.52) into Equation (3.51) produces the following result for the underdamped, unforced response of the second-order system:

$$x = e^{-\zeta\omega_n t} [\alpha \cos(\omega_d t) + \beta \, \sin(\omega_d t)]. \tag{3.53}$$

This equation describes an oscillating response that decays in time. The coefficients α and β must be determined by specifying the initial conditions of the system $x(0) = x_o$ and $\dot{x}(0) = \dot{x}_o$. At these initial conditions it can be shown that

$$x_o = \alpha \quad \text{and} \quad \dot{x}_o = \alpha \zeta \omega_n + \beta \omega_d. \tag{3.54}$$

Solving these two equations for α and β yields the following result:

$$\alpha = x_o \quad \text{and} \quad \beta = \frac{\dot{x}_o - x_o \zeta \omega_n}{\omega_d}. \tag{3.55}$$

Substituting Equation (3.55) into Equation (3.53) the following solution for the underdamped, unforced response of the second-order system may be written:

$$x = e^{-\zeta \omega_n t} \left(x_o \cos(\omega_d t) + \frac{\dot{x}_o - x_o \zeta \omega_n}{\omega_d} \sin(\omega_d t) \right). \tag{3.56}$$

Figure 3-11 shows a plot of the underdamped, unforced second-order response. This plot was generated for a system characterized by an initial velocity of zero. As the figure shows, when the damping ratio increases the amplitude of oscillation goes down and the response is finished much earlier. This is an indication of energy being dissipated through viscous friction at an increasing rate.

Critically and Overdamped Response. To consider the critically damped, unforced second-order system, we recall that this condition is characterized by $\zeta = 1$. For this case, the solution to the characteristic equation is given by a double root in which $s_1 = s_2 = -\omega_n$. Under such conditions the general solution to Equation (3.39) is given by

$$x = e^{-\omega_n t}(A_1 + A_2 t), \tag{3.57}$$

where the coefficients A_1 and A_2 depend on the initial conditions of the problem, $x(0) = x_o$ and $\dot{x}(0) = \dot{x}_o$. From the initial conditions and Equation (3.57) we can see that

$$x_o = A_1, \quad \dot{x}_o = -\omega_n A_1 + A_2. \tag{3.58}$$

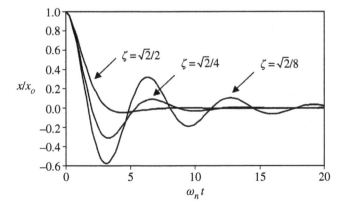

Figure 3-11. Underdamped, unforced second-order response when $\dot{x}_o = 0$.

A simultaneous solution for these two equations yields the following result for the coefficients A_1 and A_2:

$$A_1 = x_o, \quad A_2 = \dot{x}_o + \omega_n x_o. \tag{3.59}$$

Substituting these results into Equation (3.57) yields the following result for describing the critically damped, unforced second-order response:

$$x = e^{-\omega_n t}[\dot{x}_o\, t + (1 + \omega_n t)\, x_o]. \tag{3.60}$$

To consider the overdamped response of the unforced, second-order system we recall that this condition is characterized by a damping ratio given by $\zeta > 1$. Under these conditions a convenient form for Equation (3.44) is given by

$$x = e^{-\zeta\omega_n t}(A_1\, e^{\sqrt{\zeta^2-1}\,\omega_n t} + A_2\, e^{-\sqrt{\zeta^2-1}\,\omega_n t}), \tag{3.61}$$

where the coefficients A_1 and A_2 are now strictly real. This equation describes an aperiodic response that decays in time. By considering the initial conditions of the system $x(0) = x_o$ and $\dot{x}(0) = \dot{x}_o$ it may be shown that

$$x_o = A_1 + A_2, \quad \dot{x}_o = -\zeta\,\omega_n(A_1 + A_2) + \sqrt{\zeta^2 - 1}\,\omega_n(A_1 - A_2). \tag{3.62}$$

A simultaneous solution for these two equations yields the following result for the coefficients of this problem:

$$A_1 = \frac{\dot{x}_o + (\zeta + \sqrt{\zeta^2 - 1})\,\omega_n x_o}{2\sqrt{\zeta^2 - 1}\,\omega_n}, \quad A_2 = -\frac{\dot{x}_o + (\zeta - \sqrt{\zeta^2 - 1})\,\omega_n x_o}{2\sqrt{\zeta^2 - 1}\,\omega_n}. \tag{3.63}$$

Substituting these coefficients into Equation (3.61) results in the following expression for the response of the overdamped, unforced, second-order system:

$$\begin{aligned} x = {}& \frac{e^{-\zeta\omega_n t}}{2\sqrt{\zeta^2 - 1}}(e^{\sqrt{\zeta^2-1}\,\omega_n t} - e^{-\sqrt{\zeta^2-1}\,\omega_n t})\frac{\dot{x}_o}{\omega_n} \\ & + \frac{e^{-\zeta\omega_n t}}{2\sqrt{\zeta^2 - 1}}[(\zeta + \sqrt{\zeta^2 - 1})\,e^{\sqrt{\zeta^2-1}\,\omega_n t} - (\zeta - \sqrt{\zeta^2 - 1})\,e^{-\sqrt{\zeta^2-1}\,\omega_n t}]\,x_o. \end{aligned} \tag{3.64}$$

Figure 3-12 shows a plot of the critically damped and overdamped, unforced second-order response; that is, Equations (3.60) and (3.64). This plot was generated for a system characterized by an initial velocity of zero. As the damping ratio increases the response is shown to finish at a much later time.

Time Response to a Step Input.   We now turn our attention toward the forced system in which the input is a sudden step force. From Figure 3-9 the equation of motion for the system that is forced by a general input $F(t)$ is given by the following expression:

$$\ddot{x} + 2\zeta\omega_n\dot{x} + \omega_n^2 x = F(t)/m, \tag{3.65}$$

where the damping ratio and natural frequency are given in Equation (3.40). In the case of a step input, the input force may be conveniently written as

$$F(t) = kAu(t), \tag{3.66}$$

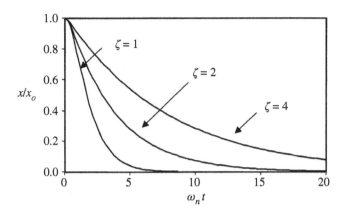

Figure 3-12. Critically and overdamped, unforced second-order response when $\dot{x}_o = 0$.

where k is the spring constant, A is the steady displacement amplitude of the system, and $u(t)$ is the unit step function defined as

$$u(t) = \begin{cases} 0 & t < 0 \\ 1 & t > 0 \end{cases}. \tag{3.67}$$

The discontinuity in this function at $t = 0$ is a bit problematic and is treated by considering the unit step response as the time integral of the unit impulse response of the system. This work has been well documented in standard dynamical textbooks, which should be consulted for further details [3]. For the purposes of this work the step response is simply presented as

$$x = A \left[1 - e^{-\zeta\omega_n t} \left(\cos(\omega_d t) + \frac{\zeta\omega_n}{\omega_d} \sin(\omega_d t) \right) \right], \tag{3.68}$$

where the damped natural frequency is given in Equation (3.46). In this result the system is assumed to be of the underdamped type $0 \leq \zeta < 1$. Figure 3-13 shows a typical plot of the step response for an underdamped, second-order system.

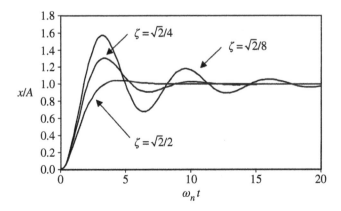

Figure 3-13. Step response of the underdamped, second-order system.

Underdamped Response Characteristics.   Figure 3-14 shows a schematic of the step response of the second-order, underdamped system with several important response characteristics being measured. These characteristics include the rise time t_r, the settling time t_s, and the maximum transient output x_{max}. The maximum transient output is discussed in terms of the maximum percent overshoot and each of these quantities are considered in the paragraphs that follow.

By definition the rise time is the amount of time it takes for the system output to reach the vicinity of the system input. Various authors define the "vicinity of the system input" differently; however, for our purposes we define the input vicinity as the input value itself and the system rise time is the amount of time required for our system to go from 0 to 100% of the system input value. Upon solving Equation (3.68) for $x(t_r) = A$ and rearranging terms we are able to conclude that the rise time for the underdamped second-order system is given by

$$t_r = \frac{\pi - \arctan(\sqrt{1 - \zeta^2}/\zeta)}{\omega_d}. \tag{3.69}$$

Since the arctan function passes through the origin when its argument is zero and asymptotically approaches $\pi/2$ when its argument goes to infinity, we are able to express the limits of the rise time as

$$\frac{\pi}{2\omega_d} \leq t_r \leq \frac{\pi}{\omega_d}. \tag{3.70}$$

Clearly the rise time may be reduced by increasing the natural frequency of the system. If the damping ratio of the system is given by $\zeta = \sqrt{2}/4 = 0.35$ the rise time may be computed using Equation (3.69) as

$$t_r \approx \frac{2}{\omega_n}. \tag{3.71}$$

This result is an approximation that may be used for estimating the rise time for most second-order, underdamped systems.

The maximum percent overshoot is defined as

$$M_p = \frac{x_{max} - A}{A} \times 100\%, \tag{3.72}$$

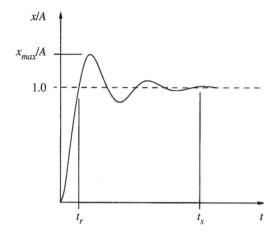

Figure 3-14. Response characteristics of the underdamped, second-order system.

where the reader will recall that A is the steady-state displacement of the system. To determine x_{max} Equation (3.68) must be differentiated with respect to time and set equal to zero. Once this result has been solved for time it may be substituted back into Equation (3.68) to determine x_{max}. Using this result in Equation (3.72) yields the following result for the maximum percent overshoot of the system:

$$M_p = e^{-\pi\zeta/\sqrt{1-\zeta^2}} \times 100\%. \tag{3.73}$$

From here it may be shown that the maximum percent overshoot may be reduced by increasing the damping ratio ζ and that this is the only parameter that the maximum percent overshoot depends on.

To discuss the settling time it is useful to consider the envelope in which the dynamic response occurs. This envelope is defined by two exponential curves, which bound the system output. The output boundaries may be written mathematically as

$$1 - e^{-\zeta\omega_n t} < x/A < 1 + e^{-\zeta\omega_n t}. \tag{3.74}$$

The time constant for each of these boundary curves is given by $1/(\zeta\omega_n)$ and it is conventional to define the settling time as four times this time constant; therefore,

$$t_s = \frac{4}{\zeta\omega_n}. \tag{3.75}$$

Clearly, to reduce the settling time one must increase the product of the damping ratio and the undamped natural frequency.

Time Response to Harmonic Excitation. To consider the response of the second-order system shown in Figure 3-9 to a harmonic excitation, we let the force acting on the system be given by

$$F(t) = kA \cos(\omega t), \tag{3.76}$$

where k is the spring constant, A is a stretched amplitude of the spring, and ω is the frequency of the harmonically applied force. Using this expression with Figure 3-9 it may be shown that the equation of motion for the second-order system is given by

$$\ddot{x} + 2\zeta\omega_n\dot{x} + \omega_n^2 x = \omega_n^2 A \cos(\omega t), \tag{3.77}$$

where the undamped natural frequency and the damping ratio are defined in Equation (3.40). To find the steady-state solution to this equation we use the method of undetermined coefficients and seek a solution of the following form:

$$x = C_1 \sin(\omega t) + C_2 \cos(\omega t), \quad \dot{x} = C_1\omega \cos(\omega t) - C_2\omega \sin(\omega t),$$
$$\ddot{x} = -C_1\omega^2 \sin(\omega t) - C_2\omega^2 \cos(\omega t). \tag{3.78}$$

Substituting these equations into Equation (3.77) and collecting terms yields the following expression:

$$\omega_n^2 A \cos(\omega t) = (C_1\omega_n^2 - C_1\omega^2 - C_2\omega 2\zeta\omega_n)\sin(\omega t)$$
$$+ (C_2\omega_n^2 - C_2\omega^2 + C_1\omega 2\zeta\omega_n)\cos(\omega t). \tag{3.79}$$

By equating the coefficients in front of the trigonometric terms it may be shown that

$$C_1(\omega_n^2 - \omega^2) - C_2 \omega 2\zeta \omega_n = 0, \quad C_2(\omega_n^2 - \omega^2) + C_1 \omega 2\zeta \omega_n = \omega_n^2 A. \tag{3.80}$$

Solving these two equations for C_1 and C_2 yields the following results:

$$C_1 = \frac{A\left(2\zeta \dfrac{\omega}{\omega_n}\right)}{\left(1 - \left[\dfrac{\omega}{\omega_n}\right]^2\right)^2 + \left(2\zeta\dfrac{\omega}{\omega_n}\right)^2}, \quad C_2 = \frac{A\left(1 - \left[\dfrac{\omega}{\omega_n}\right]^2\right)}{\left(1 - \left[\dfrac{\omega}{\omega_n}\right]^2\right)^2 + \left(2\zeta\dfrac{\omega}{\omega_n}\right)^2}. \tag{3.81}$$

Substituting these results back into Equation (3.78) yields the following solution for the harmonically excited, second-order system:

$$x = \frac{A}{\left(1 - \left[\dfrac{\omega}{\omega_n}\right]^2\right)^2 + \left(2\zeta\dfrac{\omega}{\omega_n}\right)^2} \left\{2\zeta\dfrac{\omega}{\omega_n}\sin(\omega t) + \left(1 - \left[\dfrac{\omega}{\omega_n}\right]^2\right)\cos(\omega t)\right\}. \tag{3.82}$$

Frequency Response. By using the appropriate trigonometric identities, Equation (3.82) may be written in the following form:

$$x = X \cos(\omega t - \varphi), \tag{3.83}$$

where

$$X = \frac{A}{\sqrt{\left(1 - \left[\dfrac{\omega}{\omega_n}\right]^2\right)^2 + \left(2\zeta\dfrac{\omega}{\omega_n}\right)^2}} \quad \text{and} \quad \varphi = \tan^{-1}\left(\frac{2\zeta\dfrac{\omega}{\omega_n}}{1 - \left[\dfrac{\omega}{\omega_n}\right]^2}\right). \tag{3.84}$$

In this result X is the amplitude of the harmonic response and φ is the phase angle. As shown in Equation (3.84) both of these quantities are functions of the excitation frequency ω. Figure 3-15 shows an amplitude plot of the frequency response. From this plot it may be

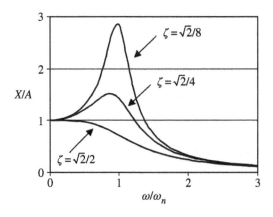

Figure 3-15. Amplitude response of the second-order system.

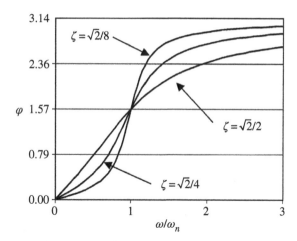

Figure 3-16. Phase response of the second-order system.

seen that the amplitude of the response gets very large for small values of damping and for operating frequencies that are near the natural frequency of the system.

These operating conditions are often troublesome to a real system and therefore it is recommended that the natural frequency and forcing frequency be as far apart from each other as possible. Figure 3-16 shows a phase plot of the frequency response. From this plot it may be seen that high frequency of oscillation tends to drive the response 180 degrees out of phase with the input oscillation.

For resonant conditions when $\omega/\omega_n = 1$ it may be shown that

$$X = \frac{A}{2\zeta} \quad \text{and} \quad \varphi = \frac{\pi}{2}. \tag{3.85}$$

As shown in this result, for lightly damped systems, the amplitude of oscillation near resonant conditions can get very large.

Similar to the first-order system one of the performance specifications for a second-order system is given in terms of frequency response. This specification is known as the bandwidth frequency, which, again, is defined as the maximum input frequency at which the output of a system will track the input sinusoid in a satisfactory manner [2]. By convention the bandwidth frequency is taken to be the input frequency ω at which the output amplitude X is attenuated by a factor of $\sqrt{2}/2$ times the input amplitude A. Solving Equation (3.84) for the case when $X/A = \sqrt{2}/2$ it may be shown that the bandwidth frequency for the second-order system is given by

$$\omega_{bw} = \omega_n \sqrt{(1 - 2\zeta^2) + \sqrt{2 - 2\,\zeta^2(1 - \zeta^2)}}, \tag{3.86}$$

where the undamped natural frequency and the damping ratio are given in Equation (3.40). Generally, large bandwidth frequencies are desirable for systems to respond quickly.

Case Study

The mass-spring-damper system shown in Figure 3-9 is characterized by the following parameters:

$$m = 0.85 \text{ kg}, \quad c = 65 \text{ Ns/m}, \quad k = 25 \text{ N/mm}.$$

Calculate the systems undamped natural frequency, the damping ratio, the damped natural frequency, the bandwidth frequency, the rise time, the maximum percent overshoot, and the settling time.

The undamped natural frequency and the damping ratio for a second-order system are given in Equation (3.40). These calculations for this problem are

$$\omega_n = \sqrt{\frac{k}{m}} = \sqrt{\frac{25 \text{ N/mm}}{0.85 \text{ kg}}} = 171.5 \frac{\text{rad}}{\text{s}} = 27.3 \text{ Hz}.$$

$$\zeta = \frac{c}{2m\omega_n} = \frac{65 \text{ N s/m}}{2 \times 0.85 \text{ kg} \times 171.5 \text{ rad/s}} = 0.223.$$

Since this system is of the underdamped type, the second-order parameters of damped natural frequency, rise time, maximum percent overshoot, and settling time may be computed. These quantities are found in Equations (3.46), (3.69), (3.73), and (3.75) and are computed as:

$$\omega_d = \omega_n \sqrt{1 - \zeta^2} = 171.5 \text{ rad/s} \times \sqrt{1 - 0.223^2} = 167.2 \frac{\text{rad}}{\text{s}} = 26.6 \text{ Hz}$$

$$t_r = \frac{\pi - \arctan \sqrt{1 - \zeta^2}/\zeta}{\omega_d} = \frac{\pi - \arctan \sqrt{1 - 0.223^2}/0.223}{167.2 \text{ rad/s}} = 10.7 \text{ ms}$$

$$M_p = e^{-\pi\zeta/\sqrt{1-\zeta^2}} \times 100\% = \text{Exp}\left(\frac{-\pi \times 0.223}{\sqrt{1 - 0.223^2}}\right) \times 100\% = 48.75\%$$

$$t_s = \frac{4}{\omega_n \zeta} = \frac{4}{171.5 \text{ rad/s} \times 0.223} = 104.6 \text{ ms}.$$

The bandwidth frequency for this system is shown in Equation (3.86). This result may be computed as 40.9 Hz.

3.4.3 Higher-Order Response

In this section we have considered first- and second-order dynamical systems. As previously mentioned, these systems are nice to work with because we have a great deal of knowledge concerning their behavior and they allow us to describe the dynamic response in very concise terms. For instance, the time constant of the first-order system is very helpful for describing how fast the response dies away. The natural frequency and the damping ratio of the second-order system are very convenient for describing the rise time, maximum percent overshoot, and the settling time of the system. For higher-order systems we tend to describe their response using similar language but without using a mathematical definition of these concepts. Within this text we model hydraulic control systems as either first- or second-order systems so as to keep the language consistent with the mathematical definitions presented here. Limits on these models are mentioned throughout the work so as to keep the models from being misapplied. However, in practice, it may be necessary to study higher-order systems using either numerical analysis or experimental methods in order to gain insight into the dynamic response of these more complex situations.

3.5   STATE SPACE ANALYSIS

3.5.1   General

State space analysis is a convenient method for describing the equations of motion for a dynamical system. This method can be used for an nth-order system and lends itself easily to a computer environment for the numerical evaluation of the equations. For instance, MATLAB® uses the state space formulation for conducting much of its analysis. Though state space methods can be used with both linear and nonlinear systems and with both time invariant and time varying systems the following discussion is concerned only with linear time invariant systems.

3.5.2   State Space Equations

The state space equations are comprised of state variables, input variables, and output variables. For linear time invariant systems, these variables are related to each other through the following state space equations:

$$\dot{\mathbf{x}} = \mathbf{A}\mathbf{x} + \mathbf{B}\mathbf{u}, \qquad \mathbf{y} = \mathbf{C}\mathbf{x} + \mathbf{D}\mathbf{u}, \tag{3.87}$$

where $\mathbf{x}$ is a vector of state variables, $\mathbf{u}$ is a vector of input variables, $\mathbf{A}$ is called the state matrix, $\mathbf{B}$ is called the input matrix, $\mathbf{C}$ is the output matrix, and $\mathbf{D}$ is called the direct transmission matrix. The first equation in Equation (3.87) is called the state equation while the second equation is called the output equation. To create the vector of state variables we utilize the fact that an nth-order differential equation can be written as a set of n first-order differential equations. This is done by defining the states of the equation properly and rewriting the equations in terms of these definitions. For instance, consider the third-order differential equation given by

$$a_3\dddot{y} + a_2\ddot{y} + a_1\dot{y} + a_o y = u(t). \tag{3.88}$$

If we define the states of this equation as

$$x_1 = y, \qquad x_2 = \dot{y}, \qquad x_3 = \ddot{y}, \tag{3.89}$$

then by substitution it may be shown that Equation (3.88) may be described using the following three first-order differential equations:

$$\begin{aligned} \dot{x}_1 &= x_2 \\ \dot{x}_2 &= x_3 \\ \dot{x}_3 &= -\frac{a_2}{a_3}x_3 - \frac{a_1}{a_3}x_2 - \frac{a_o}{a_3}x_1 + \frac{1}{a_3}u(t). \end{aligned} \tag{3.90}$$

For an nth-order system, this result may be generalized as

$$\begin{aligned} \dot{x}_1 &= x_2 \\ \dot{x}_2 &= x_3 \\ &\;\;\vdots \\ \dot{x}_{n-1} &= x_n \\ \dot{x}_n &= -\alpha_n x_n - \cdots - \alpha_2 x_2 - \alpha_1 x_1 + \alpha_o u(t). \end{aligned} \tag{3.91}$$

To express these first-order differential equations in terms of Equation (3.87) it may be shown that

$$\mathbf{x} = \begin{Bmatrix} x_1 \\ x_2 \\ x_3 \\ \vdots \\ x_n \end{Bmatrix}, \qquad \mathbf{A} = \begin{bmatrix} 0 & 1 & 0 & \cdots & 0 \\ 0 & 0 & 1 & \cdots & 0 \\ \vdots & \vdots & \vdots & & \vdots \\ 0 & 0 & 0 & \cdots & 1 \\ \alpha_1 & \alpha_2 & \alpha_3 & \cdots & \alpha_n \end{bmatrix}, \qquad \mathbf{B} = \begin{Bmatrix} 0 \\ 0 \\ 0 \\ \vdots \\ \alpha_o \end{Bmatrix}. \tag{3.92}$$

The output matrix may be tailored as desired and the direct transmission matrix is usually zero.

3.5.3 Characteristic Equation

The state space formulation allows for a convenient method of calculating the eigenvalues of the system from what is called the characteristic equation. The characteristic equation of the system may be determined as

$$\det[\mathbf{A} - s\mathbf{I}] = 0, \tag{3.93}$$

where $\mathbf{A}$ is the state matrix, $\mathbf{I}$ is the identity matrix, and "det" refers to the determinate operator. The characteristic equation is a nth polynomial in s where n is the number of first-order differential equations that are used to describe the dynamical system. By setting the characteristic equation equal to zero and solving for the roots of this polynomial the eigenvalues may be determined directly. The eigenvalues are used as time constants for determining the dynamic response of the system. An example of these results for a second-order system is given in Equation (3.43).

3.6 BLOCK DIAGRAMS AND THE LAPLACE TRANSFORM

3.6.1 General

Another convenient method for representing linear time invariant systems is by block diagrams and Laplace transforms. Block diagrams are particularly helpful for visualizing the chain of events that occur within the dynamical system and are used extensively in the chapters that follow. Since block diagrams are made possible by the use of the Laplace transform this section begins with a discussion of the Laplace transform and then presents information relative to the algebra of manipulating block diagrams.

3.6.2 Laplace Transform

Definition. The Laplace transform converts a time domain function into a Laplace domain function. This transform is defined as

$$L[f(t)] = F(s) = \int_0^\infty f(t)e^{-st}\,dt, \tag{3.94}$$

where L is the Laplace operator, s is the Laplace variable, and $f(t)$ is the time function being transformed such that $f(0^-) = 0$; that is, the initial conditions of the time function are zero. To illustrate the Laplace transform operation consider the exponential time function given by

$$f(t) = \begin{cases} 0 & t < 0 \\ Ae^{-\alpha t} & t \geq 0 \end{cases}. \tag{3.95}$$

Substituting this function into the general definition given by Equation (3.94) yields the following Laplace transform for the exponential time function:

$$F(s) = \int_0^\infty Ae^{-\alpha t}e^{-st}\,dt = \int_0^\infty Ae^{-(s+\alpha)t}\,dt = -\frac{A}{s+\alpha}\,e^{-(s+\alpha)t}\Big|_0^\infty = \frac{A}{s+\alpha}. \tag{3.96}$$

Another time function example that may be considered to illustrate the Laplace transform operation is given by the step function

$$f(t) = \begin{cases} 0 & t < 0 \\ A & t \geq 0 \end{cases}. \tag{3.97}$$

Substituting this result into Equation (3.94) yields the following Laplace transform:

$$F(s) = \int_0^\infty Ae^{-st}\,dt = -\frac{A}{s}\,e^{-st}\Big|_0^\infty = \frac{A}{s}. \tag{3.98}$$

Table 3-1 presents a number of Laplace transform pairs that may be used to convert between the time domain and the Laplace domain.

Differentiation Theorem.   The definition of the Laplace transform given in Equation (3.94) may be used with integration by parts to show that

$$L[f(t)] = -\frac{f(t)}{s}\,e^{-st}\Big|_0^\infty + \frac{1}{s}\int_0^\infty \left[\frac{d}{dt}f(t)\right]e^{-st}\,dt. \tag{3.99}$$

The right-hand term of this equation is simply the Laplace of the derivative of $f(t)$ divided by the Laplace variable s. By evaluating and rearranging Equation (3.99) it can be shown that

$$L\left[\frac{d}{dt}f(t)\right] = sF(s), \tag{3.100}$$

for the initial conditions when $f(0) = 0$. From this expression we can see that multiplication by the Laplace variable performs the operation of differentiation in the Laplace domain.

Integration Theorem.   Again, using the definition of the Laplace transform it can be shown that

$$L\left[\int\int f(t)\,dt\right] = \int_0^\infty \left[\int\int f(t)\,dt\right]e^{-st}\,dt. \tag{3.101}$$

Integration by parts shows that

$$L\left[\int\int f(t)\,dt\right] = -\frac{1}{s}\left[\int\int f(t)\,dt\right]e^{-st}\Big|_0^\infty + \frac{1}{s}\int_0^\infty \left[\frac{d}{dt}\int f(t)\,dt\right]e^{-st}\,dt. \tag{3.102}$$

Table 3-1. Selected Laplace transform pairs

Time function	Laplace transform
Unit step	$\dfrac{1}{s}$
t	$\dfrac{1}{s^2}$
$\dfrac{t^{n-1}}{(n-1)!}$ $(n = 1, 2, 3, ...)$	$\dfrac{1}{s^n}$
t^n $(n = 1, 2, 3, ...)$	$\dfrac{n!}{s^{n+1}}$
e^{-at}	$\dfrac{1}{s+a}$
te^{-at}	$\dfrac{1}{(s+a)^2}$
$\dfrac{1}{(n-1)!} t^{n-1} e^{-at}$ $(n = 1, 2, 3, ...)$	$\dfrac{1}{(s+a)^n}$
$t^n e^{-at}$ $(n = 1, 2, 3, ...)$	$\dfrac{n!}{(s+a)^{n+1}}$
$\sin(\omega t)$	$\dfrac{\omega}{s^2 + \omega^2}$
$\cos(\omega t)$	$\dfrac{s}{s^2 + \omega^2}$
$\dfrac{1}{a}(1 - e^{-at})$	$\dfrac{1}{s(s+a)}$
$\dfrac{1}{b-a}(e^{-at} - e^{-bt})$	$\dfrac{1}{(s+a)(s+b)}$
$\dfrac{-1}{b-a}(ae^{-at} - be^{-bt})$	$\dfrac{s}{(s+a)(s+b)}$
$\dfrac{1}{ab}\left[1 - \dfrac{1}{b-a}(be^{-at} - ae^{-bt})\right]$	$\dfrac{1}{s(s+a)(s+b)}$
$\dfrac{1}{a^2}[at - 1 + e^{-at}]$	$\dfrac{1}{s^2(s+a)}$
$e^{-at}\sin(\omega t)$	$\dfrac{\omega}{(s+a)^2 + \omega^2}$
$e^{-at}\cos(\omega t)$	$\dfrac{s+a}{(s+a)^2 + \omega^2}$
$\dfrac{e^{-at}}{2\omega^2}\left[(a-b)t\cos(\omega t) + \left\{\omega t - \dfrac{(a-b)}{\omega}\right\}\sin(\omega t)\right]$	$\dfrac{s+b}{[(s+a)^2 + \omega^2]^2}$

The right-hand term in this equation is simply the Laplace of the time function $f(t)$. Evaluating and rearranging Equation (3.102) shows that

$$L\left[\int f(t)\, dt\right] = \frac{1}{s} F(s), \tag{3.103}$$

for the conditions when

$$\int f(t)\, dt\Big|_0 = 0. \tag{3.104}$$

From Equation (3.103) it can be seen that dividing by the Laplace variable performs the operation of integration in the Laplace domain.

3.6.3 Partial Fraction Expansion

General. The reverse process of finding the time function from the Laplace transformation is called the inverse Laplace transformation. This transformation may be written as

$$L^{-1}[F(s)] = F^{-1}(s) = f(t). \tag{3.105}$$

A convenient method for conducting the inverse Laplace transformation is to make the Laplace transform look like a sum of known Laplace transform pairs; for example, Table 3-1, and to use these known transform pairs to determine the time function itself. In other words, if we have a Laplace function given by

$$F(s) = F_1(s) + F_2(s) + \cdots + F_n(s), \tag{3.106}$$

where the individual Laplace transforms on the right-hand side are known, then the inverse Laplace transform will be given by

$$F^{-1}(s) = F_1^{-1}(s) + F_2^{-1}(s) + \cdots + F_n^{-1}(s), \tag{3.107}$$

where the individual inverse Laplace transforms are known as well. For identifying known Laplace transform pairs the method of partial fraction expansion is useful.

Distinct and Real Roots. If the Laplace function is written as

$$F(s) = \frac{N(s)}{(s + p_1)(s + p_2)\dots(s + p_n)}, \tag{3.108}$$

where $-p_1$, $-p_2$, and so on, are the roots (poles) to the denominator of the function then the partial fraction expansion of this function may be written as

$$F(s) = \frac{C_1}{(s + p_1)} + \frac{C_2}{(s + p_2)} + \cdots + \frac{C_n}{(s + p_n)}. \tag{3.109}$$

For instance, consider the following Laplace function:

$$F(s) = \frac{100(s + 6)}{(s + 1)(s + 3)(s + 10)}. \tag{3.110}$$

Since the denominator roots are real and distinct this Laplace function may be expanded according to the form of Equation (3.109). This expansion is given by

$$F(s) = \frac{C_1}{(s + 1)} + \frac{C_2}{(s + 3)} + \frac{C_3}{(s + 10)}. \tag{3.111}$$

By equating this result with Equation (3.110) it may be shown that

$$0 = C_1 + C_2 + C_3,$$
$$100 = 13C_1 + 11C_2 + 4C_3, \tag{3.112}$$
$$600 = 30C_1 + 10C_2 + 3C_3.$$

A solution of these equations is given by

$$C_1 = \frac{250}{9}, \qquad C_2 = -\frac{150}{7}, \qquad C_3 = -\frac{400}{63}. \tag{3.113}$$

Substituting these results into Equation (3.111) and using Table 3-1 it may be shown that

$$F(s) = \frac{250}{9 \, (s+1)} - \frac{150}{7 \, (s+3)} - \frac{400}{63 \, (s+10)},$$
$$F^{-1}(s) = f(t) = \frac{250}{9} e^{-t} - \frac{150}{7} e^{-3t} - \frac{400}{63} e^{-10t}. \tag{3.114}$$

Repeated and Real Roots.   If the Laplace function is written as

$$F(s) = \frac{N(s)}{(s+p_1)^m}, \tag{3.115}$$

where $-p_1$ is the root (pole) to the denominator repeated m times then the partial fraction expansion of this function may be written as

$$F(s) = \frac{C_1}{(s+p_1)} + \frac{C_2}{(s+p_1)^2} + \cdots + \frac{C_m}{(s+p_1)^m}. \tag{3.116}$$

As an example, consider the following Laplace function:

$$F(s) = \frac{10(s+2)}{s^3}, \tag{3.117}$$

where the root to the denominator is zero, repeated three times. Using the basic form of Equation (3.116) this Laplace function may be expanded and written as

$$F(s) = \frac{C_1}{s} + \frac{C_2}{s^2} + \frac{C_3}{s^3}. \tag{3.118}$$

By equating this result with Equation (3.117) it may be shown that

$$C_1 = 0, \qquad C_2 = 10, \qquad C_3 = 20. \tag{3.119}$$

Substituting this result in Equation (3.118) and using Table 3-1 it may be shown that

$$F(s) = \frac{10}{s^2} + \frac{20}{s^3}, \qquad F^{-1}(s) = f(t) = 10t + 10t^2. \tag{3.120}$$

Distinct and Complex Roots.   When the Laplace function contains denominator roots that are complex the roots appear in complex conjugate pairs. The general form of the Laplace function in this case is given by

$$F(s) = \frac{N(s)}{(s + p_1)\,(s + p_2)(s + p_3)\,(s + p_4)\ldots(s + p_{n-1})(s + p_n)},$$

(3.121)

where $-p_1$ and $-p_2$, and so on, are complex conjugate roots to the denominator. With distinct and complex roots the Laplace function may be expanded in the following form:

$$F(s) = \frac{C_1 s + C_2}{(s + p_1)\,(s + p_2)} + \frac{C_3 s + C_4}{(s + p_3)\,(s + p_4)} + \cdots + \frac{C_{n-1} s + C_n}{(s + p_{n-1})\,(s + p_n)}.$$

(3.122)

For instance, consider the Laplace function

$$F(s) = \frac{s + 10}{(s + p_1)\,(s + p_2)(s + p_3)\,(s + p_4)},$$

(3.123)

where

$$p_1 = 2 + 3j, \qquad p_2 = 2 - 3j, \qquad p_3 = 1 + j, \qquad p_4 = 1 - j.$$

(3.124)

In this case the Laplace function may be written as

$$F(s) = \frac{C_1 s + C_2}{(s + p_1)\,(s + p_2)} + \frac{C_3 s + C_4}{(s + p_3)\,(s + p_4)}.$$

(3.125)

Setting this equation equal to Equation (3.123) it may be shown that

$$C_1 = \frac{9}{85}, \qquad C_2 = -\frac{56}{85}, \qquad C_3 = -\frac{9}{85}, \qquad C_4 = \frac{74}{85}.$$

(3.126)

Substituting this result into Equation (3.125) and using Table 3-1 the following Laplace and inverse Laplace transform may be written.

$$F(s) = \frac{9s - 56}{85(s^2 + 4s + 13)} - \frac{9s - 74}{85(s^2 + 2s + 2)}$$

$$= \frac{9}{85}\left[\frac{(s + 2)}{(s + 2)^2 + 3^2}\right] - \frac{74}{255}\left[\frac{3}{(s + 2)^2 + 3^2}\right] - \frac{9}{85}\left[\frac{(s + 1)}{(s + 1)^2 + 1^2}\right]$$

$$+ \frac{83}{85}\left[\frac{1}{(s + 1)^2 + 1^2}\right],$$

(3.127)

$$F^{-1}(s) = f(t) = e^{-2t}\left[\frac{9}{85}\cos(3t) - \frac{74}{255}\sin(3t)\right] - e^{-t}\left[\frac{9}{85}\cos(t) - \frac{83}{85}\sin(t)\right].$$

Repeated and Complex Roots.   In the case of repeated and complex roots for the denominator of the Laplace function, the Laplace equation may be written as

$$F(s) = \frac{N(s)}{[(s + p_1)\,(s + p_2)]^m},$$

(3.128)

where $-p_1$ and $-p_2$ are complex conjugate roots to the denominator. In this case the partial fraction expansion may be written as

$$F(s) = \frac{C_1 s + C_2}{[(s + p_1)\,(s + p_2)]} + \frac{C_3 s + C_4}{[(s + p_1)\,(s + p_2)]^2}$$

$$+ \cdots + \frac{C_{2m-1} s + C_{2m}}{[(s + p_1)\,(s + p_2)]^m}.$$

(3.129)

For instance, consider the Laplace function given by

$$F(s) = \frac{(s+1)(s+4)}{[(s+p_1)(s+p_2)]^2},$$ (3.130)

where

$$p_1 = 1 + 2j, \qquad p_2 = 1 - 2j.$$ (3.131)

In this case where the roots are repeated and complex the Laplace function can be written as

$$F(s) = \frac{C_1 s + C_2}{[(s+p_1)(s+p_2)]} + \frac{C_3 s + C_4}{[(s+p_1)(s+p_2)]^2}.$$ (3.132)

Setting this equation equal to Equation (3.130) it can be shown that

$$C_1 = 0, \qquad C_2 = 1, \qquad C_3 = 3, \qquad C_4 = -1.$$ (3.133)

Substituting this result into Equation (3.132) and using Table 3-1 it may be shown that

$$F(s) = \frac{1}{[(s+1)^2 + 2^2]} + \frac{3s - 1}{[(s+1)^2 + 2^2]^2},$$ (3.134)

$$F^{-1}(s) = f(t) = \frac{e^{-t}}{2} \cos(2t) + \frac{e^{-t}}{36}[2t \cos(2t) + (3t - 1) \sin(2t)].$$

Summary. In summary, the partial fraction expansions that have been presented here may be used in combination with one another when there is a mix of real and complex roots being either distinct or repeated. Again, the motivation for using the partial fraction expansion is for conducting the inverse Laplace transformation by using a transformation table similar to that of Table 3-1.

3.6.4 Block Diagrams

General. Block diagrams are used to pictorially represent dynamical systems. These diagrams are useful for understanding system causality and for thinking about adjustments that may be employed for improving the dynamic characteristics of the system. The block diagram is a schematic that represents the dynamic system using symbols of summation and multiplication. As one might imagine, the Laplace variables are very useful in such a schematic as they perform differentiation and integration by using algebraic operations of multiplication and division. In this subsection the creation of block diagrams will be described in general and illustrated using the physical systems that were presented in Section 3.2.

Components of Block Diagrams. Signals in a block diagram are represented by lines with arrows to indicate the direction of the signal "flow." Signals in a block diagram are acted on by summation and multiplication by a gain or transfer function. Summation, such as

$$D = A + B - C,$$ (3.135)

may be represented by a summation symbol with inputs, A, B, and $-C$, and output, D. Figure 3-17 represents the summation of input signals.

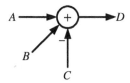

Figure 3-17. Three signals combined using summation.

Transfer functions are used to represent the relationship between the input and output of a dynamical system assuming zero initial conditions. The simplest case of a transfer function is a gain that has no dynamic behavior. For a gain, the input is multiplied by a constant to obtain an output. For example, in Figure 3-18(a), an output B is obtained by multiplying a gain, K, by the input, A,

$$B = KA. \tag{3.136}$$

A transfer function for a dynamical system is found by computing the Laplace transformation for a differential equation assuming zero initial conditions using Equation (3.100) and then solving for the ratio of output over input. As an example, the transfer function for the differential equation for the second-order system in Equation (3.65) can be found as follows. First take the Laplace transform assuming $x(0) = \dot{x}(0) = 0$ to get,

$$s^2 X(s) + s2\zeta\omega_n + \omega_n{}^2 = F(s)/m, \tag{3.137}$$

where $X(s)$ is the Laplace transform of the output (position, $x(t)$), and $F(s)$ is the Laplace transform of the input (force, $F(t)$). The final step is to find the transfer function by solving Equation (3.137) for the ratio of output over input,

$$\frac{X(s)}{F(s)} = \frac{1/m}{s^2 + s2\zeta\omega_n + \omega_n^2} = G(s). \tag{3.138}$$

If $G(s)$ is a transfer function, the relationship between input and output is identical to that of the gain, where K in Equation (3.136) can be replaced with $G(s)$, B can be replaced with $X(s)$, and A can be replaced with $F(s)$ to get $X(s) = G(s)F(s)$. The block diagram symbol for a transfer function, shown in Figure 13-18(b), is similar to that of the gain in Figure 13-18(a).

Simplification of Block Diagrams.　Interconnections of summations, gains, and transfer functions can be simplified to form one transfer function. To start, one input and one output must be chosen. Algebra is then used to find the ratio of output to input. An illustrative example is the feedback interconnection shown in Figure 3-19 with signals labeled.

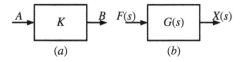

Figure 3-18. Gain with input *A* and output *B* (a) and transfer function example (b).

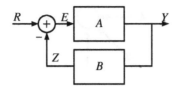

Figure 3-19. Negative feedback interconnection.

A set of equations can be formed by examining the block diagram, and then the ratio of the output, Y, to the input, R can be derived as:

$$E = R - Z$$

$$Z = BY$$

$$Y = AE$$

$$\Rightarrow \qquad\qquad\qquad\qquad\qquad (3.139)$$

$$Y = A(R - BY)$$

$$\Rightarrow \frac{Y}{R} = \frac{A}{1 + AB}$$

Mechanical System Example.   Figure 3-1 illustrates a swinging pendulum with nonlinear dynamic behavior. The linearized equation for this system is given in Equation (3.18) and is rewritten here for convenience:

$$I_o \ddot{\theta} = T - C\dot{\theta} - Mgl\theta. \qquad\qquad (3.140)$$

In this equation I_o is the mass moment of inertia of the pendulum about point O, T is a couple that is applied to the pendulum, C is a viscous drag coefficient, M is the mass of the pendulum, g is the gravitational constant, and the dimensions l and θ are shown in Figure 3-1.

To construct the block diagram of this dynamical system we begin by recognizing that the right-hand side of Equation (3.140) is comprised of terms that are added together to produce the left-hand side of the equation. For constructing the block diagram, it is convenient to consider the terms on the right-hand side as "input signals" and the term on the left-hand side as the output signal. In Figure 3-20 Step 1 shows the circular summation symbol in which the right-hand terms of Equation (3.140) are added together to produce the left-hand term. The negative sign near the input arrow indicates that a subtraction is carried out in the summation process. Before moving on to Step 2 of Figure 3-20 the student should make sure that he or she understands the relationship between Step 1 and Equation (3.140).

The next step in constructing the block diagram is to prepare the schematic for illustrating the state variables of the system. The state variables are $\dot{\theta}$ and θ. Step 2 of Figure 3-20 shows that by multiplying the output signal of the summation symbol by $1/I_o$, the highest-state derivative $\ddot{\theta}$ may be schematically determined as the output of the multiplication block. *Note*: Square symbols indicate multiplication with the input signal.

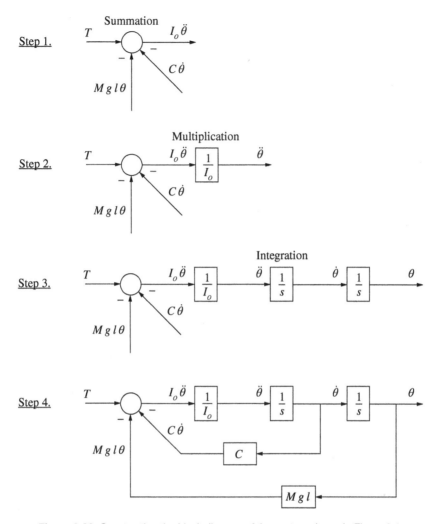

Figure 3-20. Constructing the block diagram of the system shown in Figure 3-1.

Step 3 of Figure 3-20 shows the use of the Laplace variable s for illustrating the state variables of the system. By multiplying $\ddot{\theta}$ with $1/s$ the Laplace integral is taken and the output signal is given by $\dot{\theta}$. Similarly, by multiplying $\dot{\theta}$ with $1/s$ the Laplace integral is taken and the output signal is given by θ. In this way, the state variables are represented in the block diagram as the output signals of the integral operations.

Finally, Step 4 of Figure 3-20 shows the completed block diagram for the system described in Equation (3.140). In this step the state variables are used with multiplication blocks to create the input signals for the summation symbol. The branches that are taken from the state variable lines are known as feedback signals as they point backward and contribute to the input signal for the system.

The block diagram in Figure 3-20 can be simplified using Equation (3.139) for each feedback interconnection. For example, the inner loop in Step 4 of Figure 3-22 can be simplified as,

$$\frac{\frac{1}{I_o s}}{1 + \frac{C}{I_o s}}.$$

The process can be repeated for the outer loop to get the transfer function between the input, T, and the output, θ, of the block diagram,

$$\frac{\theta(s)}{T(s)} = \frac{\frac{\frac{1}{I_o s}}{1 + \frac{C}{I_o s}} \frac{1}{s}}{1 + \frac{\frac{1}{I_o s}}{1 + \frac{C}{I_o s}} \frac{1}{s} Mgl} = \frac{\frac{1}{s(I_o s + C)}}{1 + \frac{Mgl}{s(I_o s + C)}} = \frac{1}{I_o s^2 + Cs + Mgl}.$$

This result can also be found by taking the Laplace transform of Equation (3.140), assuming zero initial conditions, and solving for the ratio of the output over the input.

Hydromechanical System Example. Figure 3-2 illustrates a hydromechanical system that utilizes pressurized fluid as the means for displacing an actuator. The linearized equations of motion for this system are given in Equation (3.21) and are rewritten here:

$$\dot{P} = \frac{\beta}{V_o}(Q - C_l P - A\dot{x}), \qquad M\ddot{x} = PA - C\dot{x} - kx - F, \tag{3.141}$$

where P is the fluid pressure, β is the fluid bulk modulus, V_o is the nominal fluid volume within the piston chamber, Q is the volumetric flow rate into the piston chamber, C_l is the coefficient of leakage, A is the cross-sectional area of the piston, and x is the actuator displacement. For the mechanical part of the system, M is the mass of the piston, C is the viscous drag coefficient, k is the spring rate, and F is an applied force by a load that may be acting on the piston.

To construct the block diagram for the system of equations shown in Equation (3.141) we begin by considering the pressure equation first. Figure 3-21 illustrates the process for constructing this block diagram by first recognizing the summation characteristics of the pressure equation shown in Equation (3.141). See Step 1 in Figure 3-21. Once the governing pressure equation is represented by the summation symbol Step 2 in Figure 3-21 may be used to prepare the schematic for representing the state variable P. This is done by multiplying the output of the summation symbol with the ratio β/V_o. Step 3 in Figure 3-20 shows the integration of the state derivative using the Laplace integration operator and Step 4 shows the block diagram for the pressure equation in its most complete form so far.

Next, we consider the equation of motion for the piston in the hydromechanical system. This is the second part of Equation (3.141). Figure 3-22 illustrates the process for constructing the block diagram of this part. As shown in this figure the first step in constructing the block diagram is to create a summation block that represents the second part of Equation (3.141). Step 2 of this process is to multiply the output of the summation symbol

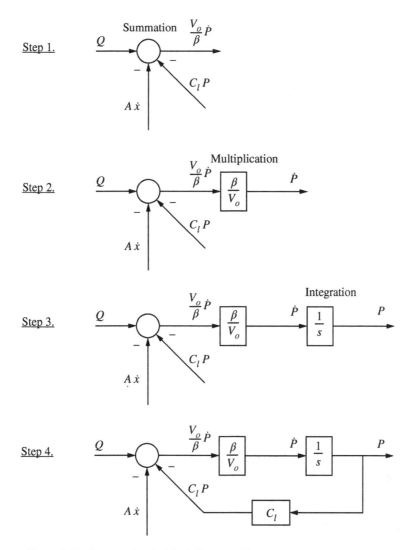

Figure 3-21. Constructing the block diagram of the pressure part of Figure 3-2.

by $1/M$ in preparation for depicting the state variables $\dot{x}$ and x. After depicting these states by using the integration operators the final step in the process is to connect branch signals with multiplication blocks to create the feedback signals to the summation symbol.

The final step in constructing the block diagram for the hydromechanical system is to join the pressure diagram with the piston diagram. These two diagrams are connected using the connecting block shown in Figure 3-23.

Electromechanical System Example. Figure 3-3 shows an electromagnetic actuator in which an electrical current is used to displace a mechanical piston. The governing equations for the motion of the piston and the electrical current are given in Equations (3.6) and (3.7) and are rewritten here for convenience:

$$L\dot{i} = v - Ri - \alpha\dot{x}, \qquad M\ddot{x} = \alpha i - C\dot{x} - kx - F, \tag{3.142}$$

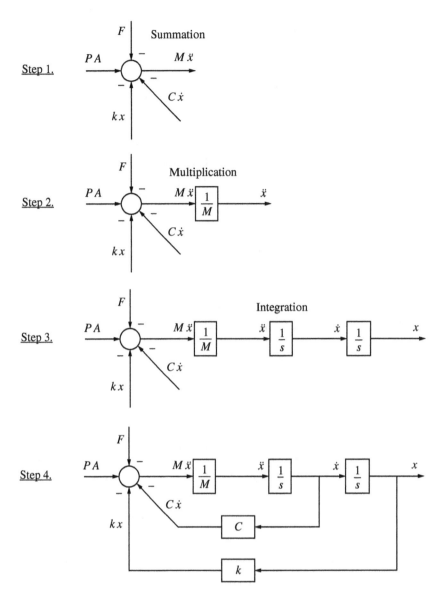

Figure 3-22. Constructing the block diagram for the piston part of Figure 3-2.

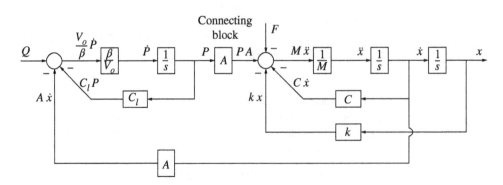

Figure 3-23. Block diagram for the hydromechanical systems of Figure 3-2.

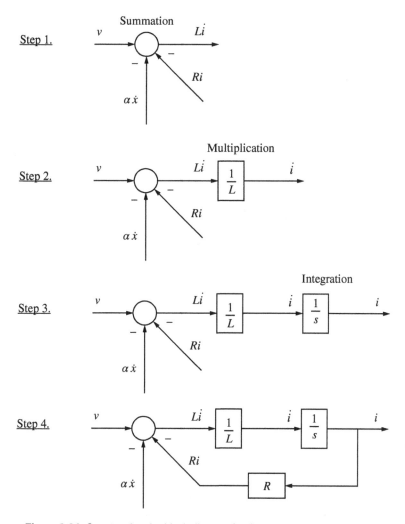

Figure 3-24. Constructing the block diagram for the current part of Figure 3-3.

where L is the coil inductance, i is the current flowing through the coil, R is the coil resistance, v is the applied voltage across the coil, α is the magnetic coupling coefficient (see Equation 3.5), and x is the displacement of the armature. Also, in Equation (3.142) M is the mass of the armature, C is the damping coefficient on the armature, k is the spring rate that resists a positive motion of the armature, and F is the applied force to the armature.

To construct the block diagram for the electromagnetic actuator we begin by considering the first part of Equation (3.142) in which the current is described dynamically. As shown in Figure 3-24, the current equation may be schematically represented using the summation symbol. By multiplying the output of the summation symbol by $1/L$ we are prepared to use the Laplace operator to represent the dynamic state of current shown by the symbol i. Step 3 in Figure 3-24 illustrates this process. Finally, by connecting the output branch of the current signal to the summation symbol through the multiplication block R the block diagram for the current dynamics is constructed in its most complete form so far.

Next, the block diagram for the armature must be constructed. This diagram is constructed in a fashion that is very similar to that of the piston for the hydromechanical system shown in Figure 3-2 except for the fact that the input to the summation symbol is now a multiple of current instead of a multiple of pressure. Compare Figures 3-22 and 3-25. Steps 1 through 4 are the same for constructing the block diagram for the armature of the electromechanical system as they have been for the previous block diagram constructions.

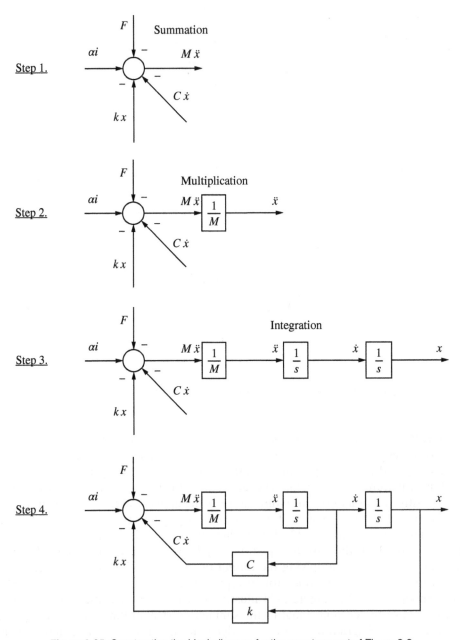

Figure 3-25. Constructing the block diagram for the armature part of Figure 3-3.

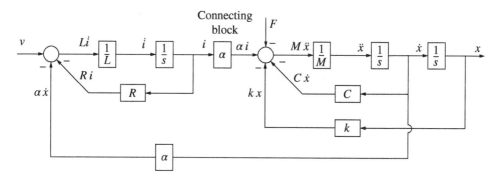

Figure 3-26. Block diagram for the electromechanical systems of Figure 3-3.

The final step in constructing the block diagram for the entire system shown in Figure 3-3 is to connect the current diagram with the armature diagram. This is done through the magnetic coupling coefficient and the final block diagram is shown in Figure 3-26.

Summary. In summary, it can be seen that the block diagram provides a graphical means for representing the governing equations of a dynamical system. This is a very useful method for visualizing the cause and effect relationship within a given system and may be used to provide insight into various adjustments to the system that can be made to alter the system's dynamic response. It is also worth mentioning that constructions like these are very common tools for the Graphical User Interface (GUI) of dynamic simulation packages such as MATLAB's Simulink® and other similar programs.

3.7 STABILITY

3.7.1 General

When a dynamical system is perturbed from a position of initial rest it will either vary from its initial condition and return to rest or it will grow in its variation until it reaches a physical limit of failure or obstruction. If the system returns to rest after being perturbed the system is considered to be stable. If the system variation continues to grow after being perturbed the system is considered to be unstable. Since an unstable control system is generally not useful, system stability is one of the most important properties of a dynamical system. In this section of our chapter we will consider the stability of linear systems using the classical Routh-Hurwitz stability criterion (1877).

3.7.2 Stability Criterion

Stability and the Complex Plane. The stability of a linear system can be determined by locating the roots of the system's characteristic equation within the complex plane. Figure 3-10 shows an example in which the roots of the characteristic equation have been located in the complex plane for a stable second-order linear system. In this figure the roots to the second-order characteristic equation are shown as complex conjugates with all real components being negative. It is the property of negative real-components that makes the system stable since these are the quantities that determine either an exponential decay or an exponential growth for the transient response. For example, see the transient response of the underdamped, unforced second-order system given in Equation (3.56). In this equation the negative sign on the exponential term comes from the negative real component of the solution to the characteristic equation. This negative sign indicates that the oscillating response of the system will decay with respect to time and in the limit as time reaches infinity this system will tend toward a steady-state condition. In other words, the system will be stable as dictated by the negative sign of the real components that are determined from the system's characteristic equation.

For a second-order system the characteristic equation is given in Equation (3.42). For higher-order systems the characteristic equation may be determined by formulating the problem in state space and carrying out the mathematical functions that are described in Equation (3.93). By setting this result to zero the following expression may be written for determining the roots to the characteristic equation of the nth-order system:

$$a_o s^n + a_1 s^{n-1} + \ldots + a_{n-1} s + a_n = 0, \tag{3.143}$$

where a_o through a_n are system coefficients and s is a system eigenvalue. By solving Equation (3.143) for the system eigenvalues the roots to the characteristic equation will have been determined and the sign of the real components to this solution will indicate whether the system is stable. Since the solution to a polynomial beyond the third degree is difficult and, in most cases, impossible to find without the use of numerical methods, it is convenient to use a systematic approach for calculating the number of eigenvalues that appear on the right-hand side of the complex plane. This approach is known as the Routh-Hurwitz stability criterion and is discussed in the following paragraphs.

The Routh-Hurwitz Stability Criterion.

The Routh-Hurwitz stability criterion is useful for determining the number of system eigenvalues that appear on the right-hand side of the complex plane. This criterion is presented here without proof; however, the reader is encouraged to examine other texts that provide a more thorough treatment of this subject [2, 4].

The first part of the Routh-Hurwitz stability criterion states that a necessary condition for system stability is that all coefficients in the characteristic equation must be positive. From Equation (3.143) it can be seen that it is possible to multiply both sides of the characteristic equation by a negative one if all coefficients appear to be negative. By doing this the first part of the Routh-Hurwitz stability criterion can be satisfied and we can move on to the second part of the Routh-Hurwitz stability criterion. If this first part cannot be satisfied then the system will be unstable and there is no need to follow the procedure further.

$$
\begin{array}{c|cccccc}
s^n & a_o & a_2 & a_4 & a_6 & \cdots \\
s^{n-1} & a_1 & a_3 & a_5 & a_7 & \cdots \\
s^{n-2} & b_1 & b_2 & b_3 & b_4 & \cdots \\
s^{n-3} & c_1 & c_2 & c_3 & c_4 & \cdots \\
\vdots & \vdots & \vdots & \vdots & & \\
s^2 & d_1 & d_2 & & & \\
s^1 & e_1 & & & & \\
s^o & f_1 & & & &
\end{array}
\tag{3.144}
$$

The second part of the Routh-Hurwitz stability criterion makes use of the Routh array, which is formed using the characteristic equation and the format of Equation (3.144).

In Equation (3.144), once the coefficients of the characteristic equation are in place, the remaining coefficients may be calculated as follows:

$$b_1 = \frac{a_1 a_2 - a_o a_3}{a_1} \quad b_2 = \frac{a_1 a_4 - a_o a_5}{a_1} \quad b_3 = \frac{a_1 a_6 - a_o a_7}{a_1} \quad \cdots$$

$$c_1 = \frac{b_1 a_3 - a_1 b_2}{b_1} \quad c_2 = \frac{b_1 a_5 - a_1 b_3}{b_1} \quad c_3 = \frac{b_1 a_7 - a_1 b_4}{b_1} \quad \cdots \qquad (3.145)$$

$$\vdots \qquad\qquad \vdots \qquad\qquad \vdots \qquad \vdots$$

The second part of the Routh-Hurwitz stability criterion states that the number of roots for the characteristic equation with positive real parts is equal to the number of sign changes that are observed in the first column of the Routh array. In other words, for absolute stability each coefficient in the first column of the Routh array must have the same sign. Let us consider a few important examples.

Second-Order Systems. For a second-order system, the characteristic equation is given by the following expression:

$$a_o s^2 + a_1 s^1 + a_2 s^o = 0. \qquad (3.146)$$

From the first part of the Routh-Hurwitz stability criterion we must show that all coefficients in the characteristic equation are positive if the system is to be stable. This part of the stability criterion is given by

$$a_o, a_1, a_2 > 0. \qquad (3.147)$$

Using the form of Equations (3.144) and (3.145), the Routh array for the second-order system may be written as

$$\begin{array}{ccc} s^2 & a_o & a_2 \\[2mm] s^1 & a_1 & \cdot \\[2mm] s^o & a_2 & \end{array} \qquad (3.148)$$

For stability, the first column of coefficients in the Routh array must have the same sign. By inspection, it can be seen that this requirement is the same stability requirement as the one given in Equation (3.147); therefore, the only stability requirement for the second-order system is that the coefficients of the characteristic equation must all be positive.

Third-Order Systems. For a third-order system the characteristic equation is given by the following expression:

$$a_o s^3 + a_1 s^2 + a_2 s^1 + a_3 s^o = 0. \qquad (3.149)$$

For stability, the first part of the Routh-Hurwitz stability criterion must be used to show that all coefficients in the characteristic equation are positive. This part of the stability criterion is given by

$$a_o, a_1, a_2, a_3 > 0. \qquad (3.150)$$

Using the form of Equations (3.144) and (3.145) the Routh array for the third-order system may be written as

$$
\begin{array}{ccc}
s^3 & a_o & a_2 \\
\\
s^2 & a_1 & a_3 \\
\\
s^1 & \dfrac{a_1 a_2 - a_o a_3}{a_1} & \\
\\
s^o & a_3 &
\end{array}
\tag{3.151}
$$

For stability the first column of coefficients in the Routh array must have the same sign. If the requirement of Equation (3.150) is satisfied it can be shown from Equation (3.151) that the remaining stability requirement for the third-order system is given by

$$a_1 a_2 > a_o a_3. \tag{3.152}$$

For the third-order system to be stable, both Equations (3.150) and (3.152) must be satisfied.

Case Study

For the hydromechanical system shown in Figure 3-2 and described by Equation (3.21) the characteristic equation may be written as

$$a_o s^3 + a_1 s^2 + a_2 s + a_3 = 0,$$

where

$$a_o = \frac{MV_o}{\beta}, \qquad a_1 = C_l M, \qquad a_2 = \frac{kV_o}{\beta} + A^2, \qquad a_3 = kC_l.$$

In this result the viscous drag coefficient C has been neglected due to its assumed small size. Using the Routh-Hurwitz stability criterion for a third-order system; that is, Equation (3.152), develop a simplified expression for the stability requirements of the hydromechanical system shown in Figure 3-2.

Substituting the above coefficients into Equation (3.152) and simplifying the result produces the following stability criterion for the system:

$$\beta A^2 > 0.$$

Obviously, this criterion is always satisfied and therefore we expect the hydromechanical system of Figure 3-2 to exhibit good stability. This method of analysis is used in Chapter 4 to discuss the stability of two-way hydraulic control valves.

Higher-Order Systems. It has just been shown that the Routh-Hurwitz stability criterion provides a useful tool for determining the stability of second- and third-order systems. The stability requirements that come from this exercise are indeed useful for designing a stable system and are employed in our analysis of hydraulic control systems; however, the number of stability requirements that must be satisfied tends to increase as the system order becomes larger. As the system order increases it becomes more and more difficult to use the Routh-Hurwitz stability criterion as a closed-form tool for designing a stable system; nevertheless, this criterion remains valid for stability assessment of higher-order systems and may be used for verification if not a priori design.

3.7.3 Summary

In summary, it has been noted that stability is one of the most important properties of a dynamical system. The Routh-Hurwitz stability criterion is a useful tool for assessing the stability of any linear system and provides a powerful means for designing stable systems that are of the second- or third-order type. Once it has been determined that the system is stable, the next most important consideration pertains to the dynamic response of the system. To adjust the dynamic response in a favorable way, feedback compensation is often employed by control engineers. In the following section feedback compensation is discussed by using the standard proportional-integral-derivative (PID) control structure. Although this is not the only control structure available to the engineer it is a standard one and provides an excellent starting place for designing a feedback control system.

3.8 FEEDBACK CONTROL

3.8.1 General

In this section we consider the important topic of feedback control. This topic seeks to show that a tuned controller design may be used to modify the output response and of a controlled system. A feedback control system includes a plant with actuators and sensors, and a controller. The plant is the system to be controlled. Actuators are used to manipulate or force the response of the plant. Sensors are used to measure the output, x, of the plant that is to be controlled. A controller is an algorithm that takes information about the desired response, and the measured output of the plant and determines the input, u, to the plant's actuator. We assume that the control system operates on the difference between the desired output and the actual output. This difference between desired and actual outputs is known as the system error, which is given by

$$\varepsilon = x_d - x, \tag{3.153}$$

where x is the actual systems output and x_d is the desired output. The error is used to quantify the performance. Another important signal to be considered is the disturbance input. The disturbance, d, is an input to the plant such as a force that causes the actual output, x, to change in an undesired way, such as moving away from the reference or desired output, x_d, which increases error, ε.

There are two main goals for a feedback control system design, to achieve stability and desired performance. Stability is of obvious importance since the controlled response should not exponentially increase as in an unstable system. Performance can be described in many ways but generally relates to specifications on how fast and accurately a feedback control system responds. For example, performance could be specified in terms of bandwidth, settling time, steady-state error, or percent overshoot. In general, when there is a disturbance or reference input signal, the error should decrease quickly to a small value to achieve good performance.

The loop transfer function, $L(s)$, is the product of all transfer functions in the feedforward and feedback parts of the closed-loop system in Figure 3-27,

$$L(s) = G(s)C(s). \tag{3.154}$$

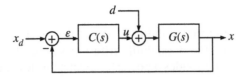

Figure 3-27. Block diagram of a feedback control system.

The transfer function between the desired output and the error can be found by block diagram algebra as

$$\frac{L[\varepsilon(t)]}{L[x_d(t)]} = \frac{1}{1 + L(s)} = S(s). \tag{3.155}$$

Similarly, the transfer function between the desired output and the actual output is given by

$$\frac{L[x(t)]}{L[x_d(t)]} = \frac{L(s)}{1 + L(s)} = T(s). \tag{3.156}$$

To illustrate the main points of this topic we select the classic second-order dynamical system as our object of control. A block diagram of this system is shown in Figure 3-28 where $u(t)$ is the controller input to the system, $d(t)$ is the disturbance input, ω_n is the undamped natural frequency of the system, ζ is the damping ratio, and x is the system output that we wish to control.

The equation of motion for this system is given as

$$\ddot{x} + 2\zeta\omega_n\,\dot{x} + \omega_n^2 x = u(t) + d(t). \tag{3.157}$$

The plant transfer function, $G(s)$, relating the Laplace transform of the input, $L[u(t) + d(t)]$ to the output, $L[x(t)]$ is ,

$$\frac{L[x(t)]}{L[u(t) + d(t)]} = \frac{X(s)}{U(s) + D(s)} = \frac{\omega_n^2}{s^2 + 2\zeta\omega_n s + \omega_n^2} = G(s). \tag{3.158}$$

Error Dynamics. Dynamical systems are characterized by time varying outputs that instantaneously differ from a desired output. If we assume that the desired system output is a constant, which describes a regulation control problem, then the system dynamics may

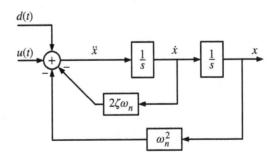

Figure 3-28. Block diagram of a classical second-order system.

be expressed in terms of the error dynamics and the desired constant system output. Using Equation (3.153) the system dynamics may be expressed as

$$x = x_d - \varepsilon, \qquad \dot{x} = -\dot{\varepsilon}, \qquad \ddot{x} = -\ddot{\varepsilon}. \tag{3.159}$$

Substituting Equation (3.159) into Equation (3.157) produces the following equation for the error dynamics of the second-order system:

$$\ddot{\varepsilon} + 2\zeta\omega_n\dot{\varepsilon} + \omega_n^2\varepsilon = \omega_n^2 x_d - u(t) + d(t). \tag{3.160}$$

3.8.2 PID Controller Design in the Time Domain

A brief glance at a classical textbook on the control of linear systems [2, 4] shows that the proportional-integral-derivative (PID) controller has been used as the basis for controlling many physical systems including space shuttle applications. These controllers are designed to adjust the input signal to a system in response to an observed output. Figure 3-29 shows a block diagram for the classical PID controller. In this figure the desired output for the system is shown by the input to the summation symbol x_d, the actual output for the dynamical system is shown by x, the calculated error between the desired and actual output is given by ε, and the adjustment input to the dynamical system is shown by the symbol $u(t)$.

In the multiplication blocks of this schematic we see the proportional controller gain K_e, the integral controller gain K_i, and the derivative controller gain K_d. The symbol s is the standard Laplace variable that indicates either a derivative operation or an integral operation depending on whether it is used to multiply or divide the gain. As shown in Figure 3-29 the controller blocks are multiplied by the error signal and summed to create the adjusted output signal $u(t)$, which is then used as the controller input signal to the dynamical system. It should be noted that this block diagram can be connected directly to the dynamical system shown in Figure 3-27 by connecting signals appropriately. Using the block diagram in Figure 3-29 the adjusted input signal from the PID controller may be written as

$$u(t) = K_e\varepsilon + K_i \int \varepsilon \, dt + K_d\dot{\varepsilon}, \tag{3.161}$$

where the controller gains are ultimately selected based on the desired output of the dynamical system. This controller is discussed in its various configurations in the following subsections.

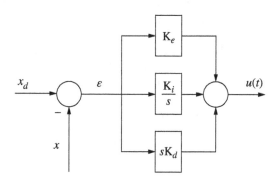

Figure 3-29. Block diagram of a classical PID controller.

P Control. The proportional control (P control) is the simplest control that can be used for adjusting the input signal to the dynamical system. By setting the integral and derivative gains equal to zero Equation (3.161) may be used to describe the control signal for the P control in the following form:

$$u(t) = K_e \varepsilon. \tag{3.162}$$

Substituting this result into Equation (3.160) shows that the error dynamics for the P-controlled system are given by

$$\ddot{\varepsilon} + 2\zeta\omega_n\dot{\varepsilon} + (\omega_n^2 + K_e)\,\varepsilon = \omega_n^2 x_d. \tag{3.163}$$

The steady-state solution to the error dynamics equation may be determined by setting the time derivatives of Equation (3.163) equal to zero and solving for ε. This result is given by the following expression:

$$\varepsilon = \frac{\omega_n^2 x_d}{\omega_n^2 + K_e}, \tag{3.164}$$

which shows that the steady-state error for the P-controlled dynamical system is nonzero. This nonzero steady-state error is a well-known characteristic of P-controlled systems and is usually one of its most undesirable features. We would usually prefer to see the system error vanish at the steady-state condition. Clearly, by increasing the proportional control-gain K_e the steady-state error may be reduced; however, this may create other dynamic problems that are discussed presently.

The error dynamics for the second-order P-controlled system may be discussed using the terminology that is germane to all second-order systems; namely, the undamped natural frequency and the damping ratio. These two quantities are defined in Equation (3.40) for the lumped parameters that are shown in Equation (3.38). Using the lumped parameter groups shown in Equation (3.163) and the general form of Equation (3.40) it may be shown that the undamped natural frequency and the damping ratio for the error dynamics of the P-controlled system are given respectively as

$$\Omega_n = \sqrt{\omega_n^2 + K_e} \quad \text{and} \quad Z = \zeta\frac{\omega_n}{\Omega_n}. \tag{3.165}$$

As noted in Equations (3.69) through (3.75) the undamped natural frequency and the damping ratio have a decisive impact on the transient response of the system. By increasing the proportional gain K_e the undamped natural frequency is increased, which tends to reduce the system rise time; however, this adjustment also tends to reduce the damping ratio, which in turn increases the maximum percent overshoot. This second effect is perhaps the most negative impact on the dynamic response. Interestingly enough it can be shown that adjustments in the proportional gain have no impact on the settling time.

PI Control. The proportional-integral control (PI control) is a commonly used control for dynamic systems. This control may be constructed by setting the derivative gain equal to zero in Equation (3.161), which yields the following result:

$$u(t) = K_e\varepsilon + K_i\int \varepsilon\, dt. \tag{3.166}$$

Substituting this result into Equation (3.160) shows that the error dynamics for the PI-controlled system are given by

$$\ddot{\varepsilon} + 2\zeta\omega_n\dot{\varepsilon} + (\omega_n^2 + K_e)\varepsilon = \omega_n^2 x_d - K_i\int \varepsilon\, dt. \tag{3.167}$$

By taking the first derivative of this equation with respect to time the derivative form of the error dynamics equation may be written as

$$\dddot{\varepsilon} + 2\zeta\omega_n\ddot{\varepsilon} + (\omega_n^2 + K_e)\dot{\varepsilon} + K_i\varepsilon = 0. \tag{3.168}$$

The first thing to notice about Equation (3.168) is that the steady-state error for this system is zero. This can be determined by setting the time derivatives equal to zero and solving for ε. This feature of having a zero error at the steady-state condition results from the integral part of the controller design and is usually the main motivation for using the integral gain in the feedback loop. The reader will recall that a simple P-controlled system exhibits a nonzero steady-state error.

To discuss the error dynamics for the PI-controlled system it will be useful to reduce the order of Equation (3.168) by one. This is possible to do when the $\dddot{\varepsilon}$ term in Equation (3.168) dies away very quickly compared to the other time derivative terms and may therefore be safely neglected when consider the dynamic response. Under these conditions the error dynamics equation may be written as

$$\ddot{\varepsilon} + \frac{\omega_n^2 + K_e}{2\zeta\omega_n}\dot{\varepsilon} + \frac{K_i}{2\zeta\omega_n}\varepsilon = 0. \tag{3.169}$$

Again, the dynamic response of this second-order equation may be discussed using the terminology that is germane to all second-order systems; namely, the undamped natural frequency and the damping ratio. Using the lumped parameter groups shown in Equation (3.169) and the general form of Equation (3.40) it may be shown that the undamped natural frequency and the damping ratio for the error dynamics of the PI-controlled system are given respectively as

$$\Omega_n = \sqrt{\frac{K_i}{2\zeta\omega_n}} \quad \text{and} \quad Z = \frac{\omega_n^2 + K_e}{4\zeta\omega_n\Omega_n}. \tag{3.170}$$

From Equations (3.69) through (3.75) of this chapter the effect of adjusting the proportional and integral gain on the transient response of the system may be determined qualitatively. For instance, the integral gain K_i may be increased to reduce the rise time while the proportional gain K_e may be increased to reduce the maximum percent overshoot and the settling time. One of the advantages of using the PI controller is that independent adjustments of the integral and proportional gains can be made to adjust noncompeting response characteristics of the second-order system. This makes the tuning of the controller very easy, especially if these adjustments can be made by means of a computer.

PID Control. The proportional-integral-derivative control (PID control) utilizes the entire control structure of Figure 3-29 and Equation (3.161). By substituting Equation (3.161) into Equation (3.160) the error dynamics for the PID-controlled system are given by

$$\ddot{\varepsilon} + (2\zeta\omega_n + K_d)\dot{\varepsilon} + (\omega_n^2 + K_e)\,\varepsilon = \omega_n^2\, x_d - K_i \int \varepsilon\, dt. \tag{3.171}$$

By taking the first derivative of this equation with respect to time, the derivative form of the error dynamics equation may be written as

$$\dddot{\varepsilon} + (2\zeta\omega_n + K_d)\ddot{\varepsilon} + (\omega_n^2 + K_e)\dot{\varepsilon} + K_i\varepsilon = 0. \tag{3.172}$$

Again, the steady-state error for this system is zero. Similar to the PI-controlled system this feature of having a zero error at the steady-state condition results from the integral part of the controller design and is usually the main motivation for using the integral gain in the feedback loop. The reader will recall that a simple P-controlled system exhibits a nonzero steady-state error.

To discuss the error dynamics for the PID-controlled system it will once again be useful to reduce the order of Equation (3.172) by one. This is possible to do when the $\dddot{\varepsilon}$ term in Equation (3.172) dies away very quickly compared to the other time derivative terms and may therefore be safely neglected when consider the dynamic response. Under these conditions the error dynamics equation may be written as

$$\ddot{\varepsilon} + \frac{\omega_n^2 + K_e}{2\zeta\omega_n + K_d}\,\dot{\varepsilon} + \frac{K_i}{2\zeta\omega_n + K_d}\varepsilon = 0. \tag{3.173}$$

Again, the dynamic response of this second-order equation may be discussed using the terminology that is germane to all second-order systems; namely, the undamped natural frequency and the damping ratio. Using the lumped parameter groups shown in Equation (3.173) and the general form of Equation (3.40) it may be shown that the undamped natural frequency and the damping ratio for the error dynamics of the PID-controlled system are given respectively as

$$\Omega_n = \sqrt{\frac{K_i}{2\zeta\omega_n + K_d}} \quad \text{and} \quad Z = \frac{\omega_n^2 + K_e}{2(2\zeta\omega_n + K_d)\,\Omega_n}. \tag{3.174}$$

From this result it can be seen that the derivative gain plays an important role in determining both the undamped natural-frequency and the damping ratio of the system. This controller design is much harder to tune since adjustments in the derivative control have competing effects on the output response of the system. The primary purpose in using the derivative control is to increase the system's stability. This is accomplished by anticipating the actuator error and introducing early corrective action into the control signal [4].

PID Control Gain Selection.   In the previous subsections the dynamic response of a compensated system was presented in terms of the system parameters and the control gains that were selected for the PID controller. As shown in these discussions the output response of the system may be designed by selecting appropriate gains for the PID controller; however, in practice there are physical saturation limits that restrict the control gain selection. This subsection presents a methodology for selecting reasonable control gains for a system that is to be controlled using a PI controller.

Figure 3-30 shows a schematic of a system error ε and the PI controller output $u(t)$ versus time. As shown in the schematic the error is modeled as though it diminishes linearly with respect to time; however, the actual error response will depend on the physics of the system being controlled. The linear model for the error shown in Figure 3-30 is only a first approximation for an error response and is used here to select reasonable gains for the PI-controlled system, which are not yet optimized. As shown in Figure 3-30 the error begins with a worst case scenario in which it is at a maximum value ε_{max}.

When time equals t_{max} the error has completely vanished. This linear approximation for the error is expressed mathematically as

$$\varepsilon = \varepsilon_{max} - \frac{\varepsilon_{max}}{t_{max}}t. \tag{3.175}$$

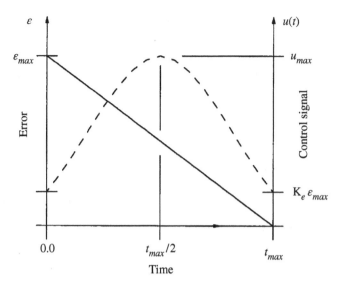

Figure 3-30. Approximate error and controller output.

Substituting this model for the error into Equation (3.166) produces the following quadratic expression for the controller output:

$$u(t) = -\frac{K_i\varepsilon_{max}}{2t_{max}}t^2 + \frac{(K_it_{max} - K_e)\varepsilon_{max}}{t_{max}}t + K_e\varepsilon_{max}. \tag{3.176}$$

By maximizing this function it may be shown that maximum controller output u_{max} occurs when

$$t = \frac{K_i t_{max} - K_e}{K_i}. \tag{3.177}$$

Substituting this result back into Equation (3.176) shows that the maximum controller output is given by

$$u_{max} = \frac{(K_e^2 + K_i^2 t_{max}^2)\varepsilon_{max}}{2K_it_{max}}. \tag{3.178}$$

This maximum controller output is shown in Figure 3-30 to occur when time equals $t_{max}/2$. The selection of this maximized location is somewhat arbitrary but has been used to produce a reasonable response for the controller. By setting t equal to $t_{max}/2$ in Equation (3.177) a simultaneous solution of Equations (3.177) and (3.178) may be used to show

$$K_e = \frac{4}{5}\frac{u_{max}}{\varepsilon_{max}} \quad \text{and} \quad K_i = \frac{2K_e}{t_{max}}. \tag{3.179}$$

As previously stated, due to the idealized model for the system error the results of Equation (3.179) are only "ballpark" or reasonable estimations for the PI control gains. The best values for these control gains may deviate from this equation somewhat depending on the actual system being controlled. If the system is being controlled electronically these gains may be adjusted online through the use of a computer; however, they will not ordinarily deviate from Equation (3.179) by an order of magnitude.

3.8.3 Control Design in the Frequency Domain

General Frequency Response.   The frequency response for first and second-order dynamical systems were derived in this chapter. The purpose of this subsection is to describe a generic method for determining the frequency response of a dynamical system described by a linear ordinary differential equation of any order to allow for control design in the frequency domain. Similar to Equation (3.88), a differential equation relating one output to one input can be written in general as,

$$\ldots + \ddot{y}a_2 + \dot{y}a_1 + ya_0 = ub_0 + \dot{u}b_1 + \ddot{u}b_2 + \ldots \tag{3.180}$$

Assuming zero initial conditions, taking the Laplace transform, and solving for the ratio of the Laplace transform of the output over the Laplace transform of the input yields,

$$\frac{\mathrm{L}[y(t)]}{\mathrm{L}[u(t)]} = \frac{Y(s)}{U(s)} = \frac{\ldots + s^2 b_2 + s b_1 + b_0}{\ldots + s^2 a_2 + s a_1 + a_0} = G(s), \tag{3.181}$$

which is a transfer function. Let the input "forcing function" be $u(t) = U(s)e^{st}$ and then the output "forced response," also called the particular solution, will have a similar form, $y(t) = Y(s)e^{st}$, where $U(s)$ and $Y(s)$ are complex functions and $s = \sigma + j\omega$ is a complex number. Also, the derivatives of the input and output are $\dot{u}(t) = U(s)se^{st}, \ddot{u}(t) = U(s)s^2 e^{st}, \ldots$ and $\dot{y}(t) = Y(s)se^{st}, \ddot{y}(t) = Y(s)s^2 e^{st}, \ldots$ If we let $s = j\omega$, then the input and its derivatives become,

$$u(t) = U(j\omega)e^{j\omega t} = U(j\omega)(\cos(\omega t) + j\sin(\omega t)),$$

$$\dot{u}(t) = U(j\omega)j\omega e^{j\omega t} = U(j\omega)j\omega(\cos(\omega t) + j\sin(\omega t)), \text{ and}$$

$$\dot{u}(t) = U(j\omega)(j\omega)^2 j\omega e^{j\omega t} = U(j\omega)(j\omega)^2(\cos(\omega t) + j\sin(\omega t)), \tag{3.182}$$

a sinusoidal form due to Euler's Formula ($e^{j\omega t} = \cos(\omega t) + j\sin(\omega t)$). Similarly, the output and its derivatives are

$$y(t) = Y(j\omega)e^{j\omega t} = Y(j\omega)(\cos(\omega t) + j\sin(\omega t)),$$

$$\dot{y}(t) = Y(j\omega)j\omega e^{j\omega t} = Y(j\omega)j\omega(\cos(\omega t) + j\sin(\omega t)), \text{ and}$$

$$\ddot{y}(t) = Y(j\omega)(j\omega)^2 e^{j\omega t} = Y(j\omega)(j\omega)^2(\cos(\omega t) + j\sin(\omega t)), \tag{3.183}$$

which are also sinusoidal functions in time. After substituting Equations (3.182) and (3.183) into Equation (3.180) and solving for the ratio of $Y(j\omega)/U(j\omega)$ the frequency response transfer function for the differential equation, Equation (3.180), can be found as,

$$\frac{Y(j\omega)}{U(j\omega)} = \frac{\ldots + b_2(j\omega)^2 + b_1(j\omega) + b_0}{\ldots + a_2(j\omega)^2 + a_1(j\omega) + a_0} = G(j\omega), \tag{3.184}$$

where $G(j\omega)$ is given as a symbol to generically represent a frequency response transfer function that can be compared to the Laplace transform transfer function, $G(s)$. When comparing Equation (3.184) to Equation (3.181) it becomes apparent that for any transfer function, the frequency response transfer function is found by substituting $s = j\omega$ into the transfer function. In general, the frequency response at any frequency, ω, can be determined for a transfer function, $G(s)$, by substituting the imaginary number, $j\omega$, for s so that the frequency response is $G(j\omega)$.

The frequency response transfer function in Equation (3.184) is the ratio of sinusoidal output to sinusoidal input. If the forcing input is sinusoidal, the sinusoidal forced output can be found as,

$$y_f(t) = Y(j\omega)e^{j\omega t} = G(j\omega)U(j\omega)e^{j\omega t}. \tag{3.185}$$

If the input is a real valued sinusoidal function, then the imaginary part of $u(t) = U(j\omega)e^{j\omega t} = U(j\omega)(\cos(\omega t) + j\sin(j\omega t))$ can be used so that $u(t) = U_0 \sin(\omega t)$. Here U_0 is the amplitude of the input. By using the polar form of the frequency response transfer function ($G(j\omega) = |G(j\omega)|e^{j\angle G(j\omega)}$), it can be shown that Equation (3.185) can be written as,

$$y_f(t) = |G(j\omega)|U(j\omega)e^{j\omega t - \angle(G(j\omega))} = |G(j\omega)|U_0 \sin(\omega t + \phi), \tag{3.186}$$

where $|G(j\omega)|$ is called the amplitude ratio and $\phi = \angle G(j\omega)$ is called the phase angle [5]. Equation (3.186) can be used to find the forced response of a system with transfer function, $G(j\omega)$, and sinusoidal input, $U_0 \sin(\omega t)$. Notice that the output is a sin function with the same frequency as the input, but with amplitude, $|G(j\omega)|U_0$, and phase angle $\angle G(j\omega)$.

In a Bode diagram for a transfer function $G(s)$, the magnitude (also referred to as gain or amplitude ratio), $|G(j\omega)|$, and phase angle, $\angle G(j\omega)$, of the complex valued frequency response of a transfer function, $G(j\omega)$, are plotted. The magnitude is usually given in decibels (dB) and the phase is usually given in degrees, although calculations involving phase angle are done using radians, unless noted. Decibels are computed as $20\log_{10}|G(j\omega)|$, making the vertical axis of the magnitude plot a logarithmic scale. The frequency axis is usually plotted on a logarithmic scale. As an example of a Bode diagram, the magnitude and phase of the transfer function $G(s) = \frac{\omega_n^2}{s^2 + 2\zeta\omega_n s + \omega_n^2}$ for the system in Figure 3-28 with $\zeta = 0.5$ are plotted versus a normalized frequency ω/ω_n in Figure 3-31. Notice that the slope of the magnitude plot decreases by 40 dB/decade and the phase angle decreases by 180 degrees around the natural frequency, where $\omega/\omega_n = 10^0 = 1$, which is a frequency equal to the magnitude of the complex conjugate pair of poles of $G(s)$. In general, for each pole, the slope of the magnitude plot decreases by 20 dB/decade and the phase angle decreases by 90 degrees around the frequency equal to the magnitude of the pole. The opposite occurs with the slope of the

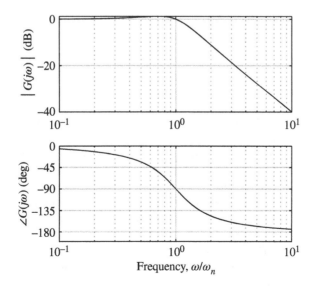

Figure 3-31. Frequency response of $G(s) = \frac{\omega_n^2}{s^2 + 2\zeta\omega_n s + \omega_n^2}$.

magnitude plot increasing by 20 dB/decade and the phase angle increasing by 90 degrees at the frequencies equal to the magnitude of each root of the numerator polynomial of a transfer function. The change in slope and phase are doubled to 40 dB and 180 degrees when two poles or roots of the numerator polynomial have the same magnitude as in a complex conjugate pair of roots.

Stability in the Frequency Domain.   In control design stability is required before consideration of performance goals since unstable systems cannot meet any of the performance objectives as we have defined them. In this subsection, we consider a way to consider stability that is different from the Routh-Hurwitz stability criterion also discussed in this chapter. Stability of the closed-loop system in Figure 3-27 may be determined using the frequency response of the open-loop transfer function, $L(j\omega)$, and Bode's stability condition. For Bode's stability condition, a few definitions are needed. The gain crossover frequency, ω_c, is the frequency at which the gain of the loop transfer function frequency response, $|L(j\omega)|$, crosses over unity gain (0 dB). The -180 frequency, ω_{180}, is the frequency at which the phase angle, $\angle L(j\omega)$, crosses over -180 deg (i.e., $\angle L(j\omega_{180}) = -180$ deg). The definition of phase margin (PM) is the difference between the phase angle, $\angle L(j\omega_c)$, and -180 deg.

$$PM = \angle L(j\omega_c) + 180° \qquad (3.187)$$

Gain margin (GM) is the largest gain that the loop transfer function could be multiplied by before the crossover frequency approaches the -180 frequency.

$$GM = \frac{1}{|L(j\omega_{180})|}. \qquad (3.188)$$

For systems that are stable with small controller gains, the closed-loop system remains stable until the gain is increased to the point that the phase margin is equal to zero, any further increases cause instability in the closed-loop system. Bode's stability condition can be stated as: A negative feedback closed-loop system with loop transfer function, $L(s)$, is stable if and only if the open-loop system, $L(s)$, is stable and,

$$\frac{1}{|L(j\omega_{180})|} > 1. \qquad (3.189)$$

Both GM and PM are considered as measures of relative stability for systems where Bode's stability condition applies. Further details and limitations of the Bode's stability condition, which is related to Nyquist's stability criterion, can be found in Reference [6] and other control systems textbooks.

Minimum values of GM and PM are often specified as design requirements for control systems. The larger these margins are, the more stability margin is built into a design. A large gain margin means that if the gain of the system is increased, the system will remain stable and that the system will be robust to uncertainty in the gain. A large phase margin means that the closed-loop system is robust to uncertainty in the phase angle. Systems will large stability margins can tolerate variations in gain and additional phase lag without jeopardizing stability and to some extent, without jeopardizing performance.

Frequency Domain Performance.   Feedback control can be used to modify the performance of the closed-loop system. Performance of feedback control systems can be specified in the frequency domain in terms of open-loop specifications and closed-loop specifications. In either case, the goal is to specify the performance of the closed-loop system in Figure 3-27.

Open-loop specifications include crossover frequency and low frequency gain. The crossover frequency is often closely associated with the closed-loop bandwidth of the control system. A large low frequency gain translates into small error at low frequencies. In addition, stability margins (GM and PM) are often given as design specifications related to performance. This is because systems with small GM and PM tend to have oscillatory behavior that is evidence of near unstable behavior, that is associated with poor performance.

The performance of closed-loop systems can be expressed in terms of closed-loop transfer functions, the transfer function between x_d and ε, called the sensitivity transfer function, $S(s)$, or the transfer function between x_d and x, called the complementary sensitivity transfer function, $T(s)$. Closed-loop performance specifications include bandwidth and gain for $S(j\omega)$ and $T(j\omega)$. For closed-loop transfer function, we can define the bandwidth frequency as the frequency that the magnitude comes within $\sqrt{2}/2$ (3 dB) of one (zero dB). To have low tracking error, it is desired to have the gain of $S(j\omega)$ as small as possible up to the highest possible bandwidth frequency. For, $T(j\omega)$, it is desired to have a gain as close as possible to one so that the desired output and the actual output have a one to one relationship up to the highest bandwidth frequency possible.

P Control in the Frequency Domain. For the second-order system in Equation (3.157) with P control, the loop transfer function found by substituting K for $C(s)$ in Equation (3.154) to get

$$L(s) = G(s)K = \frac{1}{s^2 + 2\zeta\omega_n s + \omega_n^2}K. \tag{3.190}$$

The frequency response of the loop transfer function with $\zeta = 0.5$ is plotted in Figure 3-32 for two cases, $K = 2$ and $K=5$. For $K = 2$, ω_c is about 1.5 and PM is 49 deg, and the PM decreases to 28 deg as ω_c increases to 2.3 when $K = 5$. Since in both cases, the PM is greater

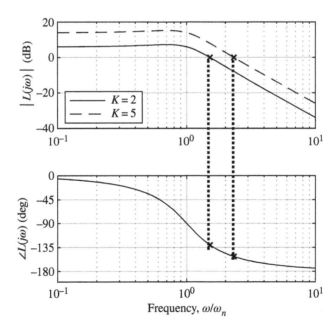

Figure 3-32. Frequency response of the loop transfer function for P control with $K = 2$ and $K = 3$. The vertical dashed lines are located at the crossover frequencies.

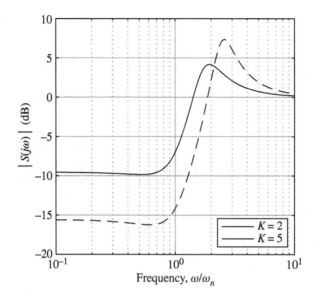

Figure 3-33. Frequency response of the closed-loop sensitivity transfer function, $S(j\omega)$, for P control with $K = 2$ and $K = 5$.

than 0 and the GM is infinite, the closed loop system is stable for both gains. The magnitude of the low frequency part of the Bode diagram increases as K is increased from 2 to 5, which indicates that tracking error is reduced as the control gain increases.

The closed-loop frequency response magnitude is given in Figure 3-33 for two control gains, K. In Figure 3-33 it can be seen that the frequency response of the sensitivity transfer function, $S(j\omega)$ (which gives the ratio of ε to x_d amplitude) has a lower gain at low frequencies when the control gain K is larger, which means that the system with a large controller gain will have a smaller error, ε, when there is a low frequency input signal, x_d. For example, at a frequency of $\omega/\omega_n = 10^{-1}$, the error gains are, -9.5 dB (0.34) and -16 dB (0.16) for the two controller gains and if the amplitude of x_d is 1.0, then the amplitude of ε is 0.34 for $K = 2$ and 0.16 for $K = 5$. The bandwidth of the closed-loop system is located at the point where the frequency response magnitude crosses -3 dB from below, which is 1.2 and 1.7 on the frequency axis for the gains $K = 2$ and $K = 5$, respectively. Since $K = 5$ results in a higher bandwidth frequency, the closed-loop system will be able to track signals with low error up to a higher frequency. The good low frequency performance for $K = 5$ is at the expense of a higher peak magnitude of $|S(j\omega)|$. In both cases of gains, K, there is a significant resonant peak in $|S(j\omega)|$ above the bandwidth frequency with the peak 4 dB higher for the system with the higher gain. Notice that the high-frequency gain in $|S(j\omega)|$ is 0 dB or 1 on the absolute scale for both control gains when $\omega/\omega_n = 10^1$. A gain of one means that the error amplitude will be 1 times the amplitude of the desired output, or in other words 100% error.

PID Control in the Frequency Domain.   In this subsection, the PID control system will be applied the second-order system in Equation (3.157). The transfer function for PID control can be derived from Equation (3.161) as

$$C(s) = \frac{K_d s^2 + K_e s + K_I}{s}. \tag{3.191}$$

The advantage of including the integral gain, K_I, is apparent by examining the magnitude of the frequency response of $C(j\omega)$. At very low frequencies, substitution of $s = j\omega \to 0$ into Equation (3.191) causes the frequency response magnitude of the controller to become very large (actually infinitely large) since there is an s in the denominator, which in turn causes the tracking error to be very small. The proportional gain, K_e, has a similar effect to that of K seen in the subsection about P control. The control gain K_d is increased to increase damping and reduce oscillations in the time response which corresponds to a reduction in a resonant peak in the closed-loop system frequency response.

The s in the denominator of $C(s)$ tends to make the phase angle more negative, contributing -90 deg of phase for all frequencies causing ω_{180} to be at a low frequency. This increase in phase lag causes the PM to be reduced significantly. If we let $C(s) = K_I/s$ (i.e., an I controller), the frequency response plot for $L(j\omega) = G(s)\frac{K_I}{s}\Big|_{s=j\omega}$ is given in Figure 3-34. Notice in this Bode diagram, that the GM is 0 dB and PM is zero degrees and therefore the system is not stable (actually just marginally stable) according to the Bode's stability condition. This stability condition is caused by the phase lag introduced by the integral control.

To counteract phase lag due to the s in the denominator of $C(s)$ and the phase lag of the plant transfer function $G(s)$, s is included in the numerator of the PID control system. The Laplace s in the numerator tends to reduce phase lag and can therefore improve stability in many cases by increasing the PM. It can be shown that the phase angle increases by 90 deg near the frequency of each root in the numerator of a transfer function and decreases by 90 deg near the frequency of each root of the denominator (pole) of a transfer function. Therefore, we choose a PID controller of the form

$$C(s) = K_d \frac{\left(s^2 + \frac{K_e}{K_d}s + \frac{K_I}{K_d}\right)}{s} = K_d \frac{(s + 2as + a^2)}{s} = K_d \frac{(s + a)^2}{s}, \qquad (3.192)$$

which has two real roots, a, in the numerator. In Equation (3.192) the values of the control gains in the PID controller can be related to a, $K_I = a^2 K_d$, $K_e = 2aK_d$. The two roots in the

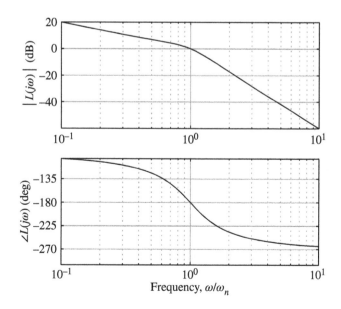

Figure 3-34. Frequency response of the open-loop transfer function, $L(j\omega)$, for I control with $K_I = 1$.

numerator will cause a beneficial increase in the phase angle of the loop transfer function at a frequency, $\omega = |a|$. In Figure 3-35, ω_c and ω_{180} both occurred at the same frequency, the natural frequency, ω_n; therefore, a can be chosen as $a = \omega_n$ so that the contribution to phase for the PID controller will occur as a reduction in phase lag near the natural frequency.

The frequency responses of the loop transfer function with PID control and two values of the derivaitve gain ($K_d = 2$ and $K_d = 5$) are plotted in Figure 3-35. The plot shows that the crossover frequency, ω_c, is 2.5 for $K_d = 2$ and $K_d = 5$. The PM is nearly 80 deg and the gain margin is infinite for both gains. The plot of the magnitude of the frequency response of $S(j\omega)$ is given in Figure 3-36. Figure 3-36 shows that the amplitude ratio of error is less than 0 dB (1.0 on the absolute scale) for all frequencies, indicating that there will be no errors greater than 100% of the desired output. Figure 3-36 also shows that the error will be small up to the closed-loop bandwidth frequency of 2.0 and 4.3 for gains $K_d = 2$ and $K_d = 5$; and the error will be zero at very low frequencies, which is a key benefit of integral control.

Plots of the time response of $x(t)$ due to a step input to $x_d(t)$, for the closed-loop system with P control and PID control are given in Figure 3-37 with each of the control gains, K_d and K, set to 5. Several characteristics of the frequency responses for these two control systems carry over to the time response. The predicted steady-state error is zero for PID control and 0.16 for P control appear in the time response simulation. The speed of response of the PID control system is much faster than the P control system, as predicted by that faster closed loop bandwidth and open loop crossover frequency, ω_c, for PID control. High PM usually translates into low overshoot, which is apparent since the PID control system had the highest PM and lower overshoot in the comparison between the P and PID control in Figure 3-37.

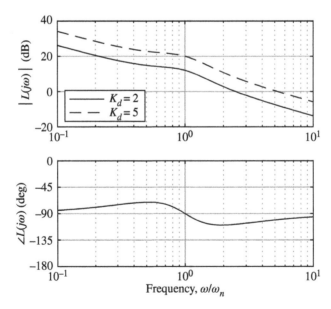

Figure 3-35. Frequency response of the open-loop transfer function, $L(j\omega)$, for PID control with $K_d = 2$ and $K_d = 5$.

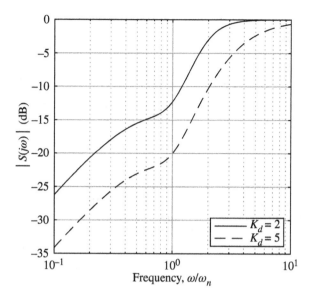

Figure 3-36. Frequency response of the closed-loop sensitivity transfer function, $S(j\omega)$, for PID control with chosen parameter, $a = \omega_n$, and $K_d = 2$ and $K_d = 5$.

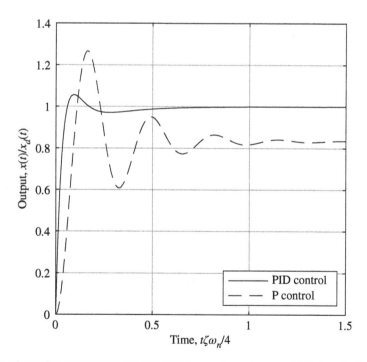

Figure 3-37. Step input time response of the closed-loop complementary sensitivity transfer function, $T(s)$, for P control and PID control with $z = \omega_n$ and $K_d = 5$ and $K = 5$.

3.8.4 Digital Control

General.   Implementation of control system is commonly achieved through the use of a microcontroller. A microcontroller is a small digital computer with an integrated capability for receiving measurement signals and producing control signals. These devices often include signal conditioning for interpreting sensor signals and the ability to amplify output signals to levels that can be used to drive electric solenoids, relay switches, and motors. Control systems like the ones considered in Section 3.8.2 and 3.8.3 are converted into a form compatible with the microcontroller, which is easily programmed, modified, and interfaced with other systems needed in machinery. These digital systems present a challenge since a sample hold effect is introduced when signals are sampled at regular time intervals, T, (discrete sampling, not continuous). To this point of the text, all signals have been considered to be sampled continuously in time. In this section the effects of discrete time on modeling and control design techniques in presented.

Discrete Time Modeling.   Since the control system (for example, the PID control system in Figure 3-29) is implemented using a microcontroller, there is a need to convert differential equations into a discrete time form. Difference equations, analogous to differential equations used to represent the dynamics in the continuous time domain, can be used to implement controllers and also approximate differential equations in general.

Consider a signal $y(t)$ in continuous time. If the signal is sampled n times starting at $t = 0$ with a constant discrete sampling rate of once every T seconds, then the sampled values are $y(0)\, y(T)\, y(2T)\, \dots\, y(nT)$. Let the $y(k)$ be the kth sample, $y(kT)$, of the signal $y(t)$, where $t = kT$ and k is a counting number starting at zero $(0, 1, 2, \dots)$. A difference equation for the time derivative of the signal $y(t)$ can be written as

$$y'(t) = \frac{d}{dt}y(t) \approx \frac{y(k) - y(k-1)}{T}, \tag{3.193}$$

by using the backward finite difference method for approximating the derivative. The term $y(k-1)$ is the value of y with one time delay. For a second derivative, two time delays are required,

$$y''(t) = \frac{d^2}{dt^2}y(t) \approx \frac{y(k) - 2y(k-1) + y(k-2)}{T^2}. \tag{3.194}$$

Higher time derivatives can be approximated in a similar manner, with each additional degree of differentiation requiring more time delays $y(k-1), y(k-2), y(k-3)$, and so on. Each derivative in a differential equation can be replaced by its discrete time equivalent to form a difference equation approximation of a differential equation. As an example, consider the second-order system in Equation (3.65), which can be converted to a difference equation,

$$x''(t) + 2\zeta\omega_n x'(t) + \omega_n^2 x(t) = F(t)/m$$

$$\Rightarrow x(k-2)\frac{1}{T^2} - x(k-1)\left(\frac{2}{T^2} + \frac{2\zeta\omega_n}{T}\right) + x(k)\left(\frac{1}{T^2} + \frac{2\zeta\omega_n}{T} + \omega_n^2\right) = F(k)/m, \tag{3.195}$$

by using Equations (3.193) and (3.194) with $x(t)$ as the variable instead of $y(t)$. This result illustrates one of the benefits of discrete time systems, which is that they are easy to solve using computers. To get the solution, $x(k)$, we can simply solve for $x(k)$ in terms of the previous values of x and F,

$$x(k) = \frac{F(k)/m - \left[x(k-2)\frac{1}{T^2} - x(k-1)\left(\frac{2}{T^2} + \frac{2\zeta\omega_n}{T} \right) \right]}{\left(\frac{1}{T^2} + \frac{2\zeta\omega_n}{T} + \omega_n^2 \right)};$$

and then evaluate the equation in a loop over and over with a few previous values of x saved and new values of the input, F, at each discrete time step. Differential equations, including those of controllers, such as the PID controller, can be solved in this way, providing useful results as long as an appropriate sample period, T, is chosen.

z-Transform. If we take the Laplace transform of the left part of Equation (3.193), $y'(t)$, and assume zero initial conditions, then we get,

$$L[\dot{y}(t)] = sY(s). \tag{3.196}$$

This relationship means that finding the Laplace transform of a differential equation can be accomplished by replacing each time derivative of a variable with the s raised to the power of the degree of the derivative, n, multiplied by the Laplace transform of the variable, that is, $L[y^{(n)}(t)] = s^n Y(s)$ for the nth derivative of $y(t)$. This method can then be used to find transfer function as the ratio of the Laplace transforms of the output signal over the input signal (See Equation (3.138) for an example). z-transforms are used for analysis of discrete time systems in a similar way that Laplace transforms are used for continuous time differential equations. The z-transform is defined as,

$$Z[y(k)] = Y(z) = \sum_{k=0}^{\infty} y(k)z^{-k}. \tag{3.197}$$

Therefore, using Equation (3.197), the z-transform of one delay, $y(k-1)$, is,

$$Z[y(k-1)] = Y(z)z^{-1}. \tag{3.198}$$

This means that the z-transform of the nth time delay in a difference equation can be found to be,

$$Z[y(k-n)] = Y(z)z^{-n}. \tag{3.199}$$

To get a z-transform transfer function we solve for the ratio of the z-transform of the output over the z-transform of the input after applying Equation (3.199) to a difference equation.

For example, a z-transform transfer function can be found for the dynamical system in Equation (3.195) by applying Equation (3.199) to get,

$$X(z)z^{-2}\frac{1}{T^2} - X(z)z^{-1}\left(\frac{2}{T^2} + \frac{2\zeta\omega_n}{T} \right) + X(z)\left(\frac{1}{T^2} + \frac{2\zeta\omega_n}{T} + \omega_n^2 \right) = F(z)/m$$

$$\Rightarrow \frac{X(z)}{F(z)} = \frac{Z[x(k)]}{Z[F(k)]} = \frac{1/m}{z^{-2}\frac{1}{T^2} - z^{-1}\left(\frac{2}{T^2} + \frac{2\zeta\omega_n}{T} \right) + \left(\frac{1}{T^2} + \frac{2\zeta\omega_n}{T} + \omega_n^2 \right)}. \tag{3.200}$$

For another example, consider the first-order dynamical system in Figure 3-5 and Equation (3.22) with a forcing function added to get $x'(t) + (k/c)x(t) = F(t)/c$. The difference equation can be found to be,

$$\frac{x(k) - x(k-1)}{T} + (k/c)x(k) = F(k)/c$$

$$\Rightarrow -x(k-1)\frac{1}{T} + x(k)\left(\frac{1}{T} + \frac{k}{c}\right) = F(k)/c. \qquad (3.201)$$

We can find the z-transform transfer function by applying Equation (3.199) to get,

$$-X(z)z^{-1}\frac{1}{T} + X(z)\left(\frac{1}{T} + \frac{k}{c}\right) = F(z)/c$$

$$\Rightarrow \frac{X(z)}{F(z)} = \frac{Z[x(k)]}{Z[F(k)]} = \frac{1/c}{-z^{-1}\frac{1}{T} + \left(\frac{1}{T} + \frac{k}{c}\right)} = \frac{z/c}{-\frac{1}{T} + z\left(\frac{1}{T} + \frac{k}{c}\right)}. \qquad (3.202)$$

Using Equation (3.199), the z-transform transfer function for the backward approximate derivative, $y(k)$, of the signal, $f(k)$, can be found to be

$$y(k) = \frac{f(k) - f(k-1)}{T}$$

$$\Rightarrow \frac{Y(z)}{F(z)} = \frac{1 - z^{-1}}{T} = \frac{z - 1}{Tz}. \qquad (3.203)$$

Similarly, starting with the function $y(k)$ equals the approximate integration of the signal $f(k)$, the z-transform transfer function of trapezoidal approximate integration of $f(k)$ can be found as,

$$y(k) = y(k-1) + T\frac{f(k-1) + f(k)}{2}$$

$$\Rightarrow \frac{Y(z)}{F(z)} = \frac{T}{2}\frac{1 + z^{-1}}{1 - z^{-1}} = \frac{T}{2}\frac{z + 1}{z - 1}. \qquad (3.204)$$

The PID controller in Equation (3.161) can be expressed as a z-transform transfer function that is similar to the Laplace transform transfer function, $C(s)$, in Equation (3.191). Starting with the continuous time PID controller equation, using Equations (3.203) and (3.203) to replace the integration and differentiation with approximates, the z-transform can be found as

$$u(t) = K_e\varepsilon(t) + K_I\int_0^t \varepsilon(t)dt + K_d\dot{\varepsilon}(t)$$

$$\Rightarrow U(z) = K_e\varepsilon(z) + K_I\frac{T(z+1)}{2(z-1)}\varepsilon(z) + K_d\frac{z-1}{zT}\varepsilon(z). \qquad (3.205)$$

To find the transfer function of the discrete-time PID controller, solve for the ratio of output over input to get,

$$
\begin{aligned}
D(z) &= \frac{U(z)}{\epsilon(z)} = K_e + K_I \frac{T(z+1)}{2(z-1)} + K_d \frac{z-1}{zT} \\
&= \frac{K_e(z^2 - z) + \frac{K_I T}{2}(z^2 + z) + \frac{K_d}{T}(z^2 - 2z + 1)}{z^2 - z} \\
&= \frac{\left(K_e + \frac{K_I T}{2} + \frac{K_d}{T}\right)z^2 + \left(-K_e + \frac{K_I T}{2} - 2\frac{K_d}{T}\right)z + \frac{K_d}{T}}{z^2 - z}.
\end{aligned}
\tag{3.206}
$$

A table of z-transforms and Laplace transforms for commonly encountered discrete-time functions is given in Table 3-2. The z-transform of a discrete time function or a Laplace transform may be found using Table 3-2. The inverse z-transform can be found using the table as well using partial fraction expansion to find z-transform terms like those in the third column and looking up corresponding discrete time functions in the middle column. Inverse z-transforms are rarely used in design; instead, simulations using software like MATLAB are used to obtain the discrete time responses of z-transforms. The z-transform may also be found by first finding the Laplace transform and then applying partial fraction expansion to get terms similar to the first column and finding the z-transform in the third column of Table 3-2.

Table 3-2. Selected z-transforms and corresponding discrete-time functions and Laplace transforms

Laplace transform	Discrete-time function	z-Transform
	$1, k = 0$; else 0 unit impulse	1
$\dfrac{1}{s}$	$1(kT)$ unit step	$\dfrac{z}{z-1}$
$\dfrac{1}{s^2}$	kT unit ramp	$T\dfrac{z}{(z-1)^2}$
$\dfrac{1}{s^3}$	$\dfrac{(kT)^2}{2!}$ parabolic	$\dfrac{T^2}{2}\dfrac{z(z+1)}{(z-1)^3}$
$\dfrac{1}{s+a}$	e^{-akT}	$\dfrac{z}{z - e^{-aT}}$
$\dfrac{\omega}{s^2 + \omega^2}$	$\sin(\omega kT)$	$\dfrac{z\sin(\omega T)}{z^2 - 2\cos(\omega T)z + 1}$
$\dfrac{s}{s^2 + \omega^2}$	$\cos(\omega kT)$	$\dfrac{z^2 - z\cos(\omega T)}{z^2 - 2\cos(\omega T)z + 1}$
$\dfrac{\omega}{(s+a)^2 + \omega^2}$	$e^{-akT}\sin(\omega kT)$	$\dfrac{ze^{-aT}\sin(\omega T)}{z^2 - 2e^{-aT}\cos(\omega T)z + e^{-2aT}}$
$\dfrac{s+a}{(s+a)^2 + \omega^2}$	$e^{-akT}\cos(\omega kT)$	$\dfrac{z(z - e^{-2aT}\cos(\omega T))}{z^2 - 2e^{-aT}\cos(\omega T)z + e^{-2aT}}$

Equivalence of s and z, Stability, and the Final Value Theorem.   The roots, z, of the denominator polynomial of a z-transform transfer function (for example, in Equation (3.200)) should not be analyzed in the same way as the roots, s, of the denominator polynomial in a Laplace transform transfer function, which are the same as the roots of the continuous time differential equation. There is however, an equivalence that can be seen between s and z. Consider the differential equation for the unforced first-order system seen in Equation (3.22), $\dot{x} - px = 0$, where $p = -k/c$ is the root of the characteristic equation. The solution seen in Equation (3.26), $x(t) = x_0 e^{pt}$, can be rewritten as a sampled form by replacing t with kT to get,

$$x(kT) = x_0 e^{pkT}, \tag{3.207}$$

where p is the root of the characteristic equation for the continuous time dynamic system.

The z-transform of Equation (3.207) can be found using Equation (3.197),

$$Z[x_0 e^{pT}] = x_0 \frac{z}{z - e^{pT}}. \tag{3.208}$$

Notice that the root of the denominator polynomial is $z = e^{pT}$. Since the Laplace transform for $x_0 e^{pt}$ is $x_0 \frac{1}{s-p}$, with root $s = p$, then z is equivalent to e^{sT}. Therefore, roots of characteristic equations (eigenvalues) in the continuous time domain are considered to be equivalent to roots of discrete z-transform characteristic equations as follows:

$$z = e^{sT}. \tag{3.209}$$

Dynamic systems can be designed by selecting desired characteristic equation roots, or eigen values, in the continuous time domain and using Equation (3.209) to convert the eigenvalues in the continuous time domain into the discrete time domain. And discrete time systems can be evaluated by converting z-transform eigenvalues into continuous time eigenvalues. Like for continuous time systems where eigenvalues are plotted on the complex plan, called the s-plane (Figure 3-10), the eigenvalues for a discrete-time system can also be plotted on the complex plane, called the z-plane shown in Figure 3-38. Using Equation (3.209), it can be clearly seen that if $s = 0$, then $z = 1$, and that if $s = j\omega$, then a plot of z for all $-\infty \le \omega \le \infty$ forms a circle with a radius of one. Indeed, all of the points on the unit circle on the z-plane correspond to points on the imaginary axis on the s-plane.

All eigenvalues outside of the unit circle on the z-plane correspond to eigenvalues in the open right-half-plane of the s-plane, which indicates unstable eigenvalues. Eigenvalues that are inside of the unit circle on the z-plane correspond to eigenvalues on the left-half-plane of the s-plane, indicating stability. Therefore, the condition for stability of discrete-time systems is,

$$\text{stable} \iff |z_n| < 1 \text{ for all eigenvalues } z_1, z_2, \dots z_n, \tag{3.210}$$

for systems with n eigenvalues. To determine the stability of a discrete time system, first find the transfer function as a ratio of polynomials in z, using z-transforms, and then find the roots of the denominator polynomial (i.e., the eigenvalues), which are generally complex numbers. Finally check to see if all of the roots are inside of the unit circle (i.e., magnitude less than one), if so, then the system is stable, if not, then the system is not stable.

Similar to the final value theorem for Laplace transforms, the final value theorem as it applies to z-transforms can be stated as follows:

$$\lim_{k \to \infty} (x(k)) = \lim_{z \to 1} (z - 1)X(z), \tag{3.211}$$

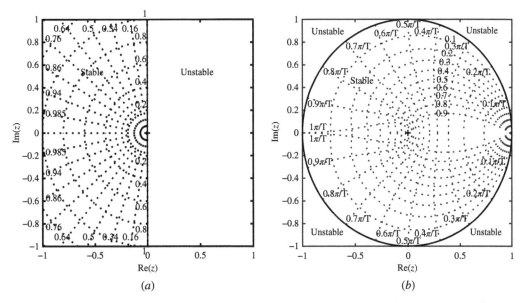

Figure 3-38. (a) s-plane with radial dotted lines showing constant damping ratio and semicircular lines for constant natural frequency. (b) z-plane with lines of constant damping ratio labeled 0.1–0.9 and lines of constant natural frequency labeled $0.1\pi/T - 0.9\pi/T$.

assuming that $(z - 1)X(z)$ is stable. This theorem can be used to analyze the steady-state behavior of discrete time systems. However, it is important to remember that it only applies to stable systems.

Interconnection Between Discrete-Time Systems and Continuous-Time Systems.

In discrete-time control systems connected to continuous-time systems, the output of the controller changes once every time step, kT, and is held for T seconds at that value until the next time step, $(k + 1)T$ in order to produce a continuous time signal. This action of updating the value and holding that value for each time step is called a zero-order hold (ZOH), which is a particular type of sample and hold. It is called "zero-order" because the function used to approximate the signal between discrete time steps is zeroth order, that is, a constant, with time. An example of the effects of the ZOH on a sinusoidal signal with frequency, ω, is given in Figure 3-39. As can be seen from the figure, selection of a sampling period that is 1/10 the period of the sine function ($T = 0.1 \times 2\pi/\omega$) creates a reasonable approximation of the original continuous time function after the ZOH is applied. A better approximation may be achieved by selecting a smaller sampling period.

The ZOH has ramifications on the dynamics of closed-loop systems. The "dynamics" of the ZOH with sample period, T, can be expressed as a Laplace transform transfer function [7],

$$G_{ZOH}(s) = \frac{(1 - e^{-sT})}{s}. \tag{3.212}$$

The ZOH transfer function in Equation (3.212) can be combined with a plant transfer function, $G(s)$, as in Figure 3-40, and the z-transform can be taken to get,

$$Z[G_{ZOH}(s)G(s)] = G(z). \tag{3.213}$$

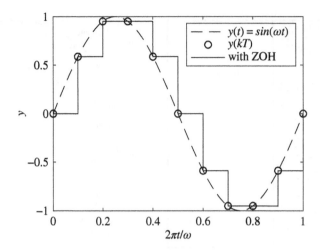

Figure 3-39. Plot of $y = \sin(\omega t)$ and y sampled with sample period $T = 0.1 \times 2\pi/\omega$ and the ZOH applied.

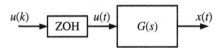

Figure 3-40. ZOH (the transfer function $G_{ZOH}(s)$) and plant transfer function, $G(s)$.

The new transfer function, $G(z)$, which includes the plant dynamics, $G(s)$, and the added dynamics due to the ZOH, $G_{ZOH}(s)$, can be used in the design and analysis of discrete-time control systems when combined with a digital controller. As an example, consider the first-order system in Equation (3.27), $c\dot{x} + kx = F(t)$, with Laplace transform transfer function,

$$G(s) = \frac{L[x(t)]}{L[F(t)]} = \frac{X(s)}{F(s)} = \frac{1/c}{s + k/c}. \tag{3.214}$$

The z-transform transfer function can be found as

$$
\begin{aligned}
Z[G_{ZOH}(s)G(s)] &\Rightarrow G(z) \\
&= Z\left[\frac{(1 - e^{-sT})}{s}\frac{1/c}{s + k/c}\right] \\
&= (1 - z^{-1})\frac{1}{k}Z\left[\frac{1}{s} - \frac{1}{s + k/c}\right] \\
&= (1 - z^{-1})\frac{1}{k}\left[\frac{z}{z - 1} - \frac{z}{z - e^{-(k/c)T}}\right] \\
&= \frac{(1/k)(z - 1)[(z - e^{-(k/c)T}) - (z - 1)]}{(z - 1)(z - e^{-(k/c)T})} \\
&= \frac{(1/k)(1 - e^{-(k/c)T})}{(z - e^{-(k/c)T})}.
\end{aligned}
\tag{3.215}
$$

Using MATLAB, there is a command, Gz = c2d(G,T,'zoh'), that allows for this transformation of a continuous time transfer function, G, to a discrete time transfer function, Gz, using stamping interval, T, using the zero-order hold method.

Discrete-Time Control Design. The closed-loop discrete-time control system is shown in Figure 3-41 with the discrete time controller expressed as a z-transform transfer function, $D(z)$, a ZOH, a continuous time plant system, $G(s)$, and a sampler in the feedback signal path indicating that the output of the plant is with is sampled with sample period, T. Control design for a discrete-time system can be accomplished by staring with a continuous-time controller, $C(s)$, and converting the design to a discrete-time controller, $D(z)$. In this method, performance specifications and control system design may proceed as in the other sections of this chapter. Then the controller and plant are converted to discrete-time models for closed-loop system analysis using discrete-time tools such as difference equations, z-transforms, and the analysis of eigenvalues on the z-plane. Other methods may be used where the design process only considers the discrete time system; however, these methods tend to be unwieldy and difficult to conceptualize since the object of control is typically a real system that operates in continuous time.

As an example, the design process for discrete-time controllers can start with a PID control design carried out using the continuous time methods as in previous sections. Consider also, the example first-order system in Figure 3-5, which has an associated discrete time transfer function found in Equation (3.215). The open-loop transfer function, $L(z)$, for the closed-loop system in Figure 3-41 with $D(z)$ from Equation (3.206) and $G(z) = Z[G_{ZOH}(s)G(s)]$ from Equation (3.215) is,

$$L(z) = D(z)Z[G_{ZOH}(s)G(s)] = D(z)G(z)$$

$$= \frac{\left(K_e + \frac{K_I T}{2} + \frac{K_d}{T}\right)z^2 + \left(-K_e + \frac{K_I T}{2} - 2\frac{K_d}{T}\right)z + \frac{K_d}{T}}{z^2 - z} \frac{(1/k)(1 - e^{-(k/c)T})}{(z - e^{-(k/c)T})}$$

$$= \frac{\left(K_e + \frac{K_I T}{2}\right)z + \left(-K_e + \frac{K_I T}{2}\right)}{z - 1} \frac{(1/k)(1 - e^{-(k/c)T})}{(z - e^{-(k/c)T})} \quad \text{(PI control, } K_d = 0\text{)}. \quad (3.216)$$

In general, the closed-loop transfer function for Figure 3-41, relating the desired output to the actual output, is

$$T(z) = \frac{L[x(t)]}{L[x_d(t)]} = \frac{X(s)}{X_d(s)} = \frac{L(z)}{1 + L(z)}, \quad (3.217)$$

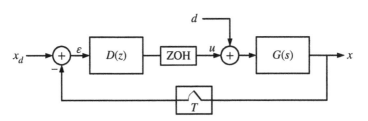

Figure 3-41. Closed-loop discrete-time control system.

where $L(z) = D(z)G(z)$. Using these equations, the closed loop transfer function, $T(z)$, can be found allowing for analysis of closed-loop eigenvalues by finding the roots of the denominator polynomial of $T(z)$. In this way closed-loop stability of the discrete time system in Figure 3-41 can be verified using Equation (3.210).

Case Study

Consider again the continuous-time PID control design example for the second-order system with closed-loop frequency response in Figure 3-36, where the closed-loop bandwidth was found to be $4.3w_n$ with the chosen PID control gains and $K_d = 5$, $K_I = a^2 K_d$, $K_e = 2aK_d$, and $a = \omega_n = 1$. A typical rule of thumb is to select a sampling interval 25 times faster than the bandwidth frequency,

$$T \le (1/25) \times \frac{2\pi}{\omega_{bw}}, \tag{3.218}$$

but sometimes smaller sampling periods are chosen for systems with complex dynamics. Since the control design resulted in a closed-loop bandwidth of $4.3w_n$, a sample period of $T = (1/25) \times 2\pi/(4.3\omega_n)$ was chosen. A sampling interval, which is chosen to be too large, will cause oscillations our instability. A sampling period, which is chosen to be too small, will unnecessarily use computing resources. The controller, $D(z)$, was found using Equation (3.206) with chosen sampling period and control gains. The discrete time plant, $G(z)$, was found using MATLAB's c2d(G,T,'zoh') function (that is a computerized implementation of Equation (3.213)) with the plant, $G(s)$, from Equation (3.158). The closed-loop transfer function was found to be,

$$T(z) = \frac{D(z)G(z)}{1 + D(z)G(z)} = \frac{0.16027(z + 0.9807)(z^2 - 1.891z + 0.894)}{(z - 0.9605)\,(z - 0.8987)\,(z - 0.6816)\,(z - 0.2388)}. \tag{3.219}$$

It can be clearly seen that the closed-loop discrete time transfer function, $T(z)$, is stable, according to Equation (3.210), since all eigenvalues (roots of denominator polynomial) are inside of the unit circle on the complex z-lane; in fact, they are all positive real numbers less than one. Since the system is stable, the final value theorem of Equation (3.211) can be applied to show that with a unit step input applied to the transfer function $T(z)$, the steady output is one, showing that the steady output exactly matches a steady input,

$$\lim_{k \to \infty} x(k) = \lim_{z \to 1}(z - 1)Z[x_d(k)]T(z)$$

$$= \lim_{z \to 1}(z - 1)\frac{z}{z-1}\frac{0.16027(z + 0.9807)(z^2 - 1.891z + 0.894)}{(z - 0.9605)\,(z - 0.8987)\,(z - 0.6816)\,(z - 0.2388)} = 1.$$

Notice that the z-transform of a unit step function can be found in Table 3-2, which can be used to get the z-transform of the input, $Z[x_d(k)] = \frac{z}{z-1}$.

A simulation comparison of the closed-loop control comparing the continuous time PID controller and the discrete time PID controller, using the transfer function in Equation (3.219), with a unit step input to the desired position is given in Figure 3-42. The response is similar for both controllers. However, a smaller sampling interval T would have allowed the discrete-time controller to nearly exactly match the continuous time control system. A larger sampling period would result in poor performance and instability in an extreme case where T is too large compared to the guidance given in Equation (3.218).

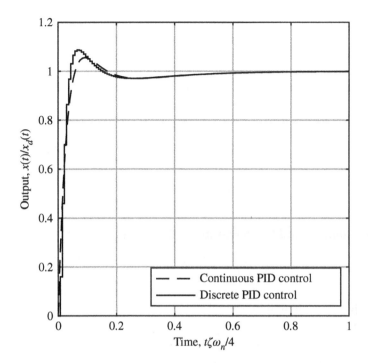

Figure 3-42. Simulation of the closed loop control system with a continuous time PID controller and a discrete time PID controller.

Control Design Using a Discrete Equivalent Approximation. In this method of control design for discrete-time systems, a controller is first designed in continuous-time and then converted to an equivalent discrete-time system. Then the closed-loop system is analyzed and simulated with the discrete time controller using an appropriate sampling period, T.

Since finding the discrete-time transfer function for the controller with sampled input and ZOH is often a cumbersome process, an approximate method has been developed. As described in Equation (3.204), the approximate integral can be expressed as a z-transform transfer function. Also, integration in terms of Laplace, given in Equation (3.103), is accomplished by multiplying the Laplace transform by $1/s$. Combining these two ideas (equating a transfer function for integration in Laplace with a transfer function for integration in z), yields the following:

$$\frac{1}{s} = \frac{T}{2}\frac{1+z^{-1}}{1-z^{-1}}. \tag{3.220}$$

This means that a discrete time approximation of continuous-time transfer function can be obtained by making the following replacement,

$$s = \frac{2}{T}\frac{1-z^{-1}}{1+z^{-1}}. \tag{3.221}$$

This method of substitution of Equation (3.221) into a continuous-time transfer function to find the discrete-time transfer function is called Tustin's method. This gives a simple way to find the discrete time transfer function $D(z)$ by substituting Equation (3.221) for s into a continuous time controller transfer function, $C(s)$. Using MATLAB, there is a command,

$D = \text{c2d}(C,T,'\text{tustin}')$ that allows for this transformation of a continuous-time transfer function, C, to a discrete time transfer function, D, using stamping interval, T, using Tustin's method.

A PD controller can be approximated by a z-transform transfer function using Tustin's method. The continuous time transfer function for the PD controller is,

$$C(s) = K_e + \frac{K_I}{s}. \tag{3.222}$$

Substituting Equation (3.221) into Equation (3.222) gives the discrete-time controller transfer function,

$$D(z) = K_e + K_I \frac{T}{2} \frac{1 + z^{-1}}{1 - z^{-1}}. \tag{3.223}$$

3.8.5 Controllability and State Feedback Controller Design

Full State Feedback. Control of a state space system may be achieved by applying full state feedback to the state space dynamic system in Equation (3.87). A control gain vector, $\mathbf{K} = \begin{bmatrix} k_1 & \cdots & k_n \end{bmatrix}$, equal in length to the state vector, $\mathbf{x}$, can be applied to the difference between the desired state and the actual state to determine the control input,

$$\mathbf{u} = \mathbf{K}(\mathbf{x}_d - \mathbf{x}). \tag{3.224}$$

After substituting Equation (3.224) into Equation (3.87), the state space representation of the closed-loop dynamic system becomes,

$$\dot{\mathbf{x}} = (\mathbf{A} - \mathbf{BK})\mathbf{x} + \mathbf{BK}\mathbf{x}_d, \quad \mathbf{y} = (\mathbf{C} - \mathbf{DK})\mathbf{x} + \mathbf{D}\mathbf{K}\mathbf{x}_d. \tag{3.225}$$

The new characteristic equation becomes,

$$\det[(\mathbf{A} - \mathbf{BK}) - s\,\mathbf{I}] = 0. \tag{3.226}$$

The roots of this characteristic equation are eigenvalues of the matrix, $\mathbf{A} - \mathbf{BK}$. These roots of the characteristic equation can be placed in any location with the appropriate choice of the matrix, $\mathbf{K}$ if the system is controllable. This means that through control design, any characteristic equation with any roots can be chosen. Let the desired roots of the closed-loop characteristic equation be, $s_1, s_2, \ldots, s_n$, which, in general, could be complex. Then the desired closed-loop characteristic equation is,

$$\alpha_{CL}(s) = (s - s_1)(s - s_2) \ldots (s - s_n) = \alpha_0 + \alpha_1 s + \ldots + \alpha_{n-1}s^{n-1} + s^n = 0. \tag{3.227}$$

The control gain matrix, $\mathbf{K}$, can be found using Akerman's formula,

$$\mathbf{K} = [0 \ldots 0\ 1]C^{-1}\alpha_{CL}(\mathbf{A}), \tag{3.228}$$

which ensures that the characteristic equation in Equation (3.226) has the same coefficients as and is equivalent to Equation (3.227). In Equation (3.228) the controllability matrix, $\mathbf{C}$, is defined as,

$$C = [\mathbf{B} \ \mathbf{AB} \ \mathbf{A}^2\mathbf{B} \ldots \mathbf{A}^{n-1}\mathbf{B}], \tag{3.229}$$

and $\alpha_{CL}(\mathbf{A})$ is a matrix defined as,

$$\alpha_{CL}(\mathbf{A}) = \mathbf{A}^n + \alpha_{n-1}\mathbf{A}^{n-1} + \ldots + \alpha_0\mathbf{I}. \tag{3.230}$$

The use of Equation (3.228) requires that C is invertible. A check of the rank of C to determine if it is full rank, and therefore invertible, is used to determine if a state space system is controllable, thus the name controllability matrix. This means that not only can the system be stabilized but also that every element of the state in $\mathbf{x}$ can be driven by the control system to approach a particular value, $\mathbf{x}_d$. The ability to place the roots in any location gives a large amount of freedom in the control design. The drawback of this method is that all states must be measured to apply Equation (3.224), which is not always feasible due to cost or other limitations. Therefore, states are often estimated if there are barriers to direct measurement.

Case Study

Consider again the second-order dynamical system described by Equation (3.157). This system can be converted into a state space system as follows. Select a state so that

$$\mathbf{x} = \begin{bmatrix} x_1 \\ x_2 \end{bmatrix} = \begin{bmatrix} x \\ \dot{x} \end{bmatrix}, \tag{3.231}$$

input,

$$\mathbf{u} = [u], \tag{3.232}$$

and an output,

$$\mathbf{y} = [x]. \tag{3.233}$$

Therefore, the state space system may be expressed as follows,

$$\dot{\mathbf{x}} = \begin{bmatrix} 0 & 1 \\ -\omega_n^2 & -2\zeta\omega_n \end{bmatrix} \mathbf{x} + \begin{bmatrix} 0 \\ 1 \end{bmatrix} \mathbf{u}, \ \mathbf{y} = [\,1\ 0\,]\mathbf{x} + [\,0\,]\mathbf{u}. \tag{3.234}$$

To start the control design process, select a desired characteristic equation,

$$\alpha_{CL}(s) = s^2 + 2Z\Omega_n s + \Omega_n^2 = 0. \tag{3.235}$$

By comparing Equation (3.235) to (3.227) we determine that $\alpha_0 = \Omega_n^2$ and $\alpha_1 = 2Z\Omega_n$. Equation (3.230) can then be applied to this problem with $\alpha_0 = \Omega_n^2$ and $\alpha_1 = 2Z\Omega_n$ to get,

$$\alpha_{CL}(\mathbf{A}) = \begin{bmatrix} 0 & 1 \\ \omega_n^2 & 2\zeta\omega_n \end{bmatrix}^2 + 2Z\Omega_n \begin{bmatrix} 0 & 1 \\ \omega_n^2 & 2\zeta\omega_n \end{bmatrix}^1 + \Omega_n^2 \begin{bmatrix} 1 & 0 \\ 0 & 1 \end{bmatrix}. \tag{3.236}$$

The controllability matrix is,

$$C = [\,\mathbf{B}\ \mathbf{AB}\,] = \begin{bmatrix} 0 & 1 \\ 1 & 2\zeta\omega_n \end{bmatrix}, \tag{3.237}$$

which is full rank (matrix rank is 2) and therefore invertible, and the system is said to be controllable. Therefore Equation (3.228) can be applied to get,

$$\mathbf{K} = [\,0\ 1\,]C^{-1}\alpha_{CL}(\mathbf{A}) = \begin{bmatrix} \Omega_n^2 - \omega_n^2 & 2Z\Omega_n - 2\zeta\omega_n \end{bmatrix}. \tag{3.238}$$

If we continue to use the plant model parameters used in previous examples, and select the desired natural frequency, $\Omega_n = 2\omega_n$ and damping ratio, $Z = \sqrt{2}/2$ to achieve a closed-loop system response with the controller from Equation (3.238), we expect a

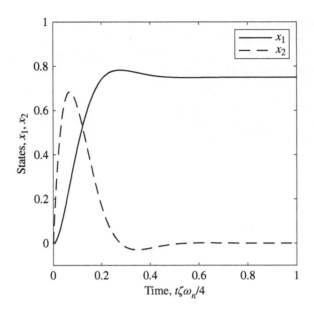

Figure 3-43. Time response of the closed-loop state space system example with full state feedback and a unit step input given as the desired position (state 1).

response with twice the speed of response and critical damping. With the state space system, it is possible to specify the desired response from all states; however, to be consistent with the previous examples in this section, a unit step input was given to the desired position, while the desired velocity was specified as zero. The simulation result given in Figure 3-43 confirms the expected performance in terms of the speed of response and damping ratio. Of interest is the steady-state error, which is not zero as highlighted by the fact that the steady response for state 1 is not close to the desired response of 1.0. This occurs for the same reasons as in the P control example in the previous subsections. The remedy is the same as in the previous sections as well; either choose higher gains to reduce steady-state error or include an integral to completely eliminate steady-state errors for the steady value of $\mathbf{x}_d$. A higher gain is achieved by selecting different desired roots of the characteristic equation that typically results in a characteristic equation with a faster response.

3.8.6 Observability and State Estimation

Every state must be measured in order to implement full state feedback control described in the previous section. However, this is often seen as a drawback of the method compared to PID control for example, which only requires measurement of the output. Measurement of any quantity adds cost to equipment or may be extremely difficult to achieve due to other design considerations. Therefore, many methods have been devised for estimating states so that only one or a few states must be measured to implement state feedback control using estimated states rather than measured states. In this section we will use the model to develop an estimator of states using limited numbers of measured signals.

To develop the estimator, we consider again the general-state space system described by Equation (3.87). Note that the matrix, **C**, called the output matrix, could better be described as a measurement matrix since the outputs of the system are considered to be measurements for the purpose of estimating states. The goal is to find a state estimate, $\hat{\mathbf{x}}$, which is as close as possible to the actual state, $\mathbf{x}$, using measurement signals, $\mathbf{y}$, and inputs, $\mathbf{u}$, to help form the estimates. The error between the actual state and estimated state is defined as $\mathbf{e} = \mathbf{x} - \hat{\mathbf{x}}$. The state estimate may be shown to have the following dynamics,

$$\dot{\hat{\mathbf{x}}} = \mathbf{A}\hat{\mathbf{x}} + \mathbf{B}\mathbf{u} + \mathbf{L}(\mathbf{y} - \mathbf{C}\hat{\mathbf{x}}), \tag{3.239}$$

where $\mathbf{L} = \begin{bmatrix} l_1 & \dots & l_n \end{bmatrix}^T$ is a vector equal in length to that of the state vector, $\mathbf{x}$. This dynamic state space system may be computed in real time requiring only the measurements, $\mathbf{y}$, and the input to the system, $\mathbf{u}$, to produce estimates of all states, $\hat{\mathbf{x}}$.

The process of finding the estimator gain, $\mathbf{L}$, is similar to that of finding $\mathbf{K}$, for full state feedback control. The time rate of change of the error can be found to be a dynamic system as,

$$\dot{\mathbf{e}} = (\mathbf{A} - \mathbf{LC})\mathbf{e}. \tag{3.240}$$

The characteristic equation of the error dynamics in Equation (3.240) is,

$$\det(\mathbf{A} - \mathbf{LC} - s\mathbf{I}) = 0. \tag{3.241}$$

Akerman's formula may be used to find $\mathbf{L}$ as it was for full state feedback control design to find $\mathbf{K}$. Let the desired roots of the estimator error dynamics characteristic equation be, $s_1, s_2, \dots, s_n$. The chosen roots of the characteristic equation of the estimator error dynamics are always chosen as stable roots and usually chosen to be real numbers that are two or three times faster than the dynamics of the state space system dynamics or of the closed-loop state space system dynamics. The desired estimator error characteristic equation is,

$$\alpha_{est}(s) = (s - s_1)(s - s_2) \dots (s - s_n) = \alpha_0 + \alpha_1 s + \dots + \alpha_{n-1} s^{n-1} + s^n = 0. \tag{3.242}$$

The estimator gain vector, $\mathbf{L}$, can be found using Akerman's formula,

$$\mathbf{L} = \alpha_{est}(\mathbf{A})O^{-1}[\, 0 \dots 0\ 1\,]^T, \tag{3.243}$$

which ensures that the characteristic equation in Equation (3.241) has the same coefficients as and is equivalent to Equation (3.242). In Equation (3.243) the observability matrix, O, is defined as,

$$O = \begin{bmatrix} \mathbf{C} \\ \mathbf{CA} \\ \vdots \\ \mathbf{CA}^{n-1} \end{bmatrix}, \tag{3.244}$$

and $\alpha_{est}(\mathbf{A})$ is a matrix defined as,

$$\alpha_{est}(\mathbf{A}) = \mathbf{A}^n + \alpha_{n-1}\mathbf{A}^{n-1} + \dots + \alpha_0\mathbf{I}. \tag{3.245}$$

The use of Equation (3.243) requires that O is invertible. A check of the rank of O to determine if it is full rank, and therefore invertible, is used to determine if a state space system is observable, thus the name observability matrix. This means that not only, can the estimation system be stabilized but also that every element of the error state in $\mathbf{e}$ can be driven by the estimation system to be arbitrarily small, but only small with an appropriate selection of the estimator gain, $\mathbf{L}$. Notice that the observability matrix depends on $\mathbf{A}$ and $\mathbf{C}$. In general, the number and the nature of the measurements, $\mathbf{y}$, determines the dimension of and the values of the elements in **C**. Therefore, if the observability matrix is rank deficient (not invertible), increase the number of measurements or try different measurements to manipulate **C** and therefore increase the rank of the observability matrix.

3.8.7 Summary

As shown in this section of the chapter, the dynamic response of a physical system may be adjusted by properly designing the gains for the PID controller or the gains of a pole placement controller. The design process for obtaining the control gains may utilize continuous and discrete time domain and frequency domain methods. There are several other classical methods for selecting control gains such as root locus, and optimization using a linear-quadratic-regulator; however, the saturation limits of the system will require that these gains be selected so that the plant input and output signals remain within some physical bound. For investigation of these optimization methods the reader is encouraged to examine classical and more advanced textbooks on control theory. Other methods of nonlinear control are also being pursued by modern day scholars but are beyond the scope of this text.

3.9 CONCLUSION

In conclusion, this chapter has been aimed at presenting the basic concepts that are germane to the subject area of dynamic systems and control. As shown throughout this chapter a great deal is known about first- and second-order dynamical systems and therefore much effort is placed on trying to reduce complex systems to one of these two models. This reduction can ordinarily be made by showing that the higher-order terms within the governing equation are associated with very small time constants and may therefore be safely neglected for the time duration of the overall transient response. This process is used throughout this text for assessing the dynamic response of various hydraulic control systems. Furthermore, the linear methods of feedback compensation are utilized extensively in Chapters 7 and 8 where the overall output response of a hydraulic control system is automated using PI and PID control structures. The objectives of these systems are to control actuator position, velocity, and effort, and the selection of controller gains are used to shape the desired output response of the system.

3.10 REFERENCES

[1] Greenwood, D.T. 1988. *Principles of Dynamics*, 2nd ed. Prentice Hall, Englewood Cliffs, NJ.

[2] Franklin, G. F., J. D. Powell, and A. Emami-Naeini. 1994. *Feedback Control of Dynamic Systems*, 3rd ed. Addison-Wesley, New York.

[3] Meirovitch, L. 1986. *Elements of Vibration Analysis*. McGraw-Hill, New York.

[4] Ogata, K. 1992. *System Dynamics*, 3rd ed. Prentice Hall, Saddle River, NJ.

[5] Kluever, C. A. 2015. *Dynamic Systems: Modeling, Simulation, and Control*. John Wiley & Sons, Hoboken, NJ.

[6] Skogestad, S., and I. Postlethwaite. 2007. *Multivariable Feedback Control: Analysis and Design*, 2nd ed. John Wiley & Sons, Hoboken, NJ.

[7] Dorf, R., and R. Bishop. 1995. *Modern Control Systems*, 7th ed. Addison-Wesley, Reading, MA.

3.11 HOMEWORK PROBLEMS

3.11.1 Modeling

3.1 Write a system of nonlinear equations to describe the dynamics of the system shown in the figure. *Note*: An equation will need to be written for the pendulum, the actuator, and the fluid pressure within the actuator chamber (three equations).

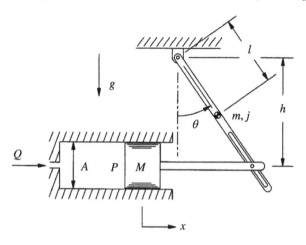

3.11.2 Linearization

3.2 Linearize the following mathematical function about the nominal value of $x = 1$:

$$f(x) = x^2 + \sec(x) - \ln(x).$$

3.3 By linearizing the system equations in Problem 3.1, reduce the number of equations from 3 to 2 by combining the pendulum equation and the actuator equation. What is the effective mass and the effective spring rate for the actuator?

3.11.3 Dynamic Behavior

3.4 The governing equation for the voltage drop across the capacitor of the RC circuit shown in the figure is given by

$$RC \frac{dv}{dt} + v = V,$$

where R is resistance, C is capacitance, and V is the voltage source. Using the form of this governing equation, write an expression for the system time constant and the bandwidth frequency. What is the settling time for the RC circuit?

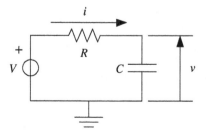

3.5   The governing equations for the RLC circuit shown in the figure are given by

$$C\frac{dv}{dt} = i, \quad L\frac{di}{dt} + Ri = V - v,$$

where R is resistance, C is capacitance, L is inductance, and V is the voltage source. Combine these equations to express the governing equation for the voltage drop across the capacitor. Write an expression for the undamped natural frequency and the damping ratio. Assuming that the inductance is 1 mH, the capacitance is 10 µF, and the resistance is 14 ohms, calculate the rise time, settling time, and the maximum percent overshoot. What is the system's bandwidth frequency?

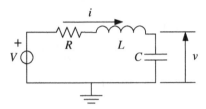

3.11.4 Block Diagrams and the Laplace Transform

3.6   Take the Laplace transform of the following function using the definition in Equation (3.94):

$$f(t) = \sin(\omega t).$$

Check your answer with Table 3-1.

3.7   Using the linearized equations for Problem 3.3, construct the block diagram for the system showing causality.

3.8   Construct a block diagram for Problem 3.4.

3.9   Construct a block diagram for Problem 3.5.

3.11.5 Feedback Control

3.10   Find the transfer function, $G(s)$, and frequency response (magnitude and phase) as a function of frequency of the system in Problem 3.5 with input V and output v. What is the open-loop bandwidth frequency in rad. / sec.? Consider a negative feedback closed-loop control system where the open-loop transfer function is $L(s) = G(s)K_e$. Select a P control (proportional control gain) so that the PM = 45 deg.

3.11   Select an appropriate sampling period, T, and find the $G(z)$ and $D(z)$ for Problem 3.10. Then find the closed-loop transfer function, $T(z)$, and eigenvalues for the discrete-time feedback control system.

3.12   Find the state space realization $(\mathbf{A}, \mathbf{B}, \mathbf{C}, \text{and } \mathbf{D})$ for Problem 3.5 with input, $\mathbf{u} = [V]$, and the states equal to the output, that is, $\mathbf{x} = \mathbf{y} = \begin{bmatrix} i \\ v \end{bmatrix}$.

3.13   For Problem 3.5 (and Problem 3.12), let $\mathbf{u} = \mathbf{K}(\mathbf{x}_d - \mathbf{x})$. Find full state feedback gains, $\mathbf{K}$, so that the roots of the closed-loop characteristic equation are two times the roots of the characteristic equation of the open-loop plant in Problem 3.5. Assuming that only the voltage, v, is measured, find the estimator gains, $\mathbf{L}$, so that the estimator characteristic equation roots are 3 times the closed-loop eigenvalues. Confirm the eigenvalues for both the closed-loop control system and the estimation system.

HYDRAULIC COMPONENTS

II

HYDRAULIC COMPONENTS

4

HYDRAULIC VALVES

4.1 INTRODUCTION

Hydraulic valves are commonly used within hydraulic control systems for the accurate modulation and control of the entire system. Within valve-controlled hydraulic circuits the valve
provides the interface between the hydraulic power element – that is, the pump – and the
hydraulic output device, which is either a linear or rotary actuator. Within these circuits, the
hydraulic control valve is the device that receives feedback from the operator or another
automatic control source and adjusts the system output accordingly. This feedback is used
to provide a controllable output for the circuit or to provide a safety function that is necessary while working with high-power devices that are typical of hydraulic control circuitry.
A good understanding of hydraulic valves is a prerequisite for understanding the control of
hydraulic pumps as well as systems; therefore, this chapter precedes our discussion of other
hydraulic components.

Hydraulic valves can be classified in a number of ways; however, the most general classification is based on the number of flow lines that are connected to the valve. For instance, a
two-way valve has a single input and a single output; that is, two flow lines. A three-way
valve has three flow lines given by a supply line, an output line, and a return line back
to the reservoir. Finally, a four-way valve utilizes a supply line, two output lines, and a
return line. More specifically, valves may also be classified either by construction type or
by function type. Since this text is focused on studying the performance characteristics of
typical valve constructions that are used within industry this chapter is ordered by construction type rather than function type. For example, we will consider spool valves, poppet
valves, and so on. Nevertheless, function type classification is useful and can be broken
down into three main categories: directional control valves, flow control valves, and pressure
control valves.

Directional control valves are used as switching devices within hydraulic circuitry. A typical example of a directional control valve is a check valve or a shuttle valve. These valves
are standard on–off devices that do not exhibit significant power losses within the system or demonstrate dynamical behavior that must be considered in the upfront design
of the circuit. Therefore, directional control valves are not considered any further within
this chapter.

Flow control valves are used to continuously modulate and direct flow within hydraulic circuitry. These valves perform the control objective of the valve-controlled circuit by implementing some form of feedback in order to adjust the amount of flow that is delivered to or away from the output device. If the feedback is generated automatically through an internal sensing and control mechanism the valve is often called a servo valve. The construction of a flow control valve is almost always of the spool valve type as these valves lend themselves nicely to feedback control through either direct or indirect actuation methods.

Pressure control valves are used to maintain or limit a specific pressure level within hydraulic circuitry. These valves can function in an on-off mode much like a pressure relief valve or they can be used to modulate and maintain pressure much like a pressure reducing valve. Pressure control valves are made in a variety of constructions. Typically, the relief valve function is accomplished by using a poppet valve construction while other pressure modulation functions may be accomplished using spool valves or flapper nozzle valves.

In this chapter we consider the function and performance characteristics of hydraulic valves. Three valve constructions are considered in this chapter: spool valves, poppet valves, and flapper nozzle valves. Spool valves are typically used for controlling volumetric flow rates while poppet valves are typically used for controlling pressure. Flapper nozzle valves are commonly used as a first stage in a two-stage flow control valve. As it turns out, each valve type must resist its own transient and steady fluid momentum effects; and, therefore, a considerable amount of time is spent in this chapter describing the fluid flow forces that are exerted on each valve. Using these flow force characteristics, mechanical and electrohydraulic valve applications are considered as they apply to certain valve types. Throughout this presentation topics of design and control are discussed in their turn. The objective of this chapter is to provide foundational material that may be used and referenced when considering the control and design of valves that are widely used in hydraulic control systems. Before entering into specific discussions on the modeling of hydraulic valves, we consider the more general topic concerning standard valve flow coefficients.

4.2 VALVE FLOW COEFFICIENTS

4.2.1 Overview

When an engineer wants to select a valve for a given application he or she will often consult a manufacturer's catalog, which contains a set of curves that characterize the steady-state performance of the valve. These curves are known as the pressure-flow curves for the valve and they are quite important for sizing and selecting a valve for a given application. Figure 4-1 shows a schematic that describes two pressure-flow curves that may be encountered in practice: one for a pressure control valve and one for a flow control valve. In this schematic the vertical axis describes the pressure drop P across the valve while the horizontal axis describes the volumetric flow rate Q through the valve.

As shown in Figure 4-1 the pressure control valve (perhaps a poppet type relief valve) is designed to operate across a wide range of flows while maintaining a fairly constant pressure drop across the valve. The flow range over which the pressure control valve is intended to operate is given by $Q_{max} - Q_{min}$. In this range the pressure-flow curve is intended to be as flat as possible to maintain the controlled pressure level of the valve. The flatter this curve is the better the pressure control valve performs. Also shown in Figure 4-1 is a pressure-flow curve for a flow control valve (most likely a spool valve construction). This valve is intended to operate over a range of pressures given by $P_{max} - P_{min}$. In this range the pressure-flow

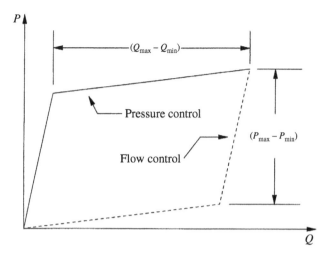

Figure 4-1. Pressure-flow curves for two valve types.

curve is intended to be as vertical as possible to maintain the controlled flow rate of the valve. The more vertical this curve is the better the flow control valve performs.

Manufacturer's data given in a form similar to that of Figure 4-1 is important for establishing the steady-state operating conditions of the valve pertaining to a change in output flow versus a change in pressure drop across the valve; however, this curve does not provide all of the help that we need for characterizing the operating performance of the valve. To consider the entire operating performance of the valve we need information that pertains to the change in flow with respect to valve displacement as well. These valve characteristics are captured in the coefficients of the linearized flow equation and, therefore, we present these coefficients in the following subsection for a valve of any construction. Throughout the remainder of this chapter these coefficients are used for describing the steady and transient behavior of the hydraulic control valve.

4.2.2 Linearized Flow Equation

The best equation that we have available for modeling the volumetric flow rate through a control valve is the classical orifice equation that was presented in Equation (2.60) of Chapter 2. The reader will recall that this equation is based on Bernoulli principles, which assume that a high Reynolds number characterizes the flow and that the flow is steady and incompressible. In the classical orifice equation viscous effects are captured in the discharge coefficient and only the fluid momentum and fluid pressure effects are modeled explicitly. The orifice equation is rewritten here for convenience as

$$Q = A C_d \sqrt{\frac{2}{\rho} P}, \tag{4.1}$$

where A is the discharge flow area of the valve, C_d is the discharge coefficient that must be determined experimentally, ρ is the fluid density, and P is the pressure drop across the valve. The parameters within this equation that change with respect to valve operation are the discharge area A and the pressure drop P across the valve. Using the Taylor series method described in Section 3.3 of Chapter 3 for linearizing Equation (4.1) we may write the linearized orifice equation as

$$Q = Q_o + \left.\frac{\partial Q}{\partial x}\right|_o (x - x_o) + \left.\frac{\partial Q}{\partial P}\right|_o (P - P_o), \tag{4.2}$$

where x is the generalized displacement of the valve that alters the discharge area and the subscript "o" identifies the nominal operating conditions of the valve. In order for Equation (4.2) to be valid the actual operating conditions of the valve cannot deviate too far from the nominal operating conditions. If it is assumed that the nominal operating position of the valve is given by $x_o = 0$ it can be shown that the linearized flow equation is given by

$$Q = \tfrac{1}{2}Q_o + \mathrm{K}_q\, x + \mathrm{K}_c P, \tag{4.3}$$

where

$$\mathrm{K}_q \equiv \left.\frac{\partial Q}{\partial A}\frac{\partial A}{\partial x}\right|_o = C_d \sqrt{\frac{2}{\rho}P}\left.\frac{\partial A}{\partial x}\right|_o \quad \text{and} \quad \mathrm{K}_c \equiv \left.\frac{\partial Q}{\partial P}\right|_o = \left.\frac{A\,C_d}{\sqrt{2\,\rho\,P}}\right|_o. \tag{4.4}$$

In Equation (4.3) Q_o is the nominal flow rate of the valve when $x = 0$ and $P = P_o$. The definitions in Equation (4.4) are known as the flow gain and the pressure-flow coefficient respectively. The pressure sensitivity of the valve is given by

$$\mathrm{K}_p \equiv \left.\frac{\partial P}{\partial x}\right|_o = \left.\frac{\partial Q/\partial x}{\partial Q/\partial P}\right|_o = \frac{\mathrm{K}_q}{\mathrm{K}_c} = \left.\frac{2P}{A}\frac{\partial A}{\partial x}\right|_o. \tag{4.5}$$

Notice: the flow gain and the pressure sensitivity may be positive or negative depending on the sign of the area gradient. Also, the pressure-flow coefficient is the basic parameter that is shown in the pressure-flow curve of Figure 4-1. For an ideal flow control valve the pressure-flow coefficient should be zero. For an ideal pressure control valve the pressure-flow coefficient should be infinite. The coefficients K_q, K_c, and K_p are known as the valve coefficients and are extremely important for determining the steady characteristics of the valve as well as the valve's stability, frequency response, and other dynamic characteristics. As is shown in subsequent chapters these coefficients have a decisive impact on the dynamic performance of hydraulic control systems and are therefore of great importance to the general theme of this text. In the following subsection these coefficients are evaluated for three flow passage or porting geometries that may be encountered within hydraulic control systems.

4.2.3 Valve Porting Geometry

As shown in Equations (4.4) and (4.5) the valve coefficients are heavily dependent on the geometry of the flow passage area A. In this part of the chapter we write expressions for the valve coefficients for the three geometry cases given by rectangular geometry, circular geometry, and triangular geometry. These three geometry cases are shown in Figure 4-2. In this figure the crosshatched area of each geometry case shows the open area of the valve port. The open distance of the valve is shown in the figure by the dimension ξ. In general, this open distance does not need to be equal to the valve displacement and may be offset by an underlapped condition of the valve. In this case the open distance of the valve is given by

$$\xi = x + u, \tag{4.6}$$

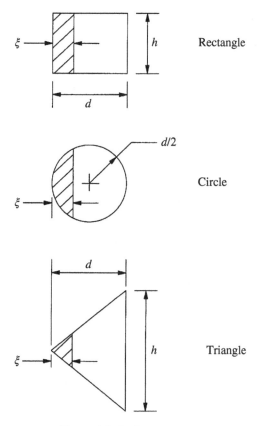

Figure 4-2. Porting geometry.

where x is the valve displacement and u is the fixed underlapped dimension of the valve. This is discussed later when we deal with critically centered and open centered valves. All other dimensions in Figure 4-2 are self-explanatory.

For the rectangular geometry shown in Figure 4-2 it is clear that the area properties are given by

$$A = h\xi \qquad \text{and} \qquad \frac{\partial A}{\partial x} = h. \tag{4.7}$$

By substituting these results into Equations (4.4) and (4.5) the valve coefficients for rectangular porting geometry may be written as

$$\mathrm{K}_q = h C_d \sqrt{\frac{2}{\rho} P_o}, \qquad \mathrm{K}_c = \frac{u h C_d}{\sqrt{2\rho P_o}}, \qquad \mathrm{K}_p = \frac{2 P_o}{u}. \tag{4.8}$$

Similarly, methods of calculus may be used to show that the area properties for circular porting geometry are given by

$$A = \frac{\pi}{8} d^2 + \left(\xi - \frac{d}{2} \right) \sqrt{(d-\xi)\xi} + \frac{d^2}{4} \arcsin \left(2\frac{\xi}{d} - 1 \right) \qquad \text{and}$$

$$\frac{\partial A}{\partial x} = 2\sqrt{(d-\xi)\,\xi}. \tag{4.9}$$

By substituting these results into Equations (4.4) and (4.5) the valve coefficients for circular porting geometry may be written as

$$K_q = C_d \sqrt{\frac{8(d-u)\,u}{\rho}\,P_o}, \qquad K_c = \frac{A_o\,C_d}{\sqrt{2\rho P_o}}, \qquad K_p = \frac{4P_o}{A_o}\sqrt{(d-u)u}, \qquad (4.10)$$

where A_o is determined by evaluating Equation (4.9) at $\xi = u$. For the triangular geometry shown in Figure 4-2 the area properties are given by

$$A = \frac{1}{2}\frac{h}{d}\,\xi^2 \qquad \text{and} \qquad \frac{\partial A}{\partial x} = \frac{h}{d}\,\xi. \qquad (4.11)$$

By substituting these results into Equations (4.4) and (4.5) the valve coefficients for triangular porting geometry may be written as

$$K_q = \frac{h}{d}uC_d\sqrt{\frac{2}{\rho}\,P_o}, \qquad K_c = \frac{h}{d}\,u^2\frac{C_d}{\sqrt{8\rho P_o}}, \qquad K_p = \frac{4P_o}{u}. \qquad (4.12)$$

It is interesting to compare the area and area gradient plots of the three geometries shown in Figure 4-2. This comparison is shown in Figure 4-3 where the heavy solid line

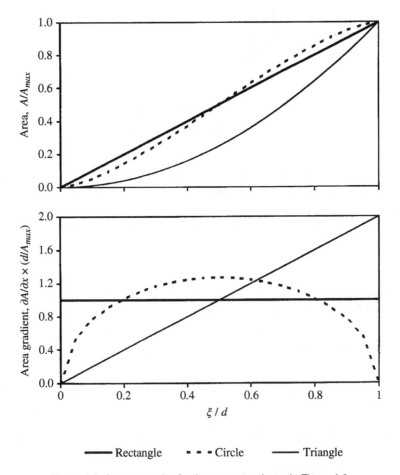

Figure 4-3. Area properties for the geometry shown in Figure 4-2.

describes the area properties of the rectangular geometry, the dashed line describes the area properties for the circular geometry, and the light solid line describes the area properties for the triangular geometry.

The area plot shows that the rectangular geometry provides a reasonable approximation for the circular geometry which demonstrates a slightly s-shape curve that varies as a function of valve opening. For the first half of the valve opening, the rectangular area is slightly larger than the circular area which will produce a larger pressure-flow coefficient when compared to a circular flow passage with the same maximum flow area. The triangular geometry will always produce the smallest pressure-flow coefficient of three geometries shown in Figure 4-2. The bottom half of Figure 4-3 shows that the area gradients for the three different geometries are substantially different over a wide range of the valve opening. The reader will recall that the area gradient determines the flow gain characteristics, which are nonzero for the rectangular geometry at the cracking position of the valve; that is, when $\xi = 0$.

From a manufacturing point of view the circular porting geometry is the easiest to make since it only requires the use of a machining drill. The rectangular and triangular porting geometry is more difficult to make due to the sharp corners at intersecting sides. It is, however, becoming more common to see sharp-edged shapes in the manufacturing environment as manufacturing processes are improving and becoming more versatile. Electrical Discharge Machining (EDM) is the most common method used for making these more intricate shapes. From an analytical point of view the circular geometry is not so attractive since the expressions for the valve coefficients are somewhat complex. For the sake of simplifying the analysis the rectangular geometry is often adopted as an approximation for the more common circular geometry which also provides a nonzero flow gain for critically centered valve designs.

4.2.4 Summary

In this section the valve coefficients have been defined and evaluated for three different porting geometries. As it has been mentioned, these coefficients are used extensively in the analysis that follows and in subsequent chapters for developing the equations that describe the performance of different valve constructions and various hydraulic control systems. These coefficients are also useful for describing the efficiency of a hydraulic control valve in the following sections.

4.3 TWO-WAY SPOOL VALVES

4.3.1 Overview

We begin our discussion of spool valves using the simplest design within this class; namely, the two-way spool valve. The two-way spool valve is important to us for the purposes of illustrating several basic spool valve properties. Though this valve is not used frequently within hydraulic control systems it is occasionally used as a pressure relief valve much like the more common application of a poppet valve, which is discussed later.

The two-way spool valve is shown in Figure 4-4 with two flow lines. The supply line is pressurized to a level given by P_s while the return line is pressurized to the level P_r. To open the valve the spool is moved in the positive x-direction thus facilitating the volumetric flow

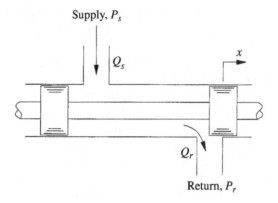

Figure 4-4. Two-way spool valve.

rate shown in Figure 4-4 by the symbol Q_r. From Equation (4.3) it can be seen that the volumetric flow rate through the valve is simply given by the expression

$$Q_r = \frac{1}{2} Q_{r_o} + K_q x + K_c (P_s - P_r), \tag{4.13}$$

where Q_{r_o} is the nominal flow rate through the valve at steady operating conditions, and K_q and K_c are the flow gain and pressure-flow coefficient defined generally in Equation (4.4). The reader will recall that the nominal flow rate is determined when $x = 0$ and when the supply and return pressures are at nominal values at which the valve coefficients have been determined. From Equation (4.13) it can be seen that the flow rate through the valve will increase as the valve displacement x increases and as the pressure drop $(P_s - P_r)$ increases. The nominal flow rate Q_{r_o} results from a nominal opening of the two-way valve that may exist when $x = 0$.

4.3.2 Efficiency

The efficiency of the two-way spool valve is determined by calculating the ratio of the useful output power to the supplied input power. From Section 2.8 of Chapter 2 it can be shown that the supplied hydraulic power to the two-way spool valve is given by

$$\dot{W}_s = P_s Q_s, \tag{4.14}$$

and the useful output power is

$$\dot{W}_o = P_r Q_r. \tag{4.15}$$

Using these two results, the overall efficiency of the two-way spool valve is given by

$$\eta = \frac{\dot{W}_o}{\dot{W}_s} = \frac{P_r\, Q_r}{P_s\, Q_s} = \eta_p\, \eta_v, \tag{4.16}$$

where the pressure efficiency of the valve is

$$\eta_p = \frac{P_r}{P_s}, \tag{4.17}$$

and the volumetric efficiency of the valve is

$$\eta_v = \frac{Q_r}{Q_s}. \tag{4.18}$$

For a valve with incompressible flow and no internal leakage the return flow Q_r and supply flow Q_s are identical, thus producing a volumetric efficiency for the valve that is equal to unity. In this case the overall efficiency of the valve is given by

$$\eta = \eta_p = \frac{P_r}{P_s}. \tag{4.19}$$

From this result we can see that a large pressure drop across the valve produces a very inefficient valve with much heat being created as a result of throttling effects. The amount of heat created from this pressure drop has been discussed in Section 2.6 of Chapter 2 where it was shown that the fluid temperature increases 0.56°C for every 1 MPa of pressure drop across the valve. This heat generation is one of the reasons that hydraulic systems require the use of convective cooling systems if they are to operate continuously for long periods of time.

4.3.3 Flow Forces

Overview. In order to design an actuation device for the two-way spool valve shown in Figure 4-4 it is necessary to have a good understanding of the forces that must be overcome to move the valve. Though it is not obvious by looking at Figure 4-4, there are significant fluid forces exerted on the valve that result from fluid momentum entering and exiting the valve. This subsection is used to derive the basic flow force equation for the two-way spool valve; however, this basic flow force equation is also used extensively in the following sections of this chapter to describe the fluid momentum effects on three-way and four-way valves, respectively. In other words, the discussion that follows is germane to all spool type valves and should be well understood in order to design adequate control devices for these valves.

Figure 4-5 shows a schematic of the control volume to be considered within the two-way spool valve. As shown in this figure, fluid enters the control volume from the top through the inlet area A_i and exits the control volume on the right through the discharge area A. The total length of the control volume is given by L_v, however, the length shown by the dimension L describes the fluid within the control volume that is flowing reasonably parallel with the axial centerline of the valve; that is, the $\hat{\mathbf{i}}$ direction. The discharge port is shown by the dimension d and the valve opening is given by the dimension ξ.

The reader will recall from Equation (4.6) that ξ includes a possible underlapped dimension of the valve as well as the actual spool displacement x. As shown by the control volume schematic the volumetric flow rate through the control volume is given by the symbol Q. Fluid enters the control volume in the negative $\hat{\mathbf{j}}$ direction and exits the control volume in the direction of the jet angle shown by the angular dimension θ. An analytical expression for this jet angle is presented later. The momentum effects of the fluid require external forces to be exerted on the control volume for static equilibrium. These forces are shown in Figure 4-4 by the components F_x and F_y. The force F_y is exerted on the control volume by the valve housing that surrounds the spool. For a discussion of forces that must be overcome, to move the valve in the axial direction the force in the vertical direction F_y is of little interest to us and is only presented here by way of completing the analysis. The force F_x is exerted on the

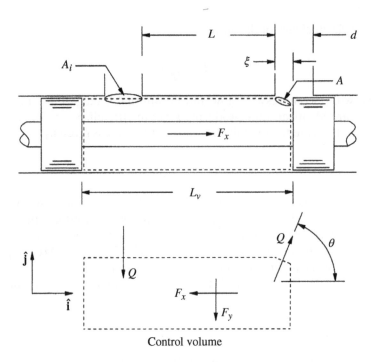

Figure 4-5. Control volume for calculating spool valve flow forces.

control volume by the spool itself. As such, an equal and opposite force is exerted on the spool as shown in Figure 4-5. This axial force is known as the spool valve flow force, which is the topic of our current discussion.

Analysis.   From the Reynolds Transport Theorem presented in Section 2.25 of Chapter 2, the conservation of fluid momentum for the control volume shown in Figure 4-5 is given by the following expression:

$$\mathbf{F} = \frac{\partial}{\partial t} \int_{c.v} \rho\mathbf{u} \, dV + \int_{c.s} \rho\mathbf{u}(\mathbf{u} \cdot \hat{\mathbf{n}}) \, dA, \tag{4.20}$$

where $\mathbf{F}$ is a vector force acting on the control volume, ρ is the fluid density, $\mathbf{u}$ is the fluid velocity vector, and $\hat{\mathbf{n}}$ is a unit vector that points normally outward from the control volume surface. The fluid forces acting on the control volume are shown in Figure 4-5 and may be written as

$$\mathbf{F} = -F_x\hat{\mathbf{i}} - F_y\hat{\mathbf{j}}. \tag{4.21}$$

The volume integral in Equation (4.20) describes the fluid momentum effects within the control volume itself. The significant fluid momentum that exists within the control volume occurs in the region of Figure 4-5 indicated by the length L. In this case the internal fluid-momentum effects may be written as

$$\frac{\partial}{\partial t} \int_{c.v} \rho\mathbf{u} \, dV = \rho L \frac{\partial Q}{\partial t}\hat{\mathbf{i}}. \tag{4.22}$$

The area integral in Equation (4.20) describes the momentum effects that enter and exit the control volume. An expression for these momentum effects is given by

$$\int_{c.s} \rho \mathbf{u}(\mathbf{u} \cdot \hat{\mathbf{n}}) \, dA = \rho \frac{Q^2}{A} \cos(\theta) \, \hat{\mathbf{i}} + \left(\rho \frac{Q^2}{A_i} + \rho \frac{Q^2}{A} \sin(\theta) \right) \hat{\mathbf{j}}. \tag{4.23}$$

Substituting Equations (4.21) through (4.23) into Equation (4.20) yields the following scalar equations for describing the conservation of momentum:

$$F_x = -\rho L \frac{\partial Q}{\partial t} - \rho \frac{Q^2}{A} \cos(\theta), \tag{4.24}$$

and

$$F_y = -\rho \frac{Q^2}{A_i} - \rho \frac{Q^2}{A} \sin(\theta). \tag{4.25}$$

As previously mentioned this subsection is only concerned with the force F_x as it is the force exerted on the spool valve that must be overcome for a control application. The force F_y is not considered further.

At this point it is convenient to write the flow force acting on the valve in the following form:

$$F_x = -F_t - F_s, \tag{4.26}$$

where

$$F_t = \rho L \frac{\partial Q}{\partial t} \qquad \text{and} \qquad F_s = \rho \frac{Q^2}{A} \cos(\theta). \tag{4.27}$$

From these equations it should be apparent that the steady flow force F_s always acts in such a way as to close the valve while the transient flow force F_t may act to either open or close the valve depending on the sign of $\partial Q / \partial t$. The transient flow force may be written more explicitly by taking the time derivative of Equation (4.13) and substituting the result into Equation (4.27). This result is given by

$$F_t = \rho L K_q \dot{x} + \rho L K_c \dot{P}, \tag{4.28}$$

where the dot notation indicates a derivative with respect to time and where P is the pressure drop across the metering land of the spool valve. *Note*: For the two-way spool valve shown in Figure 4-4, $P = P_s - P_r$. The steady flow force may be written more explicitly by substituting Equation (4.1) into Equation (4.27) to yield

$$F_s = 2APC_d^2 \cos(\theta). \tag{4.29}$$

This equation is nonlinear due to the simultaneous variation in the discharge area and the pressure. The reader will recall that the discharge area varies with the valve displacement x. By using the Taylor series method described in Section 3.3.2 of Chapter 3, Equation (4.29) may be linearized as follows:

$$F_s = F_{s_o} + \left. \frac{\partial F_s}{\partial x} \right|_o (x - x_o) + \left. \frac{\partial F_s}{\partial P} \right|_o (P - P_o), \tag{4.30}$$

where x is the valve displacement and the subscript "o" signifies the nominal operating conditions of the valve. Again, this equation is only valid as long as the actual operating conditions do not deviate too far from the nominal operating conditions. By setting $x_o = 0$ it may be shown that

$$F_s = K_{f_q} x + K_{f_c} P, \tag{4.31}$$

where

$$K_{f_q} \equiv \left.\frac{\partial F_s}{\partial A}\frac{\partial A}{\partial x}\right|_o = 2PC_d^2 \cos(\theta)\left.\frac{\partial A}{\partial x}\right|_o \quad \text{and} \quad K_{f_c} \equiv \left.\frac{\partial F_s}{\partial P}\right|_o = 2AC_d^2 \cos(\theta)|_o. \tag{4.32}$$

These definitions and results are known as the flow force gain and the pressure flow force coefficient respectively. It can be shown that these two coefficients are related to each other through the pressure sensitivity as follows:

$$K_p = \frac{K_q}{K_c} = 2\,\frac{K_{f_q}}{K_{f_c}}, \tag{4.33}$$

where the pressure sensitivity is defined in Equation (4.5). Substituting Equations (4.28) and (4.31) into Equation (4.26) produces the following expression for the total flow force:

$$F_x = -\rho L K_q \dot{x} - \rho L K_c \dot{P} - K_{f_q} x - K_{f_c} P. \tag{4.34}$$

This result provides a fundamental understanding of the flow forces that result from metering fluid across a single land of a two-way spool valve similar to what is shown in Figures 4-4 and 4-5. As shown in Equation (4.34) there are both transient and steady forcing terms that must be overcome to move the spool valve. This result is generally be applied for all spool valves that are discussed in this chapter; however, before that is done it is useful to consider the directional dependence that results from various arrangements of the two-way spool valve configuration.

Directional Dependence.   The previous analysis has been conducted for the configuration shown in Figure 4-5. This configuration shows a valve with fluid flowing past the left-hand side of the metering land away from the valve. This is called a "meter-out" condition on the left-hand side of the land. It should be noted that this is not the only configuration that is possible for the two-way valve. For instance, it is possible to have a "meter-in" condition on the left-hand side of the land in which fluid flows into the valve rather than away from the valve. In this case the transient flow force acting on the valve changes sign but the steady flow force remains in the same direction. The meter-out and meter-in conditions for the left-hand side of the land are shown in Figure 4-6.

For metering on the left-hand side of the land the flow forces are given by

$$F_x = \begin{cases} -\rho L K_q \dot{x} - \rho L K_c \dot{P} - K_{f_q} x - K_{f_c} P & \text{meter-out} \\ \rho L K_q \dot{x} + \rho L K_c \dot{P} - K_{f_q} x - K_{f_c} P & \text{meter-in} \end{cases}. \tag{4.35}$$

Similarly, there are meter-out and meter-in conditions that may exist on the right-hand side of the spool land. Under these conditions both the transient and steady flow force components change sign; however, the steady flow force still acts in such a way as to close the valve. These conditions are shown in Figure 4-7.

For metering on the right-hand side of the land the flow forces are given by

$$F_x = \begin{cases} \rho L K_q \dot{x} + \rho L K_c \dot{P} + K_{f_q} x + K_{f_c} P & \text{meter-out} \\ -\rho L K_q \dot{x} - \rho L K_c \dot{P} + K_{f_q} x + K_{f_c} P & \text{meter-in} \end{cases}. \tag{4.36}$$

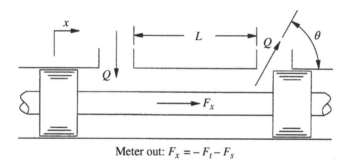

Meter out: $F_x = -F_t - F_s$

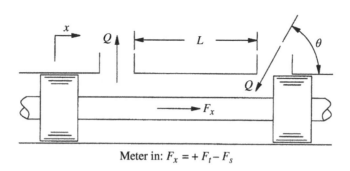

Meter in: $F_x = +F_t - F_s$

Figure 4-6. Flow forces from metering on the left-hand side of the land.

Before these results are used it should be noted that the flow gain K_q and the flow force gain K_{f_q} carry a sign with them that depends on the area gradient of the valve. See Equations (4.4) and (4.32). For flow that is metered on the left-hand side of the metering land; that is, Figures 4-4 through 4-6, these coefficients are positive. For flow that is metered on the right-hand side of the land; that is, Figure 4-7, these coefficients are negative. This sign convention is based on a positive valve displacement to the right as shown in the previous figures. It is always important to note throughout these calculations that the pressure drop P across the metering land is a positive number and should never change its sign.

Jet Angle.   Throughout the development of the basic flow force equation, the jet angle of the fluid flow has been represented by the symbol θ. See Figures 4-5 through 4-7. As shown in Equation (4.32) the jet angle is a necessary part of determining the flow force gain K_{f_q} and the pressure flow force coefficient K_{f_c}, which are the coefficients that are used to calculate the steady flow force terms. If the flow passage is assumed to be rectangular and h is large compared to ξ (see Figure 4-2) then the flow may be considered to be two-dimensional and the Laplace equation can be solved to determine the jet angle θ. This solution was carried out by Richard von Mises (1883–1953) and is presented here without its derivation:

$$\frac{\xi}{c_r} = \frac{1 + \frac{\pi}{2}\sin(\theta) - \ln\left(\tan\left[\frac{\pi-\theta}{2}\right]\right)\cos(\theta)}{1 + \frac{\pi}{2}\cos(\theta) + \ln\left(\tan\left[\frac{\pi/2-\theta}{2}\right]\right)\sin(\theta)}, \tag{4.37}$$

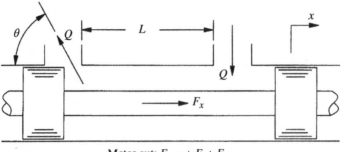

Meter out: $F_x = + F_t + F_s$

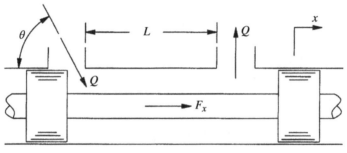

Meter in: $F_x = - F_t + F_s$

Figure 4-7. Flow forces from metering on the right-hand side of the land.

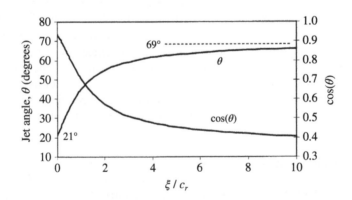

Figure 4-8. Spool valve jet-angle.

where ξ is the valve opening and c_r is the radial clearance between the valve and the valve housing. Using this result a plot of the jet angle and the cosine of the jet angle is presented in Figure 4-8. In this figure it may be seen that the jet angle varies between 21 and 69 degrees and that for most valve openings 69 degrees is a good estimation for this parameter. Only at very small valve openings do we see much of a variation in the jet angle and at these conditions the jet angle may be approximately two or three times smaller than it is at larger opened conditions.

Case Study

A two-way valve similar to the one shown in Figure 4-4 is used to meter flow away from the valve on the left-hand side of the metering land. Rectangular porting is used for the valve in which the port height h is given by 3 mm and the underlapped dimension for the valve is given by $u = 3/4$ mm. The radial clearance between the spool and the valve housing is 15 microns and the length between valve ports is 50 mm. The nominal pressure drop across the valve is 20 MPa. Using a discharge coefficient of 0.62 and a fluid density of $\rho = 850$ kg/m^3 calculate the nominal steady-state flow force exerted on the valve. What is the pressure-rise-rate that would have to occur to instantaneously cancel out the nominal steady-state flow force?

Equation (4.34) is the basic equation that must be used to calculate the flow force exerted on the two-way spool valve. Since this equation depends on the valve coefficients it will be useful to calculate the valve coefficients first. From Equations (4.4) and (4.32) it may be shown that the valve coefficients for the rectangular porting geometry of this problem are given by

$$K_q = hC_d\sqrt{\frac{2}{\rho}P_o} = 3\text{ mm} \times 0.62\sqrt{\frac{2}{850\text{ kg/m}^3} \times 20\text{ MPa}} = 0.4035\ \frac{\text{m}^2}{\text{s}},$$

$$K_c = \frac{uhC_d}{\sqrt{2\rho P_o}} = \frac{0.75\text{ mm} \times 3\text{ mm} \times 0.62}{\sqrt{2 \times 850\text{ kg/m}^3 \times 20\text{ MPa}}} = 7.565 \times 10^{-12}\ \frac{\text{m}^3}{\text{Pa s}},$$

$$K_{f_q} = 2P_oC_d^2\cos(\theta)h = 2 \times 20\text{ MPa} \times 0.62^2 \times \cos(69°) \times 3\text{ mm} = 16{,}531\ \frac{\text{N}}{\text{m}},$$

$$K_{f_c} = 2uhC_d^2\cos(\theta) = 2 \times 0.75\text{ mm} \times\ 3\text{ mm} \times 0.62^2 \times \cos(69°) = 6.2 \times 10^{-7}\text{ m}^2.$$

From Equation (4.34) it can be seen that the nominal steady-state flow force when $x = 0$ is given by

$$F_x = -K_{f_c}P_o = -6.2 \times 10^{-7}\text{ m}^2 \times 20\text{ MPa} = -12.4\text{ N}.$$

Also from Equation (4.34) it can be seen that a pressure-rise-rate effect that would cancel this force out is given by

$$\dot{P} = -\frac{K_{f_c}P_o}{\rho LK_c} = -\frac{12.4\text{ N}}{850\text{ kg/m}^3 \times 50\text{ mm} \times 7.565 \times 10^{-12}\text{ m}^3/(\text{Pa s})}$$

$$= -38.57\ \frac{\text{MPa}}{\text{ms}}.$$

Within hydraulic systems these types of pressure variations are not uncommon which illustrates the fact that pressure transient effects, while short lived, may be on the order of the steady-state flow forces themselves.

Summary.   The objective of this subsection has been to derive linearized expressions for the steady-state and transient flow forces that act on a two-way spool valve. As shown by Equations (4.35) and (4.36) these expressions depend on the valve coefficients, the direction of metered flow, and the side of the metering land that is being used relative to the axial coordinate system. The work presented in this subsection is fundamental for the consideration of flow forces that act on three-way and four-way spool valves as well; therefore, this subsection is relied on heavily for subsequent flow force discussions.

4.3.4 Pressure Relief Valves

Overview.   As previously mentioned, the two-way spool valve is occasionally used within hydraulic control systems to perform the task of relieving pressure and for providing safety for the rest of the system. Figure 4-9 shows a schematic of a two-way spool valve that is used for this function. In this figure the primary hydraulic control system is represented by the compressible fluid volume at the top. This compressible fluid volume is generally the volume of fluid within the high-pressure line of the hydraulic circuit. The nominal flow rate into the compressible fluid volume that needs to be relieved is shown in Figure 4-9 by the symbol Q_o. The volumetric flow rate exiting the compressible fluid volume through the two-way relief valve is shown by the symbol Q. The fluid bulk modulus for the fluid is given by β, the instantaneous pressuring in the fluid volume is given by P, and the volume of the compressible fluid is given by V. The two-way valve is shown in Figure 4-9 with a cross-sectional area A.

This cross-sectional area is pressurized on the left by the pressure from the fluid volume P, and is pressurized on the right by the pressure from the return line P_r. Typically the return pressure is the pressure within the hydraulic control system reservoir and, therefore, P_r is taken to be zero gauge pressure. The spool valve is shown in Figure 4-9 to have a mass given by the symbol m. A spring acts on the spool valve to force it in the negative x-direction. A region of incompressible fluid is shown within the spool valve annulus by the damping length L. When the pressure within the compressible fluid volume increases the two-way spool valve is forced to the right to facilitate the relief flow Q. This relief flow then reduces or controls the fluid pressure within the compressible fluid volume, thus providing a safety function for other components within the hydraulic control system. Thus, the basic relief valve function is accomplished.

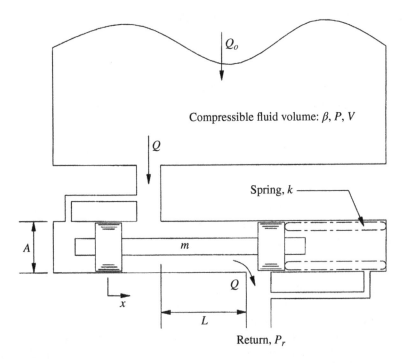

Figure 4-9. Schematic of the two-way spool type, pressure relief valve.

Analysis. The analysis of the two-way spool type, pressure relief valve shown in Figure 4-9 begins by considering the control objective of this device; namely, to control the pressure within the compressible fluid volume. This pressure may be modeled using the standard pressure-rise-rate equation, which has been presented in Equation (2.122) of Chapter 2. For the arrangement shown in Figure 4-9 this equation may be written as

$$\frac{V}{\beta}\dot{P} = Q_o - Q,$$ (4.38)

where V is the volume of compressible fluid, β is the fluid bulk modulus, P is the instantaneous pressure of the compressible fluid, Q_o is the nominal flow rate needing to be relieved by the two-way valve, and Q is the instantaneous flow rate through the valve, which will vary slightly from the nominal flow rate during the transient response of the valve. The linearized instantaneous flow rate through the two-way spool valve is given by

$$Q = \frac{1}{2}Q_o + \mathrm{K}_q x + \mathrm{K}_c P,$$ (4.39)

where x is the displacement of the two-way spool valve from its nominal position, K_q is the flow gain, and K_c is the pressure-flow coefficient. Again, these flow coefficient definitions are given in Equation (4.4). At the nominal or steady operating condition of the system Equation (4.39) may be used to show that the nominal flow rate through the valve is given by

$$Q_o = 2\mathrm{K}_c P_o,$$ (4.40)

where P_o is the nominal relief pressure of the valve. In other words, the valve is designed to provide relief flow Q_o at this pressure. From Equations (4.38) through (4.40) it may be shown that the pressure equation for the compressible volume of fluid shown in Figure 4-9 is given by

$$\frac{V}{\beta}\dot{P} + \mathrm{K}_c(P - P_o) = -\mathrm{K}_q x.$$ (4.41)

Clearly, the input for adjusting the fluid pressure within the compressible control volume is the valve displacement x. The valve displacement is modeled in the next paragraph.

By summing forces on the spool valve and setting them equal to the valve's time-rate-of-change of linear momentum, the following equation may be written to govern the axial motion of the two-way spool valve shown in Figure 4-9:

$$m\ddot{x} + c\dot{x} + kx = AP + F_x - F_o,$$ (4.42)

where m is the mass of the spool valve, c is the viscous drag coefficient, k is the spool valve spring rate, A is the spool's cross-sectional area, F_o is the nominal force or preload acting on the valve by the spring when $x = 0$, and F_x is the flow force acting on the valve in the axial direction. The flow force for the two-way spool valve is given in Equation (4.34). Substituting this result into Equation (4.42) produces the following expression for the dynamics of the two-way spool valve:

$$m\ddot{x} + (c + \rho L \mathrm{K}_q)\dot{x} + (k + \mathrm{K}_{f_q})x = (A - \mathrm{K}_{f_c})(P - P_o) - \rho L \mathrm{K}_c \dot{P},$$ (4.43)

where K_{f_q} and K_{f_c} are the flow force gain and the pressure flow force coefficient defined in Equation (4.32). In this result it has been recognized from the steady-state condition of the valve that the nominal spring force exerted on the valve is given by

$$F_o = (A - \mathrm{K}_{f_c})P_o.$$ (4.44)

In summary, Equations (4.41) and (4.43) are presented as the governing equations that describe the dynamics of the two-way spool type, pressure relief valve shown in Figure 4-9. These equations comprise a third-order system in which both the pressure and the spool valve displacement vary with respect to time. In the following paragraphs the initial design of this system from a steady-state point of view will be considered followed by an assessment of the valve's stability.

Design.   As with all design problems the steady-state governing equations must be satisfied first before any dynamic considerations become important. In the previous paragraphs the steady-state flow rate and the steady-state pressure have been used as part of the dynamic equations of motion for the system shown in Figure 4-9. These steady descriptions are given in Equations (4.40) and (4.44). From Equations (4.40) and (4.4), it can be seen that the steady flow rate through the valve must be satisfied by the following equation

$$Q_o = A_o C_d \sqrt{\frac{2}{\rho} P_o}, \tag{4.45}$$

which is simply the classical orifice equation applied at the steady-state operating conditions of the valve. For rectangular flow passages $A_o = uh$ where u is the nominal opening of the valve and h is the passage height. See Figure 4-2. If we say that $u = h/4$ it can be shown from Equation (4.45) that the rectangular porting geometry for the valve is given by

$$u = \sqrt{\frac{Q_o}{4C_d \sqrt{\frac{2}{\rho} P_o}}} \quad \text{and} \quad h = 4u. \tag{4.46}$$

These dimensions are useful for the initial sizing of the valve ports.

Another design parameter that must be specified is the spring preload F_o that is required at the nominal operating condition of the valve. This specification is given in Equation (4.44) as a function of the valve's cross-sectional area, the pressure flow force coefficient, and the nominal operating pressure of the valve. This force is also equal to the spring rate times the total spring deflection at the nominal operating condition of the valve. Therefore, the spring rate for the valve must be designed such that

$$k = \frac{(A - K_{f_c})P_o}{\delta}, \tag{4.47}$$

where δ is the total deflection of the spring at the nominal operating conditions. *Note*: The steady-state flow force of the valve $K_{f_c} P_o$ serves to reduce the spring rate requirement since this force seeks to assist the spring by trying to close the valve.

Stability.   Perhaps one of the greatest difficulties with a pressure relief valve is that of stability. In many, if not most, pressure relief valve applications the valve itself is highly unstable during operation and this instability results in audible noise problems that can be irritating to the end user. Generally, relief valve instability is tolerated in hydraulic systems due to the fact that the relief valve is designed to operate only on an intermittent basis and even though the valve is unstable it still provides an average relief effect that protects the system from being over pressurized. Along with the irritating noise that is generated by an unstable relief valve another downside is that an unstable relief valve tends to wear itself out more quickly than it might if it were designed to be stable. This wear results from the limit cycle and continual pounding against hard stops within the valve. Therefore, to avoid the adverse characteristics of a pressure relief valve it is desirable to make the valve stable. In the following paragraph stability analysis for the relief valve is developed based on the Routh-Hurwitz stability criterion.

As previously mentioned, Equations (4.41) and (4.43) comprise a third-order set of differential equations that describe the dynamics of the pressure relief valve shown in Figure 4-9. Using the Laplace transform or state space methods it can be shown that the characteristic equation for this system is given by

$$a_o s^3 + a_1 s^2 + a_2 s + a_3 = 0, \tag{4.48}$$

where

$$a_o = \frac{V}{\beta} m$$

$$a_1 = \frac{V}{\beta}(c + \rho L K_q) + m K_c$$

$$a_2 = \frac{V}{\beta}(k + K_{f_q}) + c K_c \tag{4.49}$$

$$a_3 = (k + K_{f_q})K_c + (A - K_{f_c})K_q.$$

In Chapter 3 the Routh-Hurwitz stability criterion for a third-order system was discussed and it was shown that for the linear third-order system to be stable all of the coefficients listed in Equation (4.49) need to be positive with the additional requirement that

$$a_1 a_2 > a_o a_3. \tag{4.50}$$

Since each coefficient in Equation (4.49) is always positive Equation (4.50) is the only stability requirement remaining to be satisfied. By neglecting the viscous damping coefficient c Equation (4.49) may be substituted into Equation (4.50) to produce the following stability criterion for the relief valve shown in Figure 4-9:

$$\frac{L\rho}{m}\frac{V}{\beta}\frac{(k + K_{f_q})}{(A - K_{f_c})} > 1. \tag{4.51}$$

From this criterion it can be seen that the energy storage capacity in the elastic members of the system; that is, the compressible fluid and the spring, must be sufficiently large in order to guarantee stability. Destabilizing tendencies are realized for the system when the valve inertia or the input force to the valve becomes large.

Case Study

The two-way valve in the previous case study is to be used as a pressure relief valve within an arrangement similar to the one shown in Figure 4-9. From the previous case study we learned that the following parameters apply to this relief valve application:

$$h = 3 \text{ mm}, \quad u = 0.75 \text{ mm}, \quad L = 50 \text{ mm},$$

$$P = 20 \text{ MPa}, \quad C_d = 0.62, \quad \rho = 850 \text{ kg/m}^3,$$

$$K_{f_q} = 16.5 \text{ N/mm}, \quad \text{and} \quad K_{f_c} = 0.62 \text{ mm}^2.$$

The valve is designed to fit within an 8 mm bore; therefore, its cross-sectional area is given by $A = 50.27 \text{mm}^2$. The mass of the valve is given by 0.021 kg, the fluid bulk modulus is 0.8 GPa, and the volume of compressible fluid is 0.001 m³. Calculate the nominal flow rate for this valve. Assuming that the total spring deflection in the nominal operating condition is 12 mm calculate the valve spring rate that is needed to operate the relief function at 20 MPa. Is this valve stable or not?

The nominal flow rate for the valve may be calculated by using the classical orifice equation shown in Equation (4.45). Using this equation with the rectangular porting geometry of this problem it may be shown that the nominal flow rate is given by

$$Q_o = uhC_d\sqrt{\frac{2}{\rho}P_o} = 0.75 \text{ mm} \times 3 \text{ mm} \times 0.62\sqrt{\frac{2}{850 \text{ kg/m}^3} \times 20 \text{ MPa}}$$

$$= 18.16 \text{ lpm.}$$

From Equation (4.47) it may be seen that the required spring rate for the valve is

$$k = \frac{(A - K_{f_c})P_o}{\delta} = \frac{(50.27 \text{ mm}^2 - 0.62 \text{ mm}^2) \times 20 \text{ MPa}}{12 \text{ mm}} = 82.74\frac{\text{N}}{\text{mm}}.$$

Based on this spring rate the stability criterion of Equation (4.51) may be evaluated. The left-hand side of this equation is given by

$$\frac{L\rho}{m}\frac{V}{\beta}\frac{(k + K_{f_q})}{(A - K_{f_c})} = \frac{50 \text{ mm} \times 850 \text{ kg/m}^3}{0.021 \text{ kg}} \times \frac{0.001 \text{ m}^3}{0.8 \text{ GPa}}$$

$$\times \frac{(82.74 \text{ N/mm} + 16.53 \text{ N/mm})}{(50.27 \text{ mm}^2 - 0.62 \text{ mm}^2)}$$

$$= 4.97.$$

Since this result is greater than unity the pressure relief valve will be stable.

Summary.   In this subsection the two-way spool type, pressure relief valve of Figure 4-9 has been analyzed, designed, and checked for potential instability during its normal operation. As it turns out this valve has much in common with the poppet type, pressure relief valve that is discussed in Subsection 4.6.4 and therefore much of the preceding analysis is germane to that problem as well.

4.3.5 Summary

In conclusion of this section it is important to note that our consideration of the simple two-way spool valve provides the basis for analyzing the more complex spool valves that is discussed in the following sections. Particularly, the flow force discussion of Subsection 4.3.3 is referred to and used to analyze the net flow force that is exerted on the three-way and four-way spool valve while considering the contribution of multiple metering lands that operate simultaneously for underlapped and critically lapped valve geometry.

4.4 THREE-WAY SPOOL VALVES

4.4.1 Overview

In our consideration of spool valves we now move from the simple two-way valve to the slightly more complex three-way valve. The three-way spool valve is used frequently as a control valve for pressure-controlled variable displacement pumps. It is also used as a

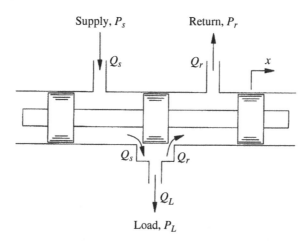

Figure 4-10. Three-way spool valve.

stand-alone control valve for hydraulic systems that utilize a single rod linear actuator for accomplishing the overall system output. These applications are discussed in the chapters that follow.

The three-way spool valve is shown in Figure 4-10 with three flow lines. The supply line is pressurized to a level given by P_s while the return line is pressurized to the level P_r. The load line indicates the working port of the control valve, which is characterized by the load pressure P_L. To direct flow into the load port the spool is moved in the positive x-direction, thus facilitating the load volumetric flow rate shown in Figure 4-10 by the symbol Q_L. The supply flow and return flow are shown in Figure 4-10 by the symbols Q_s and Q_r, respectively.

For an incompressible fluid it may be seen from Figure 4-10 that the load volumetric flow rate is given by

$$Q_L = Q_s - Q_r. \tag{4.52}$$

From the general form of Equation (4.3) it may be shown that the supply and return volumetric flow rates are given by the following expressions:

$$Q_s = \frac{1}{2}Q_{s_o} + \mathrm{K}_{q_s}x + \mathrm{K}_{c_s}(P_s - P_L)$$
$$Q_r = \frac{1}{2}Q_{r_o} + \mathrm{K}_{q_r}x + \mathrm{K}_{c_r}(P_L - P_r), \tag{4.53}$$

where Q_{s_o} and Q_{r_o} are the nominal flow rates through the valve at steady operating conditions, K_{q_s} and K_{q_r} are the flow gains, and K_{c_s} and K_{c_r} are the pressure-flow coefficients defined generally in Equation (4.4). The reader will recall that the nominal flow rate is determined when $x = 0$ and when the supply and return pressures are at nominal values at which the valve coefficients have been determined. It should also be noted that the flow gain for the return line K_{q_r} is a negative number according to its definition in Equation (4.4). This results from the fact that the return line flow area decreases with a positive displacement of the valve. At the nominal steady-state condition for the valve shown in Figure 4-10 the displacement of the valve is given by $x = 0$ and the load flow Q_L is also zero. This nominal condition requires that

$$P_{L_o} = \frac{1}{2}P_{s_o}, \qquad P_r = 0, \tag{4.54}$$

where P_{L_o} and P_{s_o} are the nominal load and supply pressure respectively. With these nominal pressure conditions it may be shown that

$$Q_{s_o} = Q_{r_o} = Q_o, \qquad K_{q_s} = K_q, \qquad K_{q_r} = -K_q, \qquad \text{and} \qquad K_{c_s} = K_{c_r} = K_c, \qquad (4.55)$$

where K_q and K_c are simply the flow gain and the pressure-flow coefficient for the entire valve. Using these results with Equations (4.52) and (4.53) it may be shown that the load flow is given by

$$Q_L = 2K_q x - 2K_c(P_L - P_s/2). \qquad (4.56)$$

From this equation it may be seen that the load flow will increase as the valve displacement x increases but will decrease as the pressure drop $(P_L - P_s/2)$ increases. Obviously, at nominal conditions of the valve the load flow vanishes as previously described.

Figure 4-10 depicts a three-way spool valve that utilizes underlapped geometry for the metering land. It must be mentioned that this is not the only design condition for this metering land and that the load flow result of Equation (4.56) changes as the design of the metering land changes. Figure 4-11 shows three possible design conditions for the metering land. The top configuration shown in this figure is the underlapped configuration that is shown in Figure 4-10, which was also used to generate the flow result of Equation (4.56). This underlapped geometry is commonly referred to as an "open centered" design since the load port is symmetrically open to both the supply and return lines when the valve is in its center position; that is, when $x = 0$. The dimension u shows this nominal opening and is usually referred to as the underlapped dimension. For the open centered design, the valve is intended to operate within a displacement regime given by $-u < x < u$. The middle configuration of Figure 4-11 shows a special case of the top configuration in which the underlapped dimension u vanishes. This condition is known as the "critically centered" design since the load port is critically blocked from both the supply and return lines when the valve is in its centered position. As the reader can see from Figure 4-11 a slight movement of the critically centered valve in either direction will open the load port to either the supply or return line while keeping the opposite line blocked. Since the edges of the metering land are coincident with the edges of the port geometry in the critically centered design, the porting for this configuration is often referred to as "line-to-line" porting.

The bottom configuration of Figure 4-11 is known as the "closed centered" configuration. The porting of this design is referred to as "overlapped" porting since the underlapped dimension u is actually negative. This configuration describes a metering land that over laps the edges of the metering port. The closed center design requires the valve to move a distance u before any flow enters or exits the load port. This initial movement is referred to as valve deadband and is generally an undesirable feature for a hydraulic control valve. Therefore, closed centered valves are not considered further in our discussion.

As previously noted, the load flow for the open centered valve design is given in Equation (4.56). For the critically centered valve the flow into the load port is achieved by moving the valve in the positive x-direction in which case the load flow is given by $Q_L = K_{q_s} x$. If the valve is moved in the negative x-direction the load flow is given by $Q_L = -K_{q_r} x$ where both K_{q_r} and x are negative numbers. *Note*: In these cases the nominal flow rate and pressure-flow coefficient are zero due to the blocked condition of the port when $x = 0$. In fact, the flow gains K_{q_s} and K_{q_r} are nonzero only if the porting geometry is rectangular. See the discussion on valve coefficients and porting geometry in Subsection 4.2.3. If the nominal pressure conditions of Equation (4.54) are satisfied

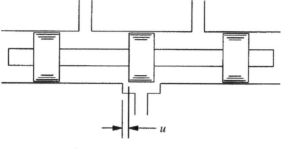

Open centered (underlapped)

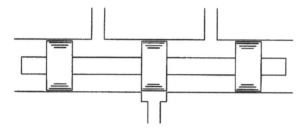

Critically centered (line-to-line)

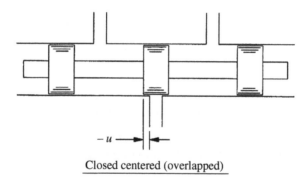

Closed centered (overlapped)

Figure 4-11. Porting geometry for the three-way spool valve.

then the valve will behave symmetrically with respect to the valve displacement and the valve flow gain will be calculated using $K_q = K_{q_s} = -K_{q_r}$. In this case the load flow may be described as:

$$Q_L = K_q x. \tag{4.57}$$

In this result the nominal pressure effects have vanished and the load flow remains proportional to the valve displacement x. The critically centered result of Equation (4.57) should be compared to the open centered result of Equation (4.56) in which the pressure effects have been retained due to the open centered configuration.

4.4.2 Efficiency

The efficiency of the three-way spool valve is determined by calculating the ratio of the useful output power to the supplied input power. For the three-way valve shown in Figure 4-10 it may be shown that the supplied hydraulic power to the valve is given by

$$\dot{W}_s = P_s Q_s, \tag{4.58}$$

and the useful output power is

$$\dot{W}_o = P_L Q_L. \tag{4.59}$$

Using these two results the overall efficiency of the three-way spool valve is given by

$$\eta = \frac{\dot{W}_o}{\dot{W}_s} = \frac{P_L Q_L}{P_s Q_s} = \eta_p\, \eta_v, \tag{4.60}$$

where the pressure efficiency of the valve is

$$\eta_p = \frac{P_L}{P_s}, \tag{4.61}$$

and the volumetric efficiency of the valve is

$$\eta_v = \frac{Q_L}{Q_s}. \tag{4.62}$$

Using Equations (4.53) and (4.55) it may be shown that the supply flow to the open centered three-way spool valve is given by

$$Q_s = K_q x - K_c(P_L - 3P_s/2), \tag{4.63}$$

where it has been recognized that the nominal supply flow is $Q_o = K_c P_s$. Using this result with Equation (4.56) the volumetric efficiency of the valve may be written as

$$\eta_v = \frac{2K_q x - 2K_c(P_L - P_s/2)}{K_q x - K_c(P_L - 3P_s/2)}. \tag{4.64}$$

A first-order Taylor series expansion of this result near the nominal operating conditions of the valve shows that the volumetric efficiency of the open center three-way valve may be approximated as

$$\eta_v = 1 + 2\frac{K_p}{P_s}x - 2\frac{P_L}{P_s}, \tag{4.65}$$

where K_p is the pressure sensitivity of the valve given in Equation (4.5). From this equation it can be seen that the volumetric efficiency of the valve is zero at the nominal operating conditions when $x = 0$ and $P_L = P_s/2$. This is because the load flow is zero in this condition and there is no useful output power from the valve. As the valve moves in the positive x-direction, however, the volumetric efficiency of the valve increases. For rectangular flow passages Equation (4.8) may be used to show that the pressure sensitivity of the open centered three-way valve is given by $K_p = P_s/u$ where u is the underlapped dimension shown in Figure 4-11. Using this result with Equation (4.65) it may be shown that the volumetric efficiency for the open centered three-way valve utilizing rectangular ports is given by

$$\eta_v = 1 + 2\frac{x}{u} - 2\frac{P_L}{P_s}. \tag{4.66}$$

From Equations (4.60), (4.61), and (4.66) the overall efficiency of the open centered three-way valve is given by

$$\eta = \eta_p\,\eta_v = \frac{P_L}{P_s}\left(1 + 2\frac{x}{u} - 2\frac{P_L}{P_s}\right), \tag{4.67}$$

for rectangular port geometry. Again, this result vanishes at the nominal operating conditions of the valve; however, when the valve is being used its overall efficiency is nonzero. For instance, when the load pressure remains at its nominal operating condition $P_L = P_s/2$ and when the valve is displaced by $x = u/4$ the overall efficiency of the valve is given by $\eta = 0.25$, which illustrates the fact that hydraulic valves create large energy losses within hydraulic systems.

If the three-way valve is a critically centered design with an underlapped dimension given by $u = 0$ then the efficiency problem reduces itself to a two-way valve problem in which the overall efficiency is given by

$$\eta = \eta_p = \frac{P_L}{P_s}. \tag{4.68}$$

See the discussion in Subsection 4.3.2 for the efficiency of a two-way valve. At the nominal operating condition of this valve the efficiency has increased to 50%, which illustrates the improvement in efficiency that may be obtained by using a critically centered design.

4.4.3 Flow Forces

The three-way spool valve in Figure 4-10 shows flow that is being metered simultaneously across the left- and right-hand side of the metering land. This simultaneous metering results from the open centered design of the valve, which creates flow forces that result from both sides. In this case, the flow forces acting on the spool valve are given by

$$F_x = F_{x_s} + F_{x_r}, \tag{4.69}$$

where the flow force F_{x_s} results from the flow being metered out of the valve across the left-hand side of the metering land while the flow force F_{x_r} results from the flow being metered into the valve across the right-hand side of the metering land. Using the general forms of Equations (4.35) and (4.36) the flow forces acting on the valve from each side of the metering land are given by

$$\begin{aligned}
F_{x_s} &= -\rho L K_{q_s}\dot{x} - \rho L K_{c_s}(\dot{P}_s - \dot{P}_L) - K_{f_{qs}}x - K_{f_{cs}}(P_s - P_L), \\
F_{x_r} &= -\rho L K_{q_r}\dot{x} - \rho L K_{c_r}(\dot{P}_L - \dot{P}_r) + K_{f_{qr}}x + K_{f_{cr}}(P_L - P_r),
\end{aligned} \tag{4.70}$$

where P_L is the fluid pressure in the load port, P_s is the fluid pressure in the supply port, and P_r is the fluid pressure in the return port. *Note*: In this result the damping length L is assumed to be the same for both sides of the valve. See Figure 4-5. As shown by the subscripts in these equations the linearized flow and flow force coefficients are generally different from each other depending on the nominal pressure drop that exists across each side of the metering land. If the nominal conditions of Equation (4.54) are used it may be shown that these coefficients are given by

$$\begin{aligned}
K_{q_s} &= K_q, & K_{c_s} &= K_c, & K_{f_{qs}} &= K_{f_q}, & K_{f_{cs}} &= K_{f_c}, \\
K_{q_r} &= -K_q, & K_{c_r} &= K_c, & K_{f_{qr}} &= -K_{f_q}, & K_{f_{cr}} &= K_{f_c},
\end{aligned} \tag{4.71}$$

where K_q, K_c, K_{f_q}, and K_{f_c} are defined in Equations (4.4) and (4.32). Using these results with Equation (4.69) it can be shown that the total axial flow force exerted on the valve is given by

$$F_x = -\rho L K_c \dot{P}_s - 2K_{f_q} x + 2K_{f_c}(P_L - P_s/2), \tag{4.72}$$

where the return pressure P_r has been assumed to be constant and zero. This result shows that the three-way spool valve in Figure 4-10 is susceptible to both transient and steady flow forces but that the nominal flow force vanishes at the steady-state conditions of the valve.

For the critically centered three-way valve the underlapped dimension is given by $u = 0$ and the flow force F_{x_r} vanishes for positive displacements of the spool valve. For negative displacements of the spool valve, the flow force F_{x_s} vanishes. Therefore, for the critically centered three-way valve design the total flow force acting on the valve is given by

$$F_x = \begin{cases} -\rho L K_q \dot{x} - K_{f_q} x & x > 0 \\ 0 & x = 0 \ . \\ \rho L K_q \dot{x} - K_{f_q} x & x < 0 \end{cases} \tag{4.73}$$

In this result the nominal conditions of Equation (4.54) have been assumed and the pressure-flow coefficient K_c and the pressure flow force coefficient K_{f_c} have vanished due to the critically centered design. The reader will recall that K_q and K_{f_q} are nonzero for the critically centered valve design only when rectangular porting geometry is used. Equation (4.73) indicates a difficulty that may be encountered when using a critically centered three-way spool valve; namely, that the damping term $\rho L K_q \dot{x}$ in the flow force equation is actually positive for negative displacements of the valve. This positive sign indicates an energy input to the valve that has destabilizing tendencies; therefore, it is not recommended that the three-way valve be used in a critically centered valve configuration if this term cannot be overcome by other damping effects acting on the valve; for example, viscous drag effects. Notice: from Equation (4.72) it can be seen that this stability issue does not exist for the open centered valve design.

Finally, it must be mentioned that if the supply and return lines are switched in Figure 4-10 the transient and steady pressure terms in Equation (4.72) will change sign. Similarly, for the critically centered valve design the transient effects in Equation (4.73) will also change sign. Notice: this change in sign for the damping term for the critically centered valve design does not eliminate the stability problem that was mentioned in the previous paragraph – it only shifts the stability problem to the positive displacement side of valve. Therefore, the recommendation for the limited use of the critically centered three-way valve remains unchanged even when the supply and return lines are switched in Figure 4-10.

4.4.4   Hydromechanical Valves

In this subsection of our chapter we present a typical design configuration for the open centered three-way spool valve. This configuration is shown in Figure 4-12 and is known as a hydromechanical three-way spool valve. The term *hydromechanical* refers to the actuation method that is used to displace the valve. Specifically, hydraulic fluid is used to directly actuate the mechanical element of the valve – thus the name "hydromechanical." As shown in Figure 4-12 the valve itself is identical to that of Figure 4-10 with underlapped porting as recommended for three-way valves in the previous subsection. The left side of the spool is pressurized by the supply pressure P_s, while the right side of the spool is pressurized by the bias pressure P_b. This bias pressure is provided by the control application of the hydraulic

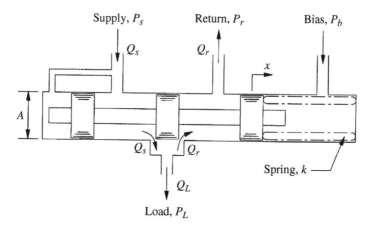

Figure 4-12. Hydromechanical three-way spool valve.

system and acts as a signal to move the spool and to generate the desired load flow Q_L. In Figure 4-12 it may be seen that a spring acts on the right-hand side of the spool to force the spool in the negative x-direction. By setting the nominal preload on the spring the null position of the valve is adjusted so that $x = 0$ in the nominal valve condition. The spring also provides a proportional relationship along with the steady flow force between the valve position x and the bias pressure P_b. This proportional relationship is shown in the analysis that follows.

For the three-way open centered spool valve shown in Figure 4-12, the spring and pressure forces acting on the spool are generally much greater than the linear momentum and viscous drag effects that result from the transient motion of the valve. Furthermore, since the valve is of the open centered type there are no stability concerns for this valve construction and the transient analysis of the valve becomes fairly unimportant for most applications of this design. This is especially true when the three-way spool valve is used to actuate a relatively slow moving device. In this case the transient characteristics of the three-way valve contribute little to the overall dynamics of the system and the valve may be modeled using steady considerations of the valve displacement itself. By summing forces on the spool valve shown in Figure 4-12 and setting them equal to zero the governing equation of motion for the spool valve is given by

$$kx = A(P_s - P_b) + F_x - F_o, \tag{4.74}$$

where k is the spring rate, A is the cross-sectional area of the valve, F_x is the flow force acting on the valve, and F_o is the spring force acting on the valve in the nominal condition when $x = 0$. The reader will recall from Equation (4.72) that the nominal flow force acting on the valve is zero. Furthermore, it is common to specify a nominal pressure difference between the supply pressure and the bias pressure. This nominal pressure difference is known as the "margin pressure" given by $P_m = (P_s - P_b)_o$ where the subscript "o" denotes the nominal pressure condition. At the nominal operating condition of the valve Equation (4.74) may be used to show that

$$F_o = AP_m, \tag{4.75}$$

where P_m is the margin pressure just mentioned. Substituting Equations (4.72) and (4.75) into Equation (4.74) it may be shown that the valve displacement is given by

$$x = \frac{A}{(k + 2K_{f_q})}(P_s - P_b - P_m) + \frac{2K_{f_c}}{(k + 2K_{f_q})}(P_L - P_s/2), \tag{4.76}$$

where the pressure transient term in the flow force equation has been neglected due to its ordinarily small size. In Equation (4.76) the difference between the supply pressure and the bias pressure $(P_s - P_b)$ is used to adjust the displacement of the spool valve and may be considered as the input to the valve design. Also shown in Equation (4.76) is the effect of the load pressure P_L. This pressure effect is generally undesirable but can often be substantial in magnitude if P_L deviates very far away from the nominal load pressure $P_{s_o}/2$. In this sense the load pressure may be considered as a disturbance effect that may need to be compensated for if accurate valve positioning is desired across a wide range of operating conditions.

Case Study

An open centered hydromechanical three-way valve similar to Figure 4-12 is used as a control valve for a load sensing variable displacement pump. The supply pressure to the valve is 40 MPa with a nominal load pressure of 20 MPa. The valve porting is rectangular with a height of 3 mm and an underlapped dimension of 0.75 mm. The flow force gain is given by $K_{f_q} = 16,531$ N/m and the pressure flow force coefficient is $K_{f_c} = 6.20 \times 10^{-7}$ m². The bore diameter for the spool valve is 8 mm, which means that the cross-sectional area $A = 50.27$ mm². The valve spring rate is 67 N/mm and the margin pressure for the design is 2 MPa. If the load pressure remains at 20 MPa calculate the change in the spool position when the supply pressure minus the bias pressure is equal to two times the margin pressure. Similarly, if the supply pressure minus the bias pressure remains equal to the margin pressure calculate the change in the spool position when the load pressure is 30 MPa.

These calculations are based on Equation (4.76). For the first calculation the change in spool position may be determined by ignoring the second term in this equation since it vanishes at a nominal load pressure of 20 MPa. Thus,

$$x = \frac{A}{(k + 2K_{f_q})}(2P_m - P_m) = \frac{50.27 \text{ mm}^2}{(67 \text{ N/mm} + 2 \times 16.531 \text{ N/mm})} (2 \text{ MPa}) = 1 \text{ mm.}$$

From this result it may be seen that an error in the supply pressure minus the bias pressure that is equal to the margin pressure will saturate the underlapped dimension of the valve. To reduce this potential for saturation the mechanical spring rate may be increased or the bore diameter of the spool valve may be decreased.

To calculate the effect of the load pressure on the spool valve position the second term in Equation (4.76) may be examined as:

$$x = \frac{2K_{f_c}}{(k + 2K_{f_q})}(P_L - P_s/2)$$

$$= \frac{2 \times 0.62 \text{ mm}^2}{(67 \text{ N/mm} + 2 \times 16.531 \text{ N/mm})} (10 \text{ MPa}) = 0.124 \text{ mm.}$$

From here it can be seen that a 10 MPa variation in the load pressure produces a significant adjustment the spool valve position relative to the underlapped dimension of 0.75 mm. This disturbance to the valve results from the steady pressure component of the flow force equation.

4.4.5 Summary

In this section of the chapter the three-way spool valve is presented and discussed. In particular it has been noted that the critically centered three-way valve exhibits a strong potential for instability due to the transient flow forces that act on the valve. Therefore, it has been recommended that three-way valves be used in open centered configurations, which are less susceptible to this problem. In this section, the example of the hydromechanical three-way valve has been used to show that the steady flow forces acting on the valve can have a significant impact on the position of the valve. In particular, the pressure component of the flow force behaves as a disturbance to the valve that can be significant. This means that if precise three-way valve control is desired something more sophisticated than the hydromechanical valve shown in Figure 4-12 will be needed to reject the steady flow force disturbance effects.

4.5 FOUR-WAY SPOOL VALVES

4.5.1 Overview

We now turn our attention to the consideration of the four-way spool valve, which is the bread-and-butter flow control valve for hydraulic circuitry. As is shown in the next chapter, this valve is used as a control valve for adjusting the displacement of the variable displacement pump. More importantly, however, this valve is often used as the main control valve for the entire hydraulic circuit as it simultaneously directs flow to and away from the working implement of the system. This application of the four-way spool valve is discussed at length in Chapter 7.

The four-way spool valve is shown in Figure 4-13 with four flow lines. The supply line is pressurized to a level given by P_s while the return line is pressurized to the level P_r. For the four-way valve there are two ports that are used for directing flow to and away from the load. These ports are indicated in Figure 4-13 by ports A and B, which are pressurized to the levels indicated by P_A and P_B, respectively. To direct flow into port A the spool valve is moved in the positive x-direction, thus facilitating the volumetric flow rate Q_A. For this same motion of the valve port B is opened to the return line, thus facilitating the volumetric flow rate Q_B.

The supply flow and return flow are shown in Figure 4-13 by the symbols Q_s and Q_r, respectively. For an incompressible fluid it may be seen from Figure 4-13 that the volumetric flow rate into port A and out of port B is given respectively by

$$Q_A = Q_2 - Q_1 \quad \text{and} \quad Q_B = Q_4 - Q_3, \tag{4.77}$$

where Q_1 through Q_4 are the volumetric flow rates across each metering land of the valve. From the general form of Equation (4.3) it can be seen that these flow rates are given by the following expressions:

$$Q_1 = \frac{1}{2}Q_{1_o} + K_{q_1} x + K_{c_1} (P_A - P_r)$$

$$Q_2 = \frac{1}{2}Q_{2_o} + K_{q_2} x + K_{c_2} (P_s - P_A)$$

$$Q_3 = \frac{1}{2}Q_{3_o} + K_{q_3} x + K_{c_3} (P_s - P_B)$$

$$Q_4 = \frac{1}{2}Q_{4_o} + K_{q_4} x + K_{c_4} (P_B - P_r), \tag{4.78}$$

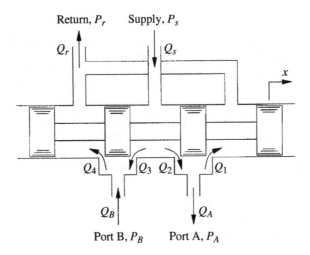

Figure 4-13. Four-way spool valve.

where Q_{1_o} through Q_{4_o} are the nominal flow rates through the valve at steady operating conditions, K_{q_1} through K_{q_4} are the flow gains, and K_{c_1} through K_{c_4} are the pressure-flow coefficients defined generally in Equation (4.4). The reader will recall that the nominal flow rate is determined when $x = 0$ and when the supply and return pressures are at nominal values at which the valve coefficients have been determined. It should also be noted that the flow gains K_{q_1} and K_{q_3} are negative numbers according to the definition in Equation (4.4). This results from the fact that the flow area at metering lands 1 and 3 decreases with a positive displacement of the valve. At the nominal steady-state condition for the valve shown in Figure 4-10 the displacement of the valve is given by $x = 0$ and the volumetric flow rates into ports A and B are also zero. This nominal condition requires that

$$P_{A_o} = P_{B_o} = \frac{1}{2}P_{s_o}, \qquad P_r = 0, \tag{4.79}$$

where P_{A_o} and P_{B_o} are the nominal pressures in ports A and B and P_{s_o} is the nominal supply pressure. With these nominal pressure conditions it may be shown that

$$Q_{1_o} = Q_{2_o} = Q_{3_o} = Q_{4_o} = Q_o,$$
$$K_{q_1} = -K_q, \quad K_{q_2} = K_q, \quad K_{q_3} = -K_q, \quad K_{q_4} = K_q, \tag{4.80}$$
$$K_{c_1} = K_{c_2} = K_{c_3} = K_{c_4} = K_c,$$

where K_q and K_c are simply the flow gain and the pressure-flow coefficient for the entire valve. Using these results with Equations (4.77) and (4.78) it may be shown that the port flows shown in Figure 4-13 are given by

$$Q_A = 2K_q x - 2K_c(P_A - P_s/2),$$
$$Q_B = 2K_q x + 2K_c(P_B - P_s/2). \tag{4.81}$$

Note: The direction of this flow is shown by the arrows in Figure 4-13. From this equation it can be seen that the port flows will increase as the valve displacement x increases and will vary with pressure. Obviously, at nominal conditions of the valve the port flows vanishes as previously described.

Figure 4-13 depicts a four-way spool valve that utilizes underlapped geometry for the metering lands. Again, it must be mentioned that this is not the only possible design for the valve and that the flow result of Equation (4.81) will change as the design of the metering lands changes. Figure 4-14 shows three possible design conditions for the metering lands of the four-way spool valve. The top configuration shown in this figure is the underlapped configuration that is shown in Figure 4-13, which was also used to generate the flow result of Equation (4.81). Again, this underlapped geometry is commonly referred to as an "open centered" design since the flow ports are symmetrically open to both the supply and return

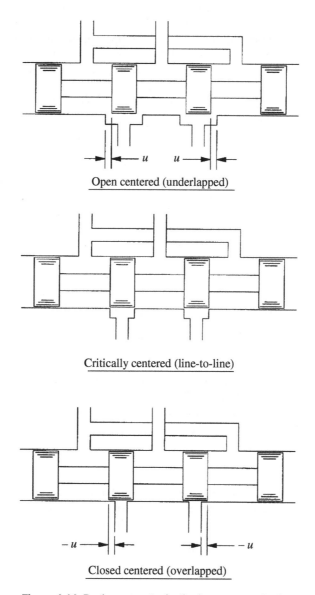

Open centered (underlapped)

Critically centered (line-to-line)

Closed centered (overlapped)

Figure 4-14. Porting geometry for the four-way spool valve.

lines when the valve is in its center position; that is, when $x = 0$. The dimension u shows this nominal opening and is referred to as the underlapped dimension. For the open centered design, the valve is intended to operate within a displacement regime given by $-u < x < u$. The middle configuration of Figure 4-14 shows a special case of the top configuration in which the underlapped dimension u vanishes. This condition is known as the "critically centered" design since the flow ports are critically blocked from both the supply and the return lines when the valve is in its centered position.

As the reader can see from Figure 4-14 a slight movement of the critically centered valve in either direction will open the flow ports to either the supply or the return lines of the valve while keeping the opposite lines blocked. Since the edges of the metering land are coincident with the edges of the port geometry in the critically centered design the porting for this configuration is often referred to as "line-to-line" porting. The bottom configuration of Figure 4-14 is known as the "closed centered" configuration. The porting of this design is referred to as overlapped porting since the underlapped dimension u is actually negative. This configuration describes a metering land that over laps the edges of the metering port. The closed center design requires the valve to move a distance u before any flow enters or exits the flow ports. Again, this initial movement is referred to as valve deadband and is generally an undesirable feature for a hydraulic control valve. Therefore, closed centered four-way valves are not considered in this chapter.

As previously noted, the load flow for the open centered four-way valve design is given in Equation (4.81). For the critically centered valve the flow into port A is achieved by moving the valve in the positive x-direction. In this case the flow into port A and out of port B is given by $Q_A = K_{q_2}x$ and $Q_B = K_{q_4}x$, respectively. If the valve is moved in the negative x-direction the flow into port A and out of port B is given by $Q_A = -K_{q_1}x$ and $Q_B = -K_{q_3}x$, respectively, where K_{q_1}, K_{q_3}, and x are negative numbers. *Note*: In these cases the nominal flow rate and pressure-flow coefficient are zero due to the blocked condition of the port when $x = 0$. In fact, the flow gains K_{q_1} through K_{q_4} are nonzero only if the porting geometry is rectangular. See the discussion on valve coefficients and porting geometry in Subsection 4.2.3. If the nominal pressure conditions of Equation (4.79) are satisfied then the valve will behave symmetrically with respect to the valve displacement and the valve flow gain will be calculated using $K_q = -K_{q_1} = K_{q_2} = -K_{q_3} = K_{q_4}$. In this case the port flows are given by:

$$Q_A = Q_B = K_q x, \tag{4.82}$$

where the positive direction of these flow is shown by the arrows in Figure 4-13. In this result the nominal pressure effects have vanished and the port flows remains proportional to the valve displacement x. The critically centered result of Equation (4.82) should be compared to the open centered result of Equation (4.81) in which the pressure effects have been retained due to the open centered configuration.

4.5.2   Efficiency

The efficiency of the four-way spool valve is determined by calculating the ratio of the useful output power to the supplied input power. For the four-way valve shown in Figure 4-13 it may be shown that the supplied hydraulic power to the valve is given by

$$\dot{W}_s = P_s Q_s, \tag{4.83}$$

and the useful output power through port A is

$$\dot{W}_o = P_A Q_A. \tag{4.84}$$

Using these two results the overall efficiency of the four-way spool valve is given by

$$\eta = \frac{\dot{W}_o}{\dot{W}_s} = \frac{P_A Q_A}{P_s Q_s} = \eta_p \eta_v, \tag{4.85}$$

where the pressure efficiency of the valve is

$$\eta_p = \frac{P_A}{P_s}, \tag{4.86}$$

and the volumetric efficiency of the valve is

$$\eta_v = \frac{Q_A}{Q_s}. \tag{4.87}$$

From Figure 4-13 it can be seen that the supply flow to the four-way hydraulic valve is given by $Q_s = Q_2 + Q_3$. Using Equation (4.78) this result is written explicitly as

$$Q_s = K_c(3P_s - P_A - P_B), \tag{4.88}$$

where it has been recognized that the nominal supply flow is $Q_o = K_c P_s$. Using this result with Equation (4.81) the volumetric efficiency of the four-way valve may be written as

$$\eta_v = \frac{2K_q x - 2K_c(P_A - P_s/2)}{K_c(3P_s - P_A - P_B)}. \tag{4.89}$$

A first-order Taylor series expansion of this result near the nominal operating conditions of the valve shows that the volumetric efficiency of the open centered four-way valve may be approximated as

$$\eta_v = \frac{1}{2} + \frac{K_p}{P_s} x - \frac{P_A}{P_s}, \tag{4.90}$$

where K_p is the pressure sensitivity of the valve given in Equation (4.5). The reader should notice that this volumetric efficiency is exactly half of that for the open center three-way valve due to the extra metering land. From this equation it can be seen that the volumetric efficiency of the valve is zero at the nominal operating conditions when $x = 0$ and $P_A = P_s/2$. Again, this is because the load flow is zero in this condition and there is no useful output power from the valve. As the valve moves in the positive x-direction, however, the volumetric efficiency of the valve increases. For rectangular flow passages Equation (4.8) may be used to show that the pressure sensitivity of the open centered four-way valve is given by $K_p = P_s/u$ where u is the underlapped dimension shown in Figure 4-14. Using this result with Equation (4.90) it may be shown that the volumetric efficiency for the open centered four-way valve utilizing rectangular ports is given by

$$\eta_v = \frac{1}{2} + \frac{x}{u} - \frac{P_A}{P_s}. \tag{4.91}$$

From Equations (4.85), (4.86), and (4.91) the overall efficiency of the open centered four-way valve is given by

$$\eta = \eta_p \eta_v = \frac{P_A}{P_s}\left(\frac{1}{2} + \frac{x}{u} - \frac{P_A}{P_s}\right), \tag{4.92}$$

for rectangular port geometry. Again, this result vanishes at the nominal operating conditions of the valve; however, when the valve is being used its overall efficiency is nonzero. For instance, when the port A pressure remains at its nominal operating condition $P_A = P_s/2$ and when the valve is displaced by $x = u/4$ the overall efficiency of the valve is given by $\eta = 0.125$, which once again illustrates the fact that hydraulic valves create large energy losses within hydraulic systems.

If the four-way valve is a critically centered design with an underlapped dimension given by $u = 0$ then the efficiency problem reduces itself to a two-way valve problem in which the overall efficiency is given by

$$\eta = \eta_p = \frac{P_A}{P_s}. \tag{4.93}$$

See the discussion in Subsection 4.3.2 for the efficiency of a two-way valve. At the nominal operating condition of this valve the efficiency has increased to 50%, which illustrates the improvement in efficiency that may be obtained by using a critically centered design.

4.5.3 Flow Forces

The four-way spool valve in Figure 4-13 shows fluid flow that is being metered simultaneously across the left- and right-hand sides of two metering lands. In this case, the flow forces acting on the spool valve are given by

$$F_x = F_{x_1} + F_{x_2} + F_{x_3} + F_{x_4}, \tag{4.94}$$

where the numerical subscripts identify the flow force associated with each flow rate shown in Figure 4-13. Using the general forms of Equations (4.35) and (4.36) the flow forces acting on the valve from each side of the metering lands are given by

$$
\begin{aligned}
F_{x_1} &= -\rho L K_{q_1}\dot{x} - \rho L K_{c_1}(\dot{P}_A - \dot{P}_r) + K_{f_{q1}}x + K_{f_{c1}}(P_A - P_r), \\
F_{x_2} &= -\rho L K_{q_2}\dot{x} - \rho L K_{c_2}(\dot{P}_s - \dot{P}_A) - K_{f_{q2}}x - K_{f_{c2}}(P_s - P_A), \\
F_{x_3} &= \rho L K_{q_3}\dot{x} + \rho L K_{c_3}(\dot{P}_s - \dot{P}_B) + K_{f_{q3}}x + K_{f_{c3}}(P_s - P_B), \\
F_{x_4} &= \rho L K_{q_4}\dot{x} + \rho L K_{c_4}(\dot{P}_B - \dot{P}_r) - K_{f_{q4}}x - K_{f_{c4}}(P_B - P_r),
\end{aligned}
\tag{4.95}
$$

where P_A is the fluid pressure in port A, P_B is the fluid pressure in port B, P_s is the fluid pressure in the supply port, and P_r is the fluid pressure in the return port. *Note*: In this result the damping length L is assumed to be the same for both sides of the valve. See Figure 4-5. As shown by the numerical subscripts in these equations the linearized coefficients are generally different from each other depending on the nominal pressure drop that exists across each side of the metering lands. If the nominal conditions of Equation (4.79) are used it may be shown that these coefficients are given by

$$
\begin{array}{llll}
K_{q_1} = -K_q, & K_{c_1} = K_c, & K_{f_{q1}} = -K_{f_q}, & K_{f_{c1}} = K_{f_c}, \\
K_{q_2} = K_q, & K_{c_2} = K_c, & K_{f_{q2}} = K_{f_q}, & K_{f_{c2}} = K_{f_c}, \\
K_{q_3} = -K_q, & K_{c_3} = K_c, & K_{f_{q3}} = -K_{f_q}, & K_{f_{c3}} = K_{f_c}, \\
K_{q_4} = K_q, & K_{c_4} = K_c, & K_{f_{q4}} = K_{f_q}, & K_{f_{c4}} = K_{f_c},
\end{array}
\tag{4.96}
$$

where K_q, K_c, K_{f_q}, and K_{f_c} are defined in Equations (4.4) and (4.32). Using these results with Equation (4.94) it can be shown that the total axial flow force exerted on the four-way spool valve is given by

$$F_x = -4K_{f_q}x + 2K_{f_c}(P_A - P_B). \tag{4.97}$$

This result shows that the four-way spool valve in Figure 4-13 is not susceptible to transient flow forces near the nominal operating conditions given in Equation (4.79). It may also be observed from this result that the nominal flow force vanishes at the steady-state conditions of the valve.

For the critically centered four-way valve the underlapped dimension is given by $u = 0$ and the flow forces F_{x_1} and F_{x_3} vanish for positive displacements of the spool valve. For negative displacements of the spool valve the flow forces F_{x_2} and F_{x_4} vanish. Therefore, for the critically centered four-way valve design it may be shown that the total flow force acting on the valve is given by

$$F_x = -2K_{f_q}x, \tag{4.98}$$

for both positive and negative displacements of the valve. In this result the nominal conditions of Equation (4.79) have been assumed, and the pressure-flow coefficient K_c and the pressure flow force coefficient K_{f_c} have vanished due to the critically centered design. The reader will recall that K_{f_q} is nonzero for the critically centered valve design only when rectangular porting geometry is used. Equation (4.98) differs from the flow force result of the critically centered three-way valve given in Equation (4.73) in that all transient terms have vanished and the potential for instability is no longer present. This is a significant difference between critically centered three-way and four-way spool valves, which removes any recommendation for strictly using open centered valve designs for the four-way spool valve. In other words, both open centered and critically centered valve designs may be safely used for the four-way spool valve without fear of introducing instability.

Last, it should be mentioned that if the supply and return lines are switched in Figure 4-13, the steady pressure term in Equation (4.97) will change sign. This switching of lines has no impact on the flow forces of the critically centered valve design since there are no pressure terms in that equation.

4.5.4 Two-Stage Electrohydraulic Valves

Overview. To achieve the most flexible control structure for a hydraulic system it is necessary to interface the system with an electronic input that will allow for closed loop feedback control. This interface is typically accomplished through the use of a two-stage electrohydraulic control valve. The first stage of the valve is an electrically actuated hydraulic valve, which then actuates the second stage or main valve of the hydraulic system. This two-stage actuation approach is necessary because electric actuators of reasonable size are not capable of exerting enough force on the main valve to overcome the opposing spring and flow forces. In this subsection of the chapter we consider the characteristics of a solenoid actuated, two-stage electrohydraulic valve.

Figure 4-15 shows a schematic of the solenoid actuated, two-stage electrohydraulic valve. This valve utilizes a four-way spool valve as the second or main stage of the valve for supplying flow through ports A and B to the hydraulic control system. Upon careful examination the reader will notice that this second-stage valve is identical to the four-way spool valve shown in Figure 4-13. The supply pressure to the valve is noted in Figure 4-15 by the symbol P_s while the return pressure is given by P_r. The flow rates through ports A and B are shown

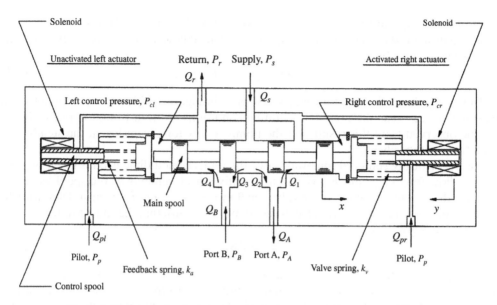

Figure 4-15. Schematic of a solenoid-actuated, two-stage electrohydraulic valve.

by the symbols Q_A and Q_B, respectively. In order for the four-way valve to direct flow into port A it must be moved in the positive x-direction by the first actuation stage of the valve. In Figure 4-15 the main spool is shown in an actuated position where the valve displacement is given by $x \neq 0$. This is done to help illustrate the working principle of the first actuation stage, which is driven by a much smaller solenoid actuated control spool.

As shown in Figure 4-15 the main spool of the two-stage control valve is controlled by the proper function of a left and right actuator. These electrohydraulic actuators are known as the first stage of the two-stage control valve since they serve the purpose of actuating the second or main stage of the valve. The left and right actuators shown in Figure 4-15 are symmetrically identical as they are each comprised of a solenoid actuated control spool, a main valve spring k_v, a lighter feedback spring for the control spool k_a, and a spring retainer with a snap ring (shown more explicitly in Figures 4-17 and 4-18). The operation of the first stage requires an auxiliary hydraulic power source to supply a pilot pressure P_p and flow rate Q_{pl} and Q_{pr} to the left and right actuators. Typically this auxiliary power source operates at very low pressures (c.a., 2 MPa) and therefore the energy loss associated with this auxiliary power supply is insignificant when compared to the power delivery of the main spool valve.

When the main spool valve is in the neutral position $x = 0$ the left- and right-hand sides of the valve are just touching the spring retainers. In this neutral position the left and right control pressures P_{cl} and P_{cr} are both pressurized to the level of half of the pilot pressure $P_p/2$ and the flow forces acting on the main spool valve are zero. In other words, the valve is perfectly centered in its nominal position with no forces acting to alter this condition. In order to move the main spool valve to the right the right-hand solenoid is activated so that it moves the right-hand control spool to the left. The motion of the control spool is indicated in Figure 4-15 by the symbol y. By moving the right-hand control spool to the left the right-hand pilot supply is choked off by the control spool and the right control pressure P_{cr} is reduced by metering a small amount of flow into the return line. During this operation the left-hand actuator shown in Figure 4-15 has remained inactive and therefore the left-hand control pressure P_{cl} is still pressured to half the level of the pilot source $P_p/2$. Now that a pressure difference exists between the left- and right-hand sides of the main spool valve a force exists that tries to move the main valve to the right. In order to accomplish this motion the pressure difference $P_{cl} - P_{cr}$ must be significant enough to overcome the preload on the spring pack that is retained by the right-hand spring retainer. Once this preload is overcome the valve moves to the right and directs fluid flow into port A and out of port B. This is, of course, the overall objective of the valve assembly. When the valve moves to the right the feedback spring between the valve retainer and the control spool is compressed, thus counteracting the electromagnetic force exerted on the control spool by the solenoid. When the forces on the control spool are perfectly balanced the control spool discontinues its motion to the left and the right control pressure P_{cr} is maintained at a level less than half of the pilot pressure $P_p/2$.

In this condition the forces on the main spool are also balanced by a combination of pressure forces, spring forces, and steady-state flow forces. It is important to notice from Figure 4-15 that there are no spring forces acting on the left-hand side of the main spool when the valve is actuated in the positive x-direction. This is due to the spring retaining mechanism, which is a critical component for setting the neutral position of the valve.

Figure 4-16 shows a schematic of the spool valve displacement x as a function of the difference in the input current between the right and left solenoid. The linearity that exists between the valve displacement and the input current is often used to describe this valve as a proportional control valve; that is, the output is proportional to the input. In the discussion that follows the solenoid actuated, two-stage electrohydraulic control valve is analyzed in greater detail. This analysis is useful for establishing design guidelines for the valve and for predicting the valve's dynamic response.

Analysis. The analysis of the electrohydraulic valve shown in Figure 4-15 begins by considering the equation of motion for the main spool valve. Similar to the discussion of the three-way valve in Subsection 4.4.4, the linear momentum and viscous damping effects on

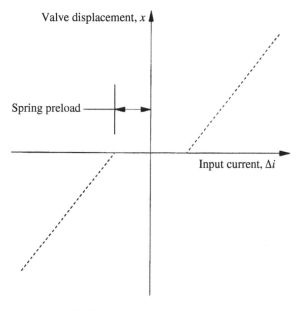

Figure 4-16. Valve displacement versus input current.

the four-way valve are considered to be negligible when compared to the pressure and spring forces exerted on the valve by the left and right actuator. Therefore, the governing equation of motion for the main spool valve is given by

$$0 = F_x + F_{al} - F_{ar}, \qquad (4.99)$$

where F_x is the flow force exerted on the valve and F_{al} and F_{ar} are the left and right actuation forces, respectively. *Note*: The flow force exerted on the valve is given either by Equation (4.97) or (4.98) depending on whether the spool valve is open centered or critically centered. The actuation forces F_{al} and F_{ar} are discussed in the following paragraphs.

Figure 4-17 shows a schematic of the left-hand actuator when the spool valve has been displaced a small distance in the positive x-direction. In this figure the control spool is now labeled as the solenoid armature since this part performs both functions.

From Figure 4-17 it may be seen that the spool valve does not touch the spring retainer; therefore, no spring forces are exerted on the left-hand side of the spool valve. In this case only a pressure force exists on the left-hand side of the spool valve and this force may be expressed as

$$F_{al} = AP_{cl}, \qquad (4.100)$$

where A is the pressurized area of the spool valve and P_{cl} is the left-hand control pressure. Since the volume of fluid on the left-hand side of the spool is extremely small the pressure transients associated with the left-hand control pressure are extremely fast and may be neglected for the purposes of this analysis. Under these incompressible flow conditions it may be shown that

$$0 = Q_{pl} - Q_{rl} - A\dot{x}, \qquad (4.101)$$

where Q_{pl} is the volumetric flow rate of fluid into the left actuator from the pilot supply and where Q_{rl} is the volumetric flow rate of fluid out of the left actuator into the return line.

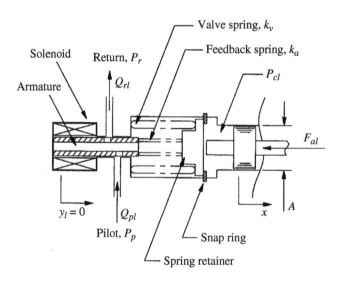

Unactivated left actuator

Figure 4-17. Left actuator of Figure 4-15.

For steady valve operation these two flow rates are identical. Since the nominal operating pressures of the left actuator are given by

$$P_{cl_o} = \tfrac{1}{2}P_{p_o} \quad \text{and} \quad P_r = 0, \tag{4.102}$$

where P_{p_o} is the nominal pilot pressure, it may be shown that the linearized volumetric flow rates in and out of the left-hand actuator are given by

$$Q_{pl} = \tfrac{1}{2}Q_o + K_{c_a}(P_p - P_{cl}), \quad Q_{rl} = \tfrac{1}{2}Q_o + K_{c_a}P_{cl}. \tag{4.103}$$

In this result it has been recognized that the left-hand control spool does not move when the main spool valve is shifted to the right; therefore, there is no control spool displacement term in this equation. In Equation (4.103) the nominal flow rate through the left-hand actuator is given by Q_o while K_{c_a} is the control spool pressure-flow coefficient generally defined in Equation (4.4). Substituting Equation (4.103) into Equation (4.101) and rearranging terms produces the following expression for the left-hand control pressure:

$$P_{cl} = \frac{P_p}{2} - \frac{A}{2K_{c_a}}\dot{x}. \tag{4.104}$$

From this result, one may see that the nominal control pressure is given by half of the pilot pressure while the motion of the main spool valve serves to reduce the left-hand control pressure in a transient fashion. Substituting Equation (4.104) into Equation (4.100) yields the following expression for the left-hand actuator force that is exerted on the main spool valve shown in Figure 4-15:

$$F_{al} = A\frac{P_p}{2} - \frac{A^2}{2K_{c_a}}\dot{x}. \tag{4.105}$$

Now that the left-hand actuator force has been determined we must develop an expression for the right-hand actuator force F_{ar}. Figure 4-18 shows a schematic of the right-hand actuator when the spool valve has been displaced a small distance in the positive x-direction.

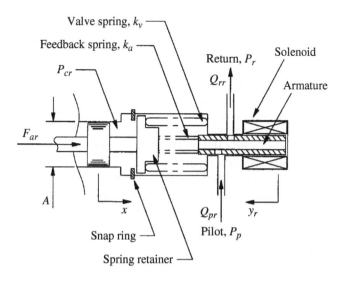

Activated right actuator

Figure 4-18. Right actuator of Figure 4-15.

From this figure it can be seen that the spool valve has exerted enough force against the spring retainer to overcome the spring pack preload. In this case pressure forces and spring forces are exerted on the valve and the following expression may be written to describe the right-hand actuator force:

$$F_{ar} = AP_{cr} + \underbrace{k_v x + F_{v_o}}_{\text{Valve spring}} + \underbrace{k_a(x + y_r) + F_{a_o}}_{\text{Feedback spring}}, \tag{4.106}$$

where P_{cr} is the right-hand control pressure, k_v is the spring rate of the main valve spring, F_{v_o} is the assembled preload of the main valve spring, k_a is the spring rate of the feedback spring, and F_{a_o} is the assembled preload of the feedback spring. Since the volume of fluid on the right-hand side of the spool is extremely small the pressure transients associated with the left-hand control pressure may be neglected for the purposes of this analysis. Under these incompressible flow conditions it may be shown that

$$0 = Q_{pr} - Q_{rr} + A\dot{x}, \tag{4.107}$$

where Q_{pr} is the volumetric flow rate of fluid into the right actuator from the pilot supply and where Q_{rr} is the volumetric flow rate of fluid out of the right actuator into the return line. Using the following nominal operating conditions of the right actuator:

$$P_{cr_o} = \tfrac{1}{2}P_{po}, \quad P_r = 0 \quad \text{and} \quad y_{r_o} = 0, \tag{4.108}$$

where P_{cr_o} is the nominal control pressure in the right actuator and y_{r_o} is the nominal displacement of the right control spool, it may be shown that the linearized volumetric flow rates in and out of the right-hand actuator are given by

$$Q_{pl} = \tfrac{1}{2}Q_o - \mathrm{K}_{q_a}y_r + \mathrm{K}_{c_a}(P_p - P_{cr}), \quad Q_{rl} = \tfrac{1}{2}Q_o + \mathrm{K}_{q_a}y_r + \mathrm{K}_{c_a}P_{cr}. \tag{4.109}$$

In this result the nominal flow rate through the right-hand actuator is given by Q_o while K_{q_a} and K_{c_a} are the control spool flow gain and the pressure-flow coefficient generally defined in Equation (4.4). Substituting Equation (4.109) into Equation (4.107) and rearranging terms the following expression for the right-hand control pressure may be written:

$$P_{cr} = \frac{P_p}{2} + \frac{A}{2\mathrm{K}_{c_a}}\dot{x} - \mathrm{K}_{p_a}y_r, \tag{4.110}$$

where K_{p_a} is the pressure sensitivity of the control spool, which is generally defined in Equation (4.5). From this result one can see that the nominal control pressure is given by half of the pilot pressure while the motion of the main spool valve serves to increase the right-hand control pressure in a transient fashion. Furthermore, displacements of the control spool also serve to reduce the right-hand control pressure, which is, of course, the purpose of the first-stage valve. Substituting Equation (4.110) into Equation (4.106) yields the following expression for the right-hand actuator force that is exerted on the main spool valve shown in Figure 4-15:

$$F_{ar} = A\frac{P_p}{2} + \frac{A^2}{2\mathrm{K}_{c_a}}\dot{x} + (k_v + k_a)x - (A\mathrm{K}_{p_a} - k_a)y_r + (F_{v_o} + F_{a_o}). \tag{4.111}$$

In order to use the result of Equation (4.111) an expression must be written that relates the motion of the right-hand control spool y_r to the input current of the right-hand solenoid.

In Figure 4-18 it may be seen that control spool is also the armature for the solenoid. By neglecting the linear momentum and viscous drag effects acting on the armature, and by neglecting the inductance and the induced electromagnetic force that results from the motion of the armature, a steady equation for the armature displacement may be written as

$$y_r = \frac{\alpha}{k_a} i - \frac{F_{a_o}}{k_a} - x,\qquad(4.112)$$

where α is the solenoid's magnetic coupling coefficient. In this result the flow forces acting on the armature have also been neglected since the construction of the armature/control spool does a fairly good job of rejecting these flow forces.

To summarize the analysis of the solenoid actuated, two-stage electrohydraulic valve shown in Figure 4-15 the preceding equations are assembled to describe the dynamics of the valve as a first-order system. If we assume that a critically centered valve is employed for the second stage, Equations (4.98), (4.105), (4.111), and (4.112) may be substituted into Equation (4.99) to yield the following expression:

$$\frac{A^2}{K_{c_a}}\dot{x} + (k_v + 2K_{f_q} + AK_{p_a})x = \frac{AK_{p_a}\alpha}{k_a}(i - i_o),\qquad(4.113)$$

where the input current required to overcome the preload of the spring pack is given by

$$i_o = \frac{F_{a_o}}{\alpha}\left[1 + \frac{k_a}{AK_{p_a}}\left(\frac{F_{v_o}}{F_{a_o}} + 1\right)\right].\qquad(4.114)$$

In Equation (4.113) it has been assumed that $AK_{p_a}/k_a \gg 1$. These results are used in the following paragraphs to discuss the steady and transient characteristics of the two-stage electrohydraulic control valve shown in Figure 4-15.

Response Characteristics. To consider both the steady and transient response characteristics of the two-stage electrohydraulic valve shown in Figure 4-15 it will be convenient to rewrite the equation of motion presented in Equation (4.113) as:

$$\tau\dot{x} + x = \psi(i - i_o),\qquad(4.115)$$

where

$$\tau = \frac{A}{K_{q_a}\left(\frac{k_v + 2K_{f_q}}{AK_{p_a}} + 1\right)} \quad\text{and}\quad \psi = \frac{\alpha}{k_a\left(\frac{k_v + 2K_{f_q}}{AK_{p_a}} + 1\right)}.\qquad(4.116)$$

All of the parameters in this equation have been previously defined. In this result, the symbol τ represents the time constant for the transient response of the system while the symbol ψ represents the constant of proportionality for the steady-state response. *Note*: ψ is the slope of the line shown in Figure 4-16. For most valve designs it may be shown that

$$\frac{k_v + 2K_{f_q}}{AK_{p_a}} \ll 1,\qquad(4.117)$$

which then means that the time constant and the constant of proportionality may be reduced to the following approximations:

$$\tau \approx \frac{A}{K_{q_a}} \quad\text{and}\quad \psi \approx \frac{\alpha}{k_a}.\qquad(4.118)$$

This simpler result can be used for initial design purposes and for close estimations of the overall valve response.

Generally, a small time constant is desired for reducing the settling time and increasing the frequency response of the valve. From Equation (4.118) it is seen that a small cross-sectional area A and a large flow gain for the actuator K_{q_a} provides the smallest time constant. To increase the flow gain for the actuator the size of the flow passage into the actuator must increase and so there is a physical limitation as to how large this parameter can get.

From a steady point of view the constant of proportionality for the valve must be sized so as to give a maximum valve displacement x_{max} for a maximum available input current i_{max}. This design requirement may be written as,

$$\psi = \frac{x_{max}}{(i_{max} - i_o)}, \tag{4.119}$$

where i_o is given in Equation (4.114). From Equation (4.118) it may be seen that to increase the maximum range of actuation for the valve the magnetic coupling coefficient for the solenoid α must be increased and the spring rate for the feedback spring k_a must be reduced. By increasing α the overall size of the solenoid will grow and therefore, once again, there are practical limitations for how large this parameter may get. All of these limitations must be considered when designing the two-stage electrohydraulic valve that is shown in Figure 4-15.

Case Study

A solenoid actuated, two-stage electrohydraulic valve similar to Figure 4-15 has been designed with a 15 mm bore diameter for the main-stage spool valve, which is a critically centered four-way valve. The control spool/armature for the actuator utilizes square flow passages that are 2 mm × 2 mm and exhibit a discharge coefficient of 0.62. The density of the hydraulic fluid is 850 kg/m³. The pilot pressure that is used to actuate the valve is 2.1 MPa. The spring rate for the feedback spring is 10 N/mm and the magnetic coupling coefficient for the solenoid is 30 N/Amp. Calculate the time constant τ and the constant of proportionality ψ for the valve. Based on these calculations estimate the settling time and the frequency response for the valve. Calculate the displacement of the main-stage valve when 2 Amps of current are applied to one solenoid. The initial amount of current required to overcome the preload of the spring pack is 1 Amp.

The time constant and constant of proportionality are shown in Equation (4.118). In order to calculate the time constant the flow gain for the control spool must be determined. Using the square flow passage geometry the flow gain for the control spool may be computed from Equation (4.8) as

$$K_{q_a} = hC_d\sqrt{\frac{2}{\rho}\frac{P_p}{2}} = 2 \text{ mm} \times 0.62 \times \sqrt{\frac{2.1 \text{ MPa}}{850 \text{ kg/m}^3}} = 6.16 \times 10^{-2} \frac{\text{m}^2}{\text{s}}.$$

From the 15 mm bore diameter given in the problem statement the cross-sectional area may be computed as $A = 176.71$ mm². Using this number with the previous calculation shows that the time constant for the electrohydraulic valve is given by

$$\tau = \frac{A}{K_{q_a}} = \frac{176.71 \text{ mm}^2}{6.16 \times 10^{-2} \text{ m}^2/\text{s}} = 2.87 \text{ ms}.$$

The constant of proportionality for the valve may be calculated directly from the problem statement as

$$\psi = \frac{\alpha}{k_a} = \frac{30\,\text{N/Amp}}{10\,\text{N/mm}} = 3\,\frac{\text{mm}}{\text{Amp}}.$$

Based on these calculations the dynamic and steady-state characteristics of the valve may be determined. In particular, the settling time for the first-order system is defined as 4 times the time constant τ; therefore, the settling time is given by $t_s = 11.48$ ms. The frequency response or bandwidth for the first-order system is defined as one over the time constant. Therefore, the frequency response is $\omega_b = 348.4$ rad/s $= 55$ Hz. The maximum spool displacement may be calculated from Equation (4.119) where it is recognized from the problem statement that the initial current $i_o = 1$ Amp. This result is given by

$$x_{\text{max}} = \psi\,(i_{\text{max}} - i_o) = 3\,\text{mm/Amp} \times (2\,\text{Amp} - 1\,\text{Amp}) = 3\,\text{mm}.$$

This result is the maximum spool displacement that are obtained on a curve much like that of Figure 4-16.

Summary.   In this subsection the two-stage electrohydraulic valve of Figure 4-15 has been analyzed and both the dynamic and steady-state response characteristics have been discussed. The response characteristics of this valve depend on the first-order time constant and the constant of proportionality that exists between the valve displacement and the input current. As it turns out, these two quantities may be closely approximated by the fairly simple expressions that are given in Equation (4.118). In this equation only four parameters are shown; namely, the pressurized area of the main spool valve, the flow gain of the control spool, the magnetic coupling coefficient of the solenoid, and the spring rate of the feedback spring. While these expressions may be used to assess the response characteristics of the valve other design parameters must be carefully specified in order to achieve the overall valve function. These other design parameters are noted throughout the discussion of this subsection.

4.5.5 Summary

In this section of the chapter the performance characteristics of the four-way spool valve are considered. As shown in this section the four-way spool valve is unique as compared to the two-way and three-way valve since it rejects all transient flow forces and only sustains steady flow forces that seek to close the valve. This means that the four-way valve does not suffer from possible instabilities as the three-way critically centered spool valve does. As the primary valve that is used for flow control within hydraulic systems, the four-way spool valve is often packaged using the electrohydraulic arrangement shown in Figure 4-15. This arrangement requires an auxiliary hydraulic power source for supplying the pilot pressure and pilot flow that is used to actuate the valve. However, the flexibility associated with the electronic control of this valve far outweighs other drawbacks that may exist within this design including the power losses that are incurred with the auxiliary pilot supply. Applications of the four-way spool valve are presented and discussed in Chapter 7 of this book.

4.6 POPPET VALVES

4.6.1 Overview

In Section 4.3 we considered the design of a two-way spool valve that is occasionally used within hydraulic circuitry for relief valve applications. In this section of our chapter we consider the more common valve construction that is used for relief type functions; namely, the two-way poppet valve. As is shown in this section there are many dynamic and steady-state similarities between the two-way spool valve and the two-way poppet valve; however, the poppet valve exhibits certain advantages over the spool valve design. These advantages may be listed as:

- Poppet valves do not leak when they are closed.
- Poppet valves do not require precise machining and are capable of adjusting themselves for wear.
- Poppet valves are self-flushing and are therefore insensitive to contamination particles.

It is for these reasons that poppet valves are used. Since pressure relief functions are needed for almost every hydraulic circuit that may be conceived of, the poppet valve is used extensively for providing safety within high-powered systems that are typical of hydraulic circuits. Furthermore, it is not unusual for a multiple number of poppet valves to be used within a single hydraulic system for the purpose of providing relief functions within different branches of the circuit. Though these valves are used on an intermittent basis their primary drawback is found in the noise that is typically generated when they are functioning. This noise results from classical poppet valve instability, which is a concept that is addressed in the next subsections. The following paragraphs discuss the flow characteristics of the two-way poppet valve.

The two-way poppet valve is shown in Figure 4-19 with two flow lines. In this figure, the supply line is shown to be pressurized to a level given by P_s while the return line is shown to be pressurized to the level P_r. To open the poppet valve the poppet is moved away from the poppet seat in the positive x-direction, thus facilitating the volumetric flow rate shown in Figure 4-19 by the symbol Q_r.

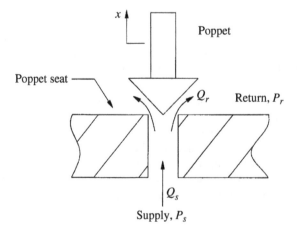

Figure 4-19. Two-way poppet valve.

From Equation (4.3), it can be seen that the volumetric flow rate through the poppet valve is simply given by the expression

$$Q_r = \tfrac{1}{2}Q_{r_o} + K_q x + K_c(P_s - P_r),$$ (4.120)

where Q_{r_o} is the nominal flow rate through the poppet valve at steady operating conditions, and K_q and K_c are the flow gain and pressure-flow coefficient defined generally in Equation (4.4). The reader will recall that the nominal flow rate is determined when $x = 0$ and when the supply and return pressures are at nominal values at which the valve coefficients have been determined. From Equation (4.120) it can be seen that the flow rate through the valve will increase as the valve displacement x increases and as the pressure drop $(P_s - P_r)$ increases. *Note:* For nominal evaluations of the poppet valve, the valve is assumed to be working with an opening of the valve determined by the steady-state flow passing through the valve. This steady-state flow rate is shown in Equation (4.120) by the symbol Q_{r_o} and the valve displacement x is measured relative to this nominal opening.

The valve coefficients in Equation (4.120) depend on the discharge area properties of the poppet valve. Figure 4-20 shows a schematic of the poppet valve discharge area as it relates to the relevant geometry features of the valve.

In this schematic A is the discharge area of the valve described by the dashed line, ξ is the open distance of the poppet in the vertical direction, θ is the poppet valve angle, and D is the diameter of the poppet valve inlet area. The discharge area A is the surface area of a cone. Using methods of calculus it may be shown that this area is given by

$$A = \pi[D \sin(\theta)\,\xi - \cos(\theta)\sin^2(\theta)\,\xi^2].$$ (4.121)

If it is assumed that ξ is small then a linear approximation for this area is

$$A = \pi D \sin(\theta)\,\xi.$$ (4.122)

By setting $\xi = u + x$, where u is the nominal open position of the valve and x is the valve displacement from this nominal position, Equation (4.122) may be used with

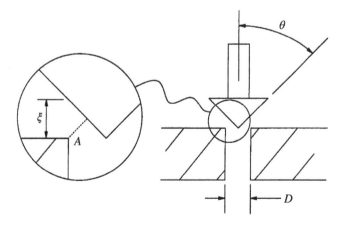

Figure 4-20. Poppet valve area geometry.

Equations (4.4) to show that the standard valve coefficients of Equation (4.120) are given by

$$K_q = \pi D \sin(\theta) C_d \sqrt{\frac{2}{\rho} P_{s_o}}, \quad K_c = \frac{\pi D \sin(\theta) u C_d}{\sqrt{2 \rho P_{s_o}}}, \tag{4.123}$$

where P_{s_o} is the nominal supply pressure to the valve and where the nominal return pressure P_r is assumed to be zero. *Note*: The poppet angle θ exhibits a decisive impact on the standard valve coefficients.

4.6.2 Efficiency

Since the poppet valve shown in Figure 4-19 is a two-way valve its efficiency characteristics are identical to that of the two-way spool valve design presented in Subsection 4.3.2. In other words, according to Equation (4.19) the efficiency of the poppet valve is given by

$$\eta = \frac{P_r}{P_s}, \tag{4.124}$$

where P_r is the return pressure and P_s is the supply pressure. It should be noted that for a poppet valve operating in a pressure relief type function the output power from the relief valve is completely lost as the fluid power is expelled into the system reservoir and never recovered. The pressure in the system reservoir is the reference pressure from which all other pressures in the hydraulic system are measured; therefore, the return pressure P_r equals zero gauge pressure and according to Equation (4.124) the efficiency of the poppet valve is zero. This means that all of the input power to the poppet relief valve turns into heat that must be dissipated by a cooling mechanism for the hydraulic system. From the specific heat discussion in Chapter 2 it can be shown that for a 40 MPa pressure drop across the relief valve the fluid crossing the relief valve will increase its temperature by 20°C. If large amounts of flow are crossing the relief valve this will necessitate a substantial cooling mechanism for the hydraulic circuit, which costs money not only from an operational point of view but also from a first cost analysis of the entire system. Obviously, pressure relief valves are a necessary evil that accompany hydraulic circuits for the purposes of providing safety at a cost.

4.6.3 Flow Forces

The fluid forces that are exerted on the two-way poppet valve may be determined by applying the Reynolds Transport Theorem to the dashed line control volume shown in Figure 4-21.

This development is very similar to what has been presented in Equations (4.20) through (4.34) for the two-way spool valve; therefore, this analysis is not repeated here. Rather, the flow force result from this analysis is simply presented as

$$F_x = -\rho L K_q \dot{x} - \rho L K_c \dot{P}_s - K_{f_q} x + (A - K_{f_c}) P_s, \tag{4.125}$$

where ρ is the fluid density, L is the damping length shown in Figure 4-21, K_q and K_c are the valve coefficients given in Equation (4.123), A is the cross-sectional area of the inlet shown in Figure 4-21, and K_{f_q} and K_{f_c} are the flow force gain and pressure flow force coefficient generally defined in Equation (4.32). *Note*: In this result the fluid momentum entering the

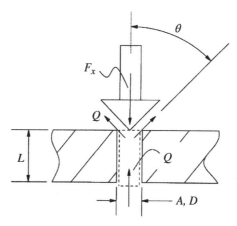

Figure 4-21. Poppet flow forces and control volume.

control volume of Figure 4-21 has been neglected due to its small size in comparison with the momentum exiting the control volume. Also, in Figure 4-21 the flow force F_x is shown as a reaction force exerted on the control volume while Equation (4.125) describes this force as it is applied to the poppet (equal and opposite). Using Equation (4.122) with Equation (4.32) the flow force gain and pressure flow force coefficient for the two-way poppet valve may be expressed as

$$\mathrm{K}_{f_q} = 2\pi D \sin(\theta) P_o C_d^2 \cos(\theta), \quad \mathrm{K}_{f_c} = 2\pi D \sin(\theta) u C_d^2 \cos(\theta), \tag{4.126}$$

where D is the diameter of the control volume shown in Figure 4-21. In this result the jet angle of the flow exiting the poppet valve is considered to be the same as the machined angle of the poppet valve itself (compare Figures 4-20 and 4-21). This is a first approximation of the jet angle that may be used to yield reasonable results for the steady-state flow forces acting on the poppet valve [1].

From Equation (4.125) the reader will observe that the poppet valve is susceptible to both transient and steady flow force effects. It should be noted that the first transient term in this equation is an effective damping term that helps to stabilize the valve. The second transient term is a disturbance effect that results from sudden changes in pressure. As shown in Equation (4.125) the steady displacement term acts to close the valve while the steady pressure term serves to open the valve. *Note*: The steady pressure term is the forcing input that is used to open the valve when the supply pressure becomes too large, thereby achieving the safety function for which the poppet valve is primarily intended to serve.

4.6.4 Pressure Relief Valves

Overview. As already noted, the primary function of the two-way poppet valve within hydraulic control systems is to perform the task of relieving pressure and for providing safety for the rest of the system. Figure 4-22 shows a schematic of a two-way poppet valve that is used for this function. In this figure the primary hydraulic control system is represented by the compressible fluid volume at the bottom. This compressible fluid volume is generally the volume of fluid within the high-pressure line of the hydraulic circuit. The nominal flow rate into the compressible fluid volume that needs to be relieved is shown in Figure 4-22 by the symbol Q_o. The volumetric flow rate exiting the compressible fluid volume through

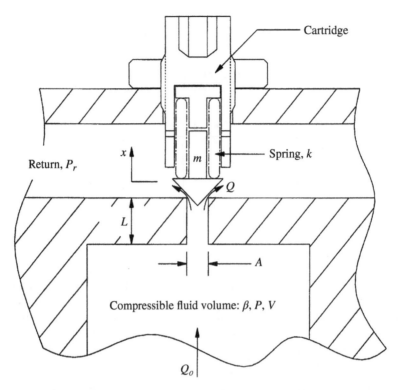

Figure 4-22. Schematic of the two-way poppet type, pressure relief valve.

the poppet valve is shown by the symbol Q. The fluid bulk modulus for the fluid is given by β, the instantaneous pressuring in the fluid volume is given by P, and the volume of the compressible fluid is given by V. The exhaust chamber of the relief valve is the return port, which is pressurized to a level given by P_r. Typically, the return pressure is the pressure within the hydraulic control system reservoir and, therefore, P_r is taken to be zero gauge pressure.

The poppet valve is shown in Figure 4-22 to have a mass given by the symbol m. A spring acts on the poppet valve to force it in the negative x-direction. A region of incompressible fluid is shown by the entry region of the poppet valve and designated by the damping length L. When the pressure within the compressible fluid volume increases the poppet valve moves upward to facilitate the relief flow Q. This relief flow then reduces or controls the fluid pressure within the compressible fluid volume, thus providing a safety function for other components within the hydraulic control system. The adjustable cartridge shown in Figure 4-22 is used to change the preload on the poppet spring.

Analysis. The analysis of the poppet type relief valve follows a similar development as that of the spool type relief valve discussed in Subsection 4.3.4. In that subsection the dynamic equation for the fluid pressure within the compressible fluid volume was given by the expression:

$$\frac{V}{\beta}\dot{P} + \mathrm{K}_c(P - P_o) = -\mathrm{K}_q x, \qquad (4.127)$$

where V is the fluid volume, β is the fluid bulk modulus, K_c and K_q are the poppet valve flow coefficients and P_o is the nominal pressure of the fluid volume. This equation is the same equation that may be used for modeling the pressure transients in the poppet type relief valve shown in Figure 4-22. When using this equation, however, it is important that the poppet valve coefficients of Equation (4.123) are used rather than spool valve coefficients. Clearly, from Equation (4.127) it may be seen that the input for adjusting the fluid pressure within the compressible fluid volume is the poppet valve displacement x. The poppet valve displacement is modeled in the next paragraph.

By summing forces on the poppet in Figure 4-22 and setting them equal to the poppet's time-rate-of-change of linear momentum the following equation may be written to govern the vertical motion of the poppet valve in the x-direction:

$$m\ddot{x} + kx = F_x - F_o, \tag{4.128}$$

where m is the mass of the poppet valve, k is the poppet valve spring rate, F_o is the nominal force or preload acting on the poppet by the spring when $x = 0$, and F_x is the fluid force acting on the poppet in the vertical direction. The fluid force for the poppet valve is given in Equation (4.125). *Note*: In Equation (4.128) viscous damping effects have been excluded from the model since they are extremely negligible for the design shown in Figure 4-22. Substituting Equation (4.125) into Equation (4.128) produces the following expression for the dynamics of the poppet valve:

$$m\ddot{x} + \rho L K_q \dot{x} + (k + K_{f_q})x = (A - K_{f_c})(P - P_o) - \rho L K_c \dot{P}, \tag{4.129}$$

where K_{f_q} and K_{f_c} are the flow force gain and the pressure flow force coefficient defined in Equation (4.126) for the poppet valve. In this result it has been recognized from the steady-state condition of the valve that the nominal spring force exerted on the valve is given by

$$F_o = (A - K_{f_c})P_o. \tag{4.130}$$

In summary, Equations (4.127) and (4.129) are presented as the governing equations that describe the dynamics of the two-way poppet type, pressure relief valve shown in Figure 4-22. The reader will notice that these equations are essentially the same as the equations for the two-way spool type relief valve presented in Subsection 4.3.4 with the exception of the viscous damping term that has been neglected for the poppet valve design in Figure 4-22. The primary differences between the two-way poppet valve design and the two-way spool valve design are buried within the flow and flow force coefficients of the analysis. For the poppet valve design it is necessary to use the Equations (4.123) and (4.126) for calculating these parameters.

Design. Our primary concern for designing the poppet type relief valve shown in Figure 4-22 is to determine what the steady-state or nominal opening of the valve u should be. Once this quantity is determined it may be used to calculate the flow and flow force coefficients of the poppet valve. To determine the nominal opening of the valve we recognize that the steady flow rate through the poppet valve must be satisfied by the classical orifice equation which is given by

$$Q_o = A_o C_d \sqrt{\frac{2}{\rho} P_o}. \tag{4.131}$$

The poppet valve flow area is generally described in Equation (4.122). For the nominal valve opening it can be shown that this area is given by $A_o = \pi D \sin(\theta)u$, where u is the nominal opening of the valve and other parameters are shown in Figure 4-20. Using this result with Equation (4.131) it may be shown that the nominal opening of the valve for a nominal flow rate Q_o and a nominal pressure drop P_o is given by

$$u = \frac{Q_o}{\pi D \sin(\theta) C_d \sqrt{\frac{2}{\rho} P_o}}. \tag{4.132}$$

Using this result the valve coefficients of Equations (4.123) and (4.126) may be determined.

Another design parameter that must be specified is the spring preload F_o that is required at the nominal operating condition of the valve. This specification is given in Equation (4.130) as a function of the valve's cross-sectional area, the pressure flow force coefficient, and the nominal operating pressure of the valve. This force is also equal to the spring rate times the total spring deflection at the nominal operating condition of the valve. Therefore, the spring rate for the valve must be designed such that

$$k = \frac{(A - K_{f_c})P_o}{\delta}, \tag{4.133}$$

where δ is the total deflection of the spring at the nominal operating conditions. *Note*: The steady-state flow force of the valve $K_{f_c} P_o$ serves to reduce the spring rate requirement since this force seeks to assist the spring by trying to close the valve.

Stability.   As previously mentioned, the dynamic equations for the spool type relief valve described in Subsection 4.3.4 are identical to the dynamic equation for the poppet type relief valve described in this subsection with the exception of the viscous damping effects. The reader will recall that the viscous damping effects were neglected in Subsection 4.3.4 for the purposes of generating a stability criterion for the valve. Therefore, this stability criterion is perfectly valid for the poppet type relief valve shown in Figure 4-22. This stability criterion is presented in Equation (4.51) and is rewritten here for convenience:

$$\frac{L\rho}{m} \frac{V}{\beta} \frac{(k + K_{f_q})}{(A - K_{f_c})} > 1. \tag{4.134}$$

From this criterion it can be seen that the energy storage capacity in the elastic members of the system; that is, the compressible fluid and the spring, must be sufficiently large in order to guarantee stability. Destabilizing tendencies are realized for the system when the valve inertia or the input force to the valve becomes large.

Case Study

A poppet type relief valve similar to what is shown in Figure 4-22 is used to limit the maximum pressure within a hydraulic circuit to 20 MPa while expelling 18 lpm of flow into the reservoir at zero gauge pressure. The following parameters shown in Figure 4-21 apply to this problem: $D = 4$ mm, $L = 25$ mm, and $\theta = 45°$. From this geometry it can be seen that the cross-sectional area of the of the poppet flow passage is given by $A = 12.57$ mm^2. The discharge coefficient of the valve is 0.62 and the fluid density is given by 850 kg/m^3. The mass of the valve is given by 0.015 kg, the fluid bulk modulus is 0.8 GPa, and the volume of compressible fluid is 0.001 m^3. Assuming that the total spring deflection in the nominal operating condition is 10 mm, calculate the valve spring rate that is needed to operate the relief function at 20 MPa. Is this valve stable or not?

To answer the questions posed in this problem the flow force valve coefficients of Equation (4.126) need to be evaluated first. Using Equation (4.132) it can be shown that the nominal valve opening for this design is given by $u = 0.25$ mm. Therefore, the flow force valve coefficients are given by

$$K_{f_q} = 2\pi D \sin(\theta) P_o C_d^2 \cos(\theta) = \pi \times 4 \text{ mm} \times 20 \text{ MPa} \times 0.62^2 = 96.61 \frac{\text{N}}{\text{mm}}$$

$$K_{f_c} = 2\pi D \sin(\theta) u C_d^2 \cos(\theta) = \pi \times 4 \text{ mm} \times 0.25 \text{ mm} \times 0.62^2 = 1.208 \text{ mm}^2.$$

Using Equation (4.133) and the pressure flow force coefficient just calculated the required spring rate for this design is given by

$$k = \frac{(A - K_{f_c})P_o}{\delta} = \frac{(12.57 \text{ mm}^2 - 1.208 \text{ mm}^2)20 \text{ MPa}}{10 \text{ mm}} = 22.72 \quad \frac{\text{N}}{\text{mm}}.$$

To examine the valve's stability the stability criterion in Equation (4.134) is be evaluated. The left-hand side of this equation may be calculated as

$$\frac{L\rho}{m} \frac{V}{\beta} \frac{(k + K_{f_q})}{(A - K_{f_c})} = \frac{25 \text{ mm} \times 850 \text{ kg/m}^3}{0.015 \text{ kg}} \frac{0.001 \text{ m}^3}{0.8 \text{ GPa}}$$

$$\frac{(22.72 \text{ N/mm} + 96.61 \text{ N/mm})}{(12.57 \text{ mm}^2 - 1.208 \text{ mm}^2)} = 18.6.$$

Since this result is greater than unity the poppet valve will be stable.

Summary. In this subsection the steady and dynamic characteristics of the poppet type relief valve shown in Figure 4-22 have been considered. Perhaps the most important part of this analysis has been the stability criterion that is shared with the two-way spool valve discussed in Subsection 4.3.4. This result is presented in Equation (4.134) and has been used by way of example in the previous case study. For this case study the poppet valve was shown to be stable; however, it should be noted that the poppet valve stability criterion is not always satisfied in practice and as a result many poppet valves create a great deal of noise when they operate. This unstable operation is characterized by a limit cycle in which the poppet slams back and forth between hard stops in the poppet valve design. Though many poppet valves operate in an unstable fashion their frequency of operation is on the order of 1000 Hz and other components within the hydraulic system are unable to respond to this instability. Therefore, the relief function of the unstable poppet valve is carried out on an average basis and since the unstable valve is normally closed during the operation of the hydraulic circuit it is deemed somewhat acceptable for various applications. Even so, it must be mentioned that unstable valves wear out much faster than stable valves do and therefore it is highly desirable from both a noise and machine design point of view to satisfy the stability criterion in Equation (4.134).

4.6.5 Summary

In this section the two-way poppet valve construction has been considered. As previously mentioned, this valve design is widely used as a pressure relief valve within hydraulic circuitry to provide safety functions for the entire system and in many hydraulic circuits a multiple number of these valves may be used. As a pressure relief valve the poppet valve is used only on an intermittent basis in an on–off mode and does not controllably meter flow as the four-way spool valve does that was previously discussed in Section 4.5. Even so, the advantages of poppet valves over spool valves that have been mentioned in Subsection 4.6.1 are real and they provide much motivation for developing poppet valve constructions that may be used to controllably meter fluid flow even as the four-way spool valve does. The challenge of this task lies in the mode of actuation for the poppet valve and in guaranteeing stability throughout the entire range of operation. Since these obstacles are not easily overcome the flow control poppet valve remains an item of research, which may only appear in practice years from now. But, until then, it seems that poppet valves will primarily function as pressure relief devices.

4.7 FLAPPER NOZZLE VALVES

4.7.1 Overview

The two-way flapper nozzle valve is rarely used as a standalone valve. Typically the flapper nozzle valve is used as the first stage of a two-stage electrohydraulic valve in which a controlled pressure force is supplied to a main flow control valve for the purposes of serving the larger hydraulic system. In this section of the chapter the two-stage electrohydraulic valve is considered as an application for the flapper nozzle valve; however, before we are able to discuss this application a few fundamental concepts for the flapper nozzle valve must be presented and discussed.

A two-way flapper nozzle valve is shown in Figure 4-23 with two flow lines. In this figure the control line is shown to be pressurized to a level given by P_c while the return line is shown to be pressurized to the level P_r. This valve facilitates the return flow Q_r by altering the flapper angle φ as shown in Figure 4-23. The flapper is a mechanical device that pivots about a pivot point located a distance r away from the centerline of the nozzle. By rotating the flapper in the positive direction the valve opens and the volumetric flow rate through the valve increases. By rotating the flapper in the negative direction the valve closes and

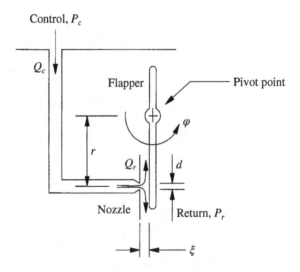

Figure 4-23. A two-way flapper nozzle valve.

the volumetric flow rate through the valve decreases. In Figure 4-23, the nozzle diameter is shown by the symbol d.

Using the standard form of Equation (4.3) it can be shown that the linearized volumetric flow rate through the flapper nozzle valve is simply given by the expression

$$Q_r = \tfrac{1}{2}Q_{r_o} + K_q r\varphi + K_c(P_c - P_r), \tag{4.135}$$

where Q_{r_o} is the nominal flow rate through the flapper nozzle valve at steady operating conditions and K_q and K_c are the flow gain and pressure-flow coefficient defined generally in Equation (4.4). In this result the opening of the flapper nozzle valve shown in Figure 4-23 is given by $\xi = u + x$ where x is taken to be $r\,\varphi$ for small values of φ and where u is the nominal opening of the valve. The reader will recall that the nominal flow rate Q_{r_o} is determined when $\varphi = 0$ and when the control and return pressures are at nominal values at which the valve coefficients have been determined. From Equation (4.135) it can be seen that the flow rate through the valve will increase as the valve displacement φ increases and as the pressure drop $(P_c - P_r)$ increases.

The flow coefficients in Equation (4.135) depend on the discharge area of the nozzle. For the flapper nozzle valve it is not the nozzle area that is the restricted flow area of the valve; rather, it is the curtain area that restricts the flow. From Figure 4-23 it can be shown that the curtain area is given by

$$A = \pi d\xi, \tag{4.136}$$

where ξ is the curtain height and d is the nozzle diameter. Again, if we let $\xi = u + x$ Equations (4.4) and (4.136) may be used to show that the flow coefficients in Equation (4.135) are given by

$$K_q = \pi d C_d \sqrt{\frac{2}{\rho}P_{c_o}}, \quad K_c = \frac{\pi d u C_d}{\sqrt{2\rho P_{c_o}}}, \tag{4.137}$$

where P_{c_o} is the nominal control pressure to the valve and where the nominal return pressure to the valve P_r is assumed to be zero. In practice the nominal opening of the valve u is on the order of tenths of millimeters [2], which illustrates more clearly that the curtain area is the restricting flow area and that the flapper rotates very small amounts during operation.

4.7.2 Efficiency

As previously mentioned, the typical application of the flapper nozzle valve is to perform as the first stage of a two-stage electrohydraulic valve. Since the power to operate the first stage is negligible when compared to the transmitted power of the second stage, issues of efficiency are rarely discussed for the flapper nozzle valve. In any event, the flapper nozzle valve shown in Figure 4-23 is a two-way valve and its overall efficiency is simply given by the ratio of the return pressure to the control pressure

$$\eta = \frac{P_r}{P_c}. \tag{4.138}$$

Since the return pressure P_r is typically zero the efficiency of the flapper nozzle valve is also zero, which illustrates the point that this valve is not used to transmit power; rather, it is used to generate a control signal for another valve. Therefore, the energy used to operate this valve is a parasitic drain on the overall hydraulic system.

4.7.3 Flow Forces

Figure 4-24 shows a schematic of a flapper nozzle valve with a single nozzle. The right-hand side of this figure shows an enlarged view of the nozzle with a dashed control volume that is used for studying the flow forces that act on the flapper. The volumetric flow rate through the control volume is given by the symbol Q. The nozzle length is shown in Figure 4-24 by the dimension L while the nozzle diameter is given by d. In Figure 4-24 the dimension x describes the linear displacement of the flapper while u is a nominal open dimension of the valve. As previously mentioned, by rotating the flapper about its pivot point the open area is effectively altered for the purposes of changing the flow rate through the nozzle.

Using the Reynolds Transport Theorem the conservation of fluid momentum for the dashed line control volume shown in Figure 4-24 is given by

$$\mathbf{F} = \frac{\partial}{\partial t} \int_{c.v} \rho \mathbf{u} dV + \int_{c.s} \rho \mathbf{u}(\mathbf{u} \cdot \hat{\mathbf{n}}) \, dA, \tag{4.139}$$

where $\mathbf{F}$ is a vector force acting on the control volume, ρ is the fluid density, $\mathbf{u}$ is the fluid velocity vector, and $\hat{\mathbf{n}}$ is a unit vector that points normally outward from the control volume surface. The forces acting on the control volume are given by

$$\mathbf{F} = (A_n P_c - F_x)\hat{\mathbf{i}}, \tag{4.140}$$

where F_x is the horizontal flow force acting on the flapper, A_n is the cross-sectional area of the nozzle given by $\pi d^2/4$, and where the vertical flow force component has vanished due to the symmetry of the flow field about the centerline of the nozzle. The volume integral in Equation (4.139) describes the fluid momentum effects within the control volume itself. In this case the internal fluid-momentum effects may be written as

$$\frac{\partial}{\partial t} \int_{c.v} \rho \mathbf{u} dV = \rho L \frac{\partial Q}{\partial t} \hat{\mathbf{i}}. \tag{4.141}$$

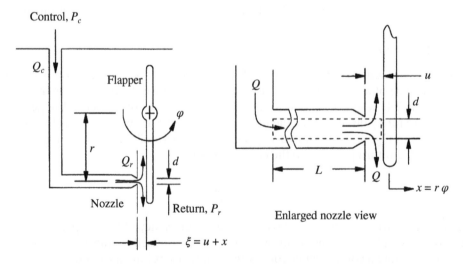

Figure 4-24. Flow force control volume for a two-way flapper nozzle valve.

The area integral in Equation (4.139) describes the momentum effects that enter and exit the control volume. An expression for these momentum effects is given by

$$\int_{c.s} \rho\mathbf{u}(\mathbf{u} \cdot \hat{\mathbf{n}})dA = -\rho\frac{Q^2}{A_n}\hat{\mathbf{i}}, \tag{4.142}$$

where the fluid momentum leaving the control volume in the horizontal direction is zero due to the obstruction of the flapper and where the vertical components cancel each other out due to symmetry. In Equation (4.142), $A_n = \pi\, d^2/4$. Substituting Equations (4.140) through (4.142) into Equation (4.139) yields the following scalar equation for the fluid momentum force acting on the flapper in the horizontal direction:

$$F_x = -\rho L\frac{\partial Q}{\partial t} + \rho\frac{Q^2}{A_n} + A_n P_c. \tag{4.143}$$

As in the spool valve case it is convenient to express the flow force acting on the flapper in the following form:

$$F_x = -F_t + F_s, \tag{4.144}$$

where F_t is the transient flow force and F_s is the steady flow force. These two quantities are given by

$$F_t = \rho L\frac{\partial Q}{\partial t} \quad \text{and} \quad F_s = \rho\frac{Q^2}{A_n} + A_n P_c. \tag{4.145}$$

From these equations it should be apparent that the steady flow force F_s always acts in such a way as to open the valve while the transient flow force F_t may act to either open or close the valve depending on the sign of $\partial Q/\partial t$. The transient flow force may be written more explicitly by taking the time derivative of Equation (4.135) and substituting the result into Equation (4.145). This result is given by

$$F_t = \rho L K_q r\dot{\varphi} + \rho L K_c \dot{P}_c, \tag{4.146}$$

where the dot notation indicates a derivative with respect to time and where the flow gain K_q and pressure-flow coefficient K_c are given in Equation (4.137). In this result it has been assumed that the return pressure P_r is a constant. The steady flow force may be written more explicitly by substituting the classical orifice equation of Equation (4.1) into Equation (4.145) to yield

$$F_s = \frac{2A^2 P_c C_d^2}{A_n} + A_n P_c, \tag{4.147}$$

where A is the curtain area of the flapper nozzle valve given in Equation (4.136) and P_c is the control pressure shown in Figure 4-24. *Note*: In this result the return pressure is assumed to be zero. This equation is nonlinear due to the simultaneous variation in the discharge area A and the control pressure P_c. By using the Taylor series method, Equation (4.147) may be linearized as

$$F_s = K_{f_q} r\varphi + (A_n + K_{f_c})\, P_c, \tag{4.148}$$

where K_{f_q} and K_{f_c} are the flow force gain and the pressure flow force coefficient given by

$$K_{f_q} = 16\pi u P_{c_o} C_d^2 \quad \text{and} \quad K_{f_c} = 8\pi u^2 C_d^2, \tag{4.149}$$

and where P_{c_o} is the nominal control pressure for the valve. Substituting Equations (4.146) and (4.148) into Equation (4.144) produces the following expression for the flow forces acting on the flapper along the centerline of the nozzle shown in Figure 4-24:

$$F_x = -\rho L K_q r\dot{\varphi} - \rho L K_c \dot{P}_c + K_{f_q} r\varphi + (A_n + K_{f_c})P_c. \tag{4.150}$$

From this result it may be seen that the transient flow force associated with angular velocity of the flapper $\dot{\varphi}$ is stabilizing while the steady flow force associated with the angular position of the flapper φ is destabilizing.

Equation (4.150) describes the flow force acting on the flapper of a single nozzle, flapper nozzle valve shown in Figure 4-23. In practice, this valve configuration is rarely encountered as flapper nozzle valves almost always utilize two matched nozzles – one on each side of the flapper. Figure 4-25 shows a typical configuration for this valve.

Using similar analysis that was used to generate Equation (4.150) it may be shown that the total fluid force acting on the flapper of the double nozzle valve shown in Figure 4-25 is given by

$$F_x = -2\rho L K_q r\dot{\varphi} - \rho L K_c(\dot{P}_{cl} - \dot{P}_{cr}) + 2K_{f_q} r\varphi$$
$$+ (A_n + K_{f_c})(P_{cl} - P_{cr}), \tag{4.151}$$

where P_{cl} is the left control pressure and P_{cr} is the right control pressure. In this result the flow coefficients and the flow force coefficients are given in Equations (4.137) and (4.149), respectively. Furthermore, the nominal operating pressure for the left and right control pressures is assumed to be identical. This pressure assumption allows the valve coefficients to be matched on both sides of the valve. It may still be shown from this result that the overall transient flow force associated with angular velocity of the flapper $\dot{\varphi}$ is stabilizing while the overall steady flow force associated with the angular position of the flapper φ is destabilizing. This destabilizing effect of the angular position term requires the use of a spring mechanism to regain valve stability for the flapper nozzle valve. The flow force result of

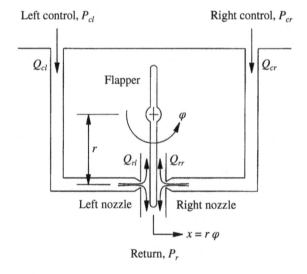

Figure 4-25. Double flapper nozzle valve.

Equation (4.151) is used to model a two-stage electrohydraulic valve in the following sub-section of this chapter.

4.7.4 Two-Stage Electrohydraulic Valves

Overview.   In Subsection 4.5.4 a solenoid actuated, two-stage electrohydraulic valve was presented and discussed. In this subsection, it was noted that electrohydraulic valves are used to achieve the most flexible control structure for a hydraulic system by interfacing the system with an electronic input that allows for closed loop feedback control. In this present subsection another type of electrohydraulic valve is presented; namely, a torque motor actuated valve that utilizes a flapper nozzle valve for the first stage. The second stage or main stage for this valve is the standard four-way spool valve that is presented in Section 4.5. Again, this two-stage actuation approach is necessary because electric actuators of rea-sonable size are not capable of exerting enough force on the main valve to overcome the opposing spring and flow forces.

Figure 4-26 shows a schematic of the torque motor–actuated, two-stage electrohydraulic valve. This valve utilizes a four-way spool valve as the second or main stage of the valve for supplying flow through ports A and B to the hydraulic system. Upon careful examina-tion, the reader will notice that this second-stage valve is identical to the four-way spool valve shown in Figure 4-13 except for the fact that the supply and return ports have been switched. The reader will recall that switching these ports has little impact on the dynamic characteristics of the four-way spool valve since only the pressure flow forces change sign for the open centered design. The supply pressure to the valve is noted in Figure 4-26 by the symbol P_s while the return pressure is given by P_r. The flow rates through ports A and B

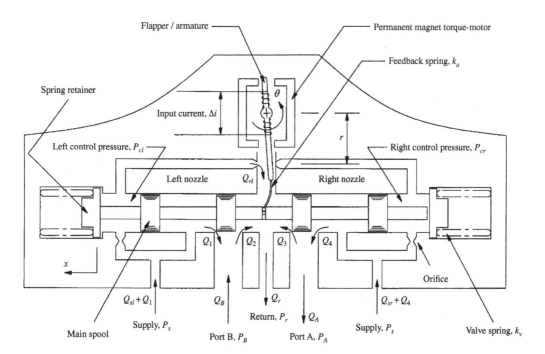

Figure 4-26. Schematic of a torque motor–actuated, two-stage electrohydraulic valve.

are shown by the symbols Q_A and Q_B, respectively. In order for the four-way valve to direct flow into port A it must be moved in the positive x-direction by the first actuation stage of the valve. In Figure 4-26 the main spool is shown in an actuated position where the valve displacement is given by $x \neq 0$. This is done to help illustrate the working principle of the second actuation stage, which is driven by a torque motor–actuated, flapper nozzle valve.

As shown in Figure 4-26, the main spool of the two-stage control valve is controlled by properly adjusting the right and left control pressure using the double flapper nozzle valve. The nozzle valve is known as the first stage of the two-stage control valve, as it serves the purpose of actuating the second stage. The left- and right-hand sides of the valve shown in Figure 4-26 are symmetrically identical as they each utilize a captured centering spring k_v and a control pressure for actuation. As shown in Figure 4-26 the first and second stage of the electrohydraulic valve utilize the same supply pressure P_s, which is typically a low pressure of about 2 MPa. This low supply pressure for the main spool valve indicates that the entire valve package is not a power transmitting valve; but rather, that this valve is used for controlling another power transmitting hydraulic device. For instance, the two-stage electrohydraulic valve shown in Figure 4-26 is typically used to control the displacement of an axial-piston pump, which then serves to supply the main flow of the larger hydraulic system. The supply pressure for the electrohydraulic valve shown in Figure 4-26 is usually provided by an auxiliary hydraulic power source, which expels very low amounts of energy when compared to the power delivery of the overall hydraulic system.

When the main spool valve is in the neutral position $x = 0$ the left- and right-hand sides of the valve are just touching the spring retainers. In this neutral position the left and right control pressures P_{cl} and P_{cr} are both pressurized to the level of half of the supply pressure $P_s/2$ and the flow forces acting on the main spool valve are zero. In other words, the valve is perfectly centered in its nominal position with no forces acting to alter this condition. In order to move the main spool valve to the left (the positive x-direction) the flapper for the flapper nozzle valve is rotated in a positive direction in order to block the right nozzle and to open the left nozzle. By blocking the right nozzle the right control pressure P_{cr} is increased to a level greater than $P_s/2$. By simultaneously opening the left nozzle the left control pressure P_{cl} is decreased to a level less than $P_s/2$. Now that a pressure difference exists between the left- and right-hand sides of the main spool valve a force exists that tries to move the main valve to the left. In order to accomplish this motion the pressure difference $P_{cr} - P_{cl}$ must be significant enough to overcome the preload on the spring pack that is retained by the left-hand spring retainer. Once this preload is overcome, the valve moves to the left and directs fluid flow into port A and out of port B. This is, of course, the overall objective of the valve assembly. *Note*: During this motion of the main spool valve the right-hand spring does not act on the main spool due to the spring retaining mechanism. When the second-stage valve moves to the left the feedback spring k_a shown in Figure 4-26 exerts a restoring force on the flapper that serves to stabilize the overall valve design. This stabilizing effect is shown in the analysis that follows.

As shown in Figure 4-26, the flapper of the flapper nozzle valve is also the armature of a permanent magnet torque motor that exerts a torque on the flapper for the purposes of actuating the valve. The torque on the flapper/armature is proportional to the input current to the torque motor. This proportional relationship provides a linear relationship between the displacement of the main spool valve x and the input current to the torque motor Δi similar to what is shown in Figure 4-16 of this chapter. The linearity that exists between the valve displacement and the input current is used to describe this valve as a proportional control valve; that is, the output displacement x is proportional to the input current Δi.

In the discussion that follows the torque motor–actuated, two-stage electrohydraulic control valve of Figure 4-26 is analyzed in greater detail. This analysis is useful for establishing design guidelines for the valve and for predicting the valve's dynamic response.

Analysis.   The analysis of the electrohydraulic valve shown in Figure 4-26 begins by considering the equation of motion for the main spool valve. Similar to the analysis of the two-stage electrohydraulic valve of Subsection 4.5.4 the linear momentum and viscous damping effects on the four-way valve are considered to be negligible when compared to the pressure and spring forces exerted on the valve by the left and right actuator. Therefore, the governing equation of motion for the main spool valve is given by

$$0 = F_x + F_{ar} - F_{al}, \tag{4.152}$$

where F_x is the flow force exerted on the valve and F_{al} and F_{ar} are the left and right actuation forces respectively. *Note*: The flow force exerted on the valve is given either by Equation (4.97) or (4.98) depending on whether the spool valve is open centered or critically centered. In Equation (4.152) the feedback spring force between the flapper and the main spool is considered to be negligible when compared to the other terms in this equation. Therefore, the feedback spring effects have been excluded at this stage. The actuation forces F_{al} and F_{ar} will be discussed in the following paragraphs.

Figure 4-27 shows a schematic of the right-hand actuator when the spool valve has been displaced a small distance in the positive x-direction.

From this figure, the spool valve does not touch the spring retainer; therefore, no spring forces are exerted on the right-hand side of the spool valve. In this case only a pressure force exists on the right-hand side of the spool valve and this force may be expressed as

$$F_{ar} = A P_{cr}, \tag{4.153}$$

where A is the pressurized area of the spool valve and P_{cr} is the right-hand control pressure. Since the volume of fluid on the right-hand side of the spool is extremely small, the pressure transients associated with the right-hand control pressure are extremely fast and may be

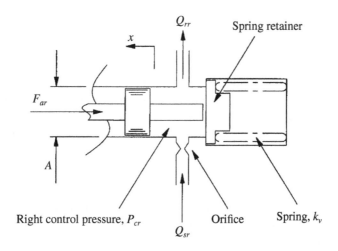

Figure 4-27. Right actuator of Figure 4-26.

neglected for the purposes of this analysis. Under these incompressible flow conditions it may be shown that

$$0 = Q_{sr} - Q_{rr} - A\dot{x}, \tag{4.154}$$

where Q_{sr} is the volumetric flow rate of fluid into the right actuator from the supply line and where Q_{rr} is the volumetric flow rate of fluid out of the right actuator into the return line through the flapper nozzle valve. For steady valve operation these two flow rates are identical. Since the nominal operating pressures of the right actuator are given by

$$P_{cr_o} = \tfrac{1}{2}P_{s_o} \quad \text{and} \quad P_r = 0, \tag{4.155}$$

where P_{s_o} is the nominal supply pressure, it may be shown that the linearized volumetric flow rates in and out of the right-hand actuator are

$$Q_{sr} = \tfrac{1}{2}Q_o + K_{c_a}(P_s - P_{cr}), \quad Q_{rr} = \tfrac{1}{2}Q_o - K_{q_a}r\varphi + K_{c_a}P_{cr}. \tag{4.156}$$

In this result Q_o is the nominal flow rate through the actuator, K_{c_a} is the pressure-flow coefficient that is designed to be identical for the actuator orifice and the flapper nozzle valve, K_{q_a} is the flow gain for the flapper nozzle valve, and φ is the angular displacement of the flapper. Substituting Equation (4.156) into Equation (4.154) and rearranging terms produces the following expression for the right-hand control pressure:

$$P_{cr} = \frac{P_s}{2} + \frac{K_{p_a}r}{2}\varphi - \frac{A}{2K_{c_a}}\dot{x}, \tag{4.157}$$

where K_{p_a} is the pressure sensitivity of the flapper nozzle valve given by K_{q_a}/K_{c_a}. From this result one may see that the nominal control pressure is given by half of the supply pressure while the adjustment of the angular position of the flapper φ may be used to increase or decrease the right-hand control pressure. *Note*: In Equation (4.157), the linear velocity of the main spool valve serves to alter the right-hand control pressure in a transient fashion thereby producing a favorable damping effect for the valve. Substituting Equation (4.157) into Equation (4.153) yields the following expression for the right-hand actuator force that is exerted on the main spool valve shown in Figure 4-26:

$$F_{ar} = A\frac{P_s}{2} + \frac{AK_{p_a}r}{2}\varphi - \frac{A^2}{2K_{c_a}}\dot{x}. \tag{4.158}$$

Now that the right-hand actuator force has been determined we must develop an expression for the left-hand actuator force F_{al}. Figure 4-28 shows a schematic of the left-hand actuator when the spool valve has been displaced a small distance in the positive x-direction.

From this figure it can be seen that the spool valve has exerted enough force against the spring retainer to overcome the spring pack preload. In this case, pressure forces and spring forces are exerted on the valve and the following expression may be written to describe the left-hand actuator force:

$$F_{al} = AP_{cl} + k_v x + F_{v_o}, \tag{4.159}$$

where P_{cl} is the left-hand control pressure, k_v is the spring rate of the main valve spring, and F_{v_o} is the assembled preload of the main valve spring. Since the volume of fluid on the left-hand side of the spool is extremely small, the pressure transients associated with the

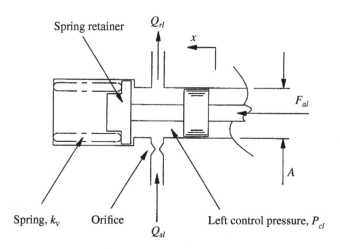

Figure 4-28. Left actuator of Figure 4-26.

left-hand control pressure may be neglected for the purposes of this analysis. Under these incompressible flow conditions it may be shown that

$$0 = Q_{sl} - Q_{rl} + A\dot{x}, \tag{4.160}$$

where Q_{sl} is the volumetric flow rate of fluid into the left actuator from the supply line and where Q_{rl} is the volumetric flow rate of fluid out of the left actuator into the return line through the flapper nozzle valve. Using the following nominal operating conditions for the left actuator:

$$P_{cl_o} = \tfrac{1}{2}P_{s_o} \quad \text{and} \quad P_r = 0, \tag{4.161}$$

where P_{cl_o} is the nominal control pressure in the left actuator, it may be shown that the linearized volumetric flow rates in and out of the left-hand actuator are given by

$$Q_{sl} = \tfrac{1}{2}Q_o + K_{c_a}(P_s - P_{cl}), \quad Q_{rl} = \tfrac{1}{2}Q_o + K_{q_a}r\varphi + K_{c_a}P_{cl}. \tag{4.162}$$

In this result the nominal flow rate through the left-hand actuator is given by Q_o while K_{q_a} and K_{c_a} are the flapper nozzle valve flow gain and the pressure-flow coefficient generally defined in Equation (4.137). Substituting Equation (4.162) into Equation (4.160) and rearranging terms the following expression for the left-hand control pressure may be written:

$$P_{cl} = \frac{P_s}{2} - \frac{K_{p_a}r}{2}\varphi + \frac{A}{2K_{c_a}}\dot{x}, \tag{4.163}$$

where K_{p_a} is the pressure sensitivity of the flapper nozzle valve. Again, one may see that the nominal control pressure is given by half of the supply pressure while the adjustment of the angular position of the flapper φ may be used to increase or decrease the left-hand control pressure. *Note*: In Equation (4.163) the linear velocity of the main spool valve serves to alter the left-hand control pressure in a transient fashion thereby producing a favorable damping effect for the valve. Substituting Equation (4.163) into Equation (4.159) yields the following expression for the left-hand actuator force that is exerted on the main spool valve shown in Figure 4-26:

$$F_{al} = A\,\frac{P_s}{2} - \frac{A\,\mathrm{K}_{p_a}r}{2}\,\varphi + \frac{A^2}{2\mathrm{K}_{c_a}}\,\dot{x} + k_v x + F_{v_o}. \tag{4.164}$$

At this point it is useful to summarize the equation of motion for the two-stage electrohydraulic valve thus far. If we assume that the main-stage spool valve is of the critically centered type, then Equations (4.98), (4.158), and (4.164) may be used with Equation (4.152) to show that

$$\frac{A^2}{\mathrm{K}_{c_a}}\dot{x} + (k_v + 2\mathrm{K}_{f_q})\,x = A\,\mathrm{K}_{p_a}r\varphi - F_{v_o}, \tag{4.165}$$

where K_{f_q} is the flow force gain for the main spool valve generally defined in Equation (4.32). In order to use this result an expression must be written, which relates the angular motion of the flapper φ to the input current of the torque motor Δi. This relationship is developed in the following paragraphs.

By neglecting viscous damping and the time-rate-of-change of angular momentum for the flapper nozzle valve a static equation may be written for the moments exerted on the flapper. This result is given by

$$0 = \underbrace{\mathrm{K}_t\,\Delta i + \mathrm{K}_m\,\varphi}_{\text{Torque } motor} + F_x r - T_a - k_t\varphi, \tag{4.166}$$

where K_t is the torque motor's torque constant, K_m is the torque motor's magnetic spring constant, F_x is the flow force exerted on the flapper in the horizontal direction and given in Equation (4.151), T_a is the torque exerted on the flapper by the feedback spring, and k_t is the spring rate of a torsional spring that may exist at the pivot point of the flapper nozzle valve. *Note*: This torsional spring is not explicitly shown in any of the previous figures and may result from an elastic/semi-rigid connection at the flapper pivot point as opposed to a well-lubricated pin joint. The torque motor constants K_t and K_m are well-known quantities that may be provided to the engineer by any torque motor manufacturer. These quantities are also well documented in previous literature [3] and are therefore not discussed at length in this chapter. For the purposes of our discussion it must be mentioned that the magnetic spring constant of the torque motor K_m produces a destabilizing effect for the flapper/armature and therefore the torsional spring rate of the flapper's pivot connection is usually designed to cancel the magnetic spring effect [4]. This means that for Equation (4.166) the magnetic spring constant K_m is usually designed to be equal to the torsional spring rate k_t. For the purposes of generality this equality is not assumed for the analysis that follows but may be employed for designs in which this design feature applies.

Figure 4-29 shows a schematic of the feedback spring that exists between the flapper and the main-stage spool of the electrohydraulic valve. The length of the feedback spring is given by the symbol l and the diameter of the spring is given by d_w (not shown in Figure 4-29). The deflection of the spring in the vertical direction is given by z and the distance along the spring is shown by the coordinate s. The reaction force between the spring and the main spool valve is shown in Figure 4-29 by the symbol F. Using the classical strength-of-materials equation for evaluating beam deflection the force and deflection characteristics of the feedback spring may be determined.

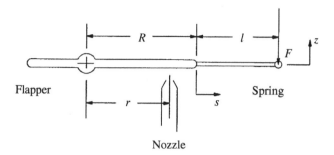

Figure 4-29. Flapper and feedback spring geometry.

The deflection equation is given by

$$\frac{d^2z}{ds^2} = \frac{M}{EI},$$ (4.167)

where M is the internal bending moment of the spring, E is the spring's modulus of elasticity, and I is the spring's second area moment of inertia given by $\pi d_w^4/64$. The solution to this equation at the tip of the spring is given by

$$z = -\frac{Fl^3}{3EI},$$ (4.168)

where, as previously mentioned, F is the reaction force between the tip of the spring and the main spool valve. From geometry, it may be shown that this deflection is also given by

$$z = -x - R\varphi,$$ (4.169)

where R is the distance from the tip of the flapper to the pivot point, φ is the angular position of the flapper, and x is the displacement of the main spool valve. Using Equations (4.168) and (4.169) it may be shown that the reaction force between the spring and the spool valve is

$$F = k_a(x + R\varphi),$$ (4.170)

where the spring rate of the feedback spring is

$$k_a = \frac{3EI}{l^3}.$$ (4.171)

The opposing torque exerted on the flapper by this force is given by

$$T_a = F(R + l) = k_a(R + l)(x + R\varphi).$$ (4.172)

This result is used with Equation (4.166) to determine the linear relationship between the input current of the torque motor Δi and the angular position of the flapper φ.

Using the steady form of Equation (4.151) with Equations (4.157), (4.163) and (4.172), Equation (4.166) may be written more explicitly as

$$k_{eff}\varphi = \frac{K_t}{r^2}\Delta i - \frac{k_a(R+l)}{r^2}x + \frac{(A_n + K_{f_{ca}})A}{K_{c_a}r}\dot{x},$$ (4.173)

where the effective spring constant for the flapper is given by

$$k_{eff} = A_n K_{p_a} + \frac{k_a(R+l)R}{r^2} + \frac{(k_t - K_m)}{r^2} - K_{f_{qa}}.$$ (4.174)

To generate this result it has been recognized that $K_{f_{qa}} = K_{f_{ca}} K_{p_a}$. Equation (4.173) shows that a positive input current Δi applies a positive torque on the flapper that tries to displace the flapper in the positive angular direction. This second term in this equation shows that the feedback spring tries to restore the flapper to its original position as the main spool valve moves in the positive x-direction. Finally the linear velocity of the spool valve $\dot{x}$ creates a pressure rise in the left-hand nozzle and a pressure drop in the right-hand nozzle. This pressure difference results in a transient pressure force that seeks to destabilize the valve. This effect is observed more fully in the subsequent discussion. As it turns out, all of the effective spring rate terms shown in Equation (4.174) have the potential of being important and therefore none of them can be generally eliminated due to their consistently small size. The reader is reminded that the torque motor will occasionally be designed so that $(k_t - K_m) = 0$ in which case this term in Equation (4.174) will vanish.

By combining Equations (4.165) and (4.173) the governing equation of motion for the two-stage electrohydraulic valve of Figure 4-26 may be written as

$$\frac{A}{K_{q_a}} \left(1 - \frac{A_n K_{p_a} + K_{f_{qa}}}{k_{eff}} \right) \dot{x} + \left(\frac{(k_v + 2K_{f_q})}{A K_{p_a}} + \frac{k_a(R+l)}{k_{eff} r} \right) x = \frac{K_t}{k_{eff} r}(\Delta i - \Delta i_o), \quad (4.175)$$

where the required input current to overcome the main valve spring preload F_{v_o} is given by

$$\Delta i_o = \frac{F_{v_o} k_{eff} r}{K_t A K_{p_a}}.$$ (4.176)

From this result a stability criterion for the valve may be determined. This stability criterion requires the leading coefficient of the velocity term on the left-hand side of Equation (4.175) to be positive. Using Equation (4.175), it may be seen that this stability criterion is given by

$$\frac{k_{eff}}{A_n K_{p_a} + K_{f_{qa}}} > 1,$$ (4.177)

where the effective spring constant for the flapper nozzle valve k_{eff} is given in Equation (4.174). All other coefficients in Equation (4.175) are positive by definition. A quick study of Equation (4.177) shows that some form of stiffness, either in the form of a feedback spring or a torsional spring, must be added to the flapper if the valve is to be strictly stable.

Response Characteristics.   To consider both the steady and transient response characteristics of the two-stage electrohydraulic valve shown in Figure 4-26 it is convenient to rewrite the equation of motion presented in Equation (4.175) as:

$$\tau \dot{x} + x = \psi(\Delta i - \Delta i_o),$$ (4.178)

where

$$\tau = \frac{A}{K_{q_a}} \left(1 - \frac{A_n K_{p_a} + K_{f_{qa}}}{k_{eff}} \right) \left(\frac{(k_v + 2K_{f_q})}{AK_{p_a}} + \frac{k_a(R+l)}{k_{eff}\, r} \right)^{-1}$$

and   (4.179)

$$\psi = \frac{K_t}{k_{eff}\, r} \left(\frac{(k_v + 2K_{f_q})}{AK_{p_a}} + \frac{k_a(R+l)}{k_{eff}\, r} \right)^{-1}.$$

All of the parameters in this equation have been previously defined. In this result the symbol τ represents the time constant for the transient response of the system while the symbol ψ represents the constant of proportionality for the steady-state response. *Note:* ψ is the slope of the line shown in Figure 4-16. Unfortunately, Equation (4.179) does not simplify nicely as it did for the solenoid actuated electrohydraulic valve of Subsection 4.5.4. All of the terms in this equation are important and must generally be considered when seeking to evaluate the valve's time constant and the constant of proportionality.

As previously mentioned, a small time constant is desired for reducing the settling time and increasing the frequency response of the valve. Due to the destabilizing tendency of the flapper nozzle valve its time constant tends to be very small and the response time of the valve is very fast. Similar to the electrohydraulic valve of Subsection 4.5.4 Equation (4.179) shows that a smaller cross-sectional area A and a larger flow gain for the nozzle K_{q_a} will provide the smallest time constant.

From a steady point of view the constant of proportionality for the valve must be sized so as to give a maximum valve displacement x_{max} for a maximum available input current Δi_{max}. This design requirement may be written as

$$\psi = \frac{x_{max}}{(\Delta i_{max} - \Delta i_o)},$$ (4.180)

where Δi_o is given in Equation (4.176). From Equation (4.179) it may be seen that to increase the maximum range of actuation for the valve the torque constant for the motor K_t must be increased and the effective spring rate k_{eff} must be reduced. By increasing K_t the overall size of the torque motor will grow and by reducing k_{eff} there is a potential for introducing instability into the valve design. Once again, a tradeoff between these two parameters must be considered when designing the two-stage electrohydraulic valve that is shown in Figure 4-26.

Summary.   This subsection has been used to study the two-stage electrohydraulic valve shown in Figure 4-26. This valve is frequently used to provide control functions for hydraulic devices that are used to transmit large amounts of power to the overall hydraulic system; however, this valve is not typically used to transmit large amounts of power itself. As previously mentioned, for example, the flapper nozzle valve design is often employed to control the displacement of a variable displacement pump. It should be mentioned that while the flapper nozzle valve tends to package itself better than the two solenoid design shown in Figure 4-15, it is a more difficult valve to design and stability is less certain. Even so, many flapper nozzle valves are unwittingly designed to operate in an unstable manner in which the flapper will slap back and forth between nozzles in a limit cycle while providing an average output that is acceptable from an end user's point of view. Since a large amount of power is not being consumed by the unstable flapper nozzle valve this valve does not tend to make excessive noise as it operates; therefore, its instability often goes unnoticed. The process, however, of operating in an unstable manner is not good from a flapper wear point of view and therefore a stable valve is more desirable than an unstable one.

4.7.5 Summary

This section of our chapter studies the characteristics of flapper nozzle valves. By way of application the flapper nozzle valve has been applied in its most common use as a first stage within a two-stage electrohydraulic control valve as shown in Figure 4-26. In order to adequately design the flapper nozzle valve a thorough understanding is needed of the interactions between the torque motor, the nozzle flow forces, and the main-stage spool valve that is being controlled. This section of our chapter brings these issues together and presents them in Equation (4.178). This result should be consulted when designing a flapper nozzle valve similar to Figure 4-26.

4.8 CONCLUSION

This chapter has attempted to present the governing equations that describe the performance characteristics of spool valves, poppet valves, and flapper nozzle valves. Though this chapter has not been exhaustive in its coverage of material it has provided a basis for analyzing common valve constructions and for applying similar methods to unique valve designs that have not been addressed in this chapter. Perhaps the most useful exercise employed in this chapter has been the repetitive use of the Reynolds Transport Theorem for calculating the fluid flow forces that are exerted on the mechanical components of the valve. These forces are the primary obstacles that must be overcome to actuate hydraulic valves; and therefore, they must be understood if the actuation system for the valve is to be designed properly. Since hydraulic valves are used to control both main and auxiliary hydraulic systems the material of this chapter are referred to regularly in subsequent chapters of this book.

4.9 REFERENCES

[1] Stone, J. A. 1960. Discharge coefficients and steady-state flow forces for hydraulic poppet valves. ASME *Journal of Basic Engineering*. March issue: 144–154.

[2] Anderson, W. R. 1988. *Controlling Electrohydraulic Systems. Marcel Dekker*, New York.

[3] Merritt, H. E. 1967. *Hydraulic Control Systems*. John Wiley & Sons, New York.

[4] Lin, S. J., and A. Akers. 1991. Dynamic analysis of a flapper nozzle valve. ASME *Journal of Dynamic Systems, Measurement, and Control*. 113:163–67.

4.10 HOMEWORK PROBLEMS

4.10.1 Valve Flow Coefficients

4.1   For a fixed nominal flow rate through a rectangular flow passage, determine the flow gain for a valve with its nominal opening given by 1/10, 1/4, 1/2, 3/4, and 9/10 of the maximum opening of the valve.

4.2   Repeat Problem 4.1 for a circular and triangular flow passage.

4.3 For a fixed nominal flow rate through a valve what is the pressure-flow coefficient for a rectangular, circular, and triangular flow passage?

4.10.2 Spool Valves

4.4 A two-way spool valve uses a 3 mm × 3 mm square flow passage to relieve a pressure line at 40 MPa using a flow rate of 4 lpm. The jet angle of the fluid exiting the valve is 68°. The fluid density is 850 kg/m^3 and the discharge coefficient for the valve is 0.62. Calculate the nominal pressure-flow coefficient for the valve. What is the nominal opening of the valve? How much power is lost across this valve?

4.5 The two-way spool valve in Problem 4.4 is used as a pressure relief valve similar to the configuration of Figure 4-9. The spool mass is given by 0.01 kg, the spool bore diameter is 6 mm, the spring is compressed 10 mm in the nominal operating mode, the damping length is 25 mm, the fluid bulk modulus is 1 GPa, and the compressible fluid volume is 5 × 10^{-4} m^3. Is the two-way spool valve stable? Why or why not?

4.6 A three-way spool valve similar to Figure 4-12 uses 3 mm × 3 mm square flow passages. The mass of the spool is given by 0.0151 kg, the viscous drag coefficient is 1.1562 Ns/m, and the mechanical spring rate is 42 N/mm. Assume that the jet angle is 60°, the discharge coefficient is 0.62, and the fluid density is 850 kg/m^3. The supply pressure to the valve is 10 MPa. Calculate the natural frequency, damping ratio, rise time, maximum percent overshoot, and settling time for the open centered valve with an underlapped dimension of 0.4 mm. The length between ports is 10 mm. Is this valve sensitive to load disturbances? Why or why not?

4.7 Repeat Problem 4.6 for a critically centered valve. What is the impact of switching the supply and return ports of Figure 4-12?

4.8 A four-way spool valve similar to Figure 4-13 utilizes 3 mm × 3 mm square flow passages. The mass of the spool valve is 0.0156 kg, the viscous drag coefficient is 1.307 Ns/m, and the mechanical spring rate is 65 N/mm. Assume that the jet angle is 65°, the discharge coefficient is 0.62, and the fluid density is 850 kg/m^3. The supply pressure to the valve is 20 MPa. Calculate the natural frequency, damping ratio, rise time, maximum percent overshoot, and settling time for the open centered valve with an underlapped dimension of 0.5 mm.

4.9 Repeat Problem 4.8 for a critically centered valve.

4.10 The four-way valve of Problem 4.9 is used as the main stage of a two-stage electro-hydraulic valve similar to what is shown in Figure 4-15. The main stage has been designed with a 10 mm bore diameter. The control spool/armature for the actuator utilizes square flow passages that are 1 mm × 1 mm and exhibit a discharge coefficient of 0.62. The pilot pressure that is used to actuate the valve is 2.1 MPa. The spring rate for the feedback spring is 7 N/mm and the magnetic coupling coefficient for the solenoid is 25 N/Amp. Calculate the time constant τ and the constant of proportionality ψ for the valve. Based on these calculations estimate the settling time and the frequency response for the valve. Calculate the displacement of the main-stage valve when 1.5 Amps of current are applied to one solenoid. The initial amount of current required to overcome the preload of the spring pack is 0.75 Amp.

4.10.3 Poppet Valves

4.11 A hydraulic poppet valve is used to relieve high-pressure spikes within a hydraulic circuit. The pressure drop across the relief valve is 6000 psi. What is the expected temperature rise in the hydraulic fluid?

4.12   A poppet valve is designed similar to Figure 4-22 to relieve a pressure line at 40 MPa using a flow rate of 4 lpm. The diameter of the inlet passage to the poppet valve is 2 mm and the jet angle of the fluid is 45°. The fluid density is 850 kg/m^3 and the discharge coefficient for the valve is 0.62. Calculate the nominal pressure-flow coefficient for the valve. What is the nominal opening of the valve? What does the spring preload on the valve need to be to achieve this steady-state operating condition? How much power is lost across this relief valve?

4.13   The poppet mass in Problem 4.12 is given by 0.01 kg, the spring is compressed 10 mm in the nominal operating mode, the damping length is 25 mm, the fluid bulk modulus is 1 GPa, and the compressible fluid volume is 5×10^{-4} m^3. Is the poppet valve stable? Why or why not?

4.10.4 Flapper Nozzle Valves

4.14   A flapper nozzle valve utilizes a nozzle that is located 19 mm away from the flapper pivot point. The nozzle diameter is 3.65 mm and the nominal gap between the flapper and the nozzle is 0.375 mm. The pilot pressure supply is 2.1 MPa, the fluid density is 850 kg/m^3, and the discharge coefficient for the nozzle is 0.62. Calculate the flow gain and the pressure-flow coefficient for the valve. Using the same information, calculate the flow force gain and the pressure flow force coefficient.

4.15   The flapper nozzle valve in Problem 4.14 is used as the first stage in a two-stage electrohydraulic valve similar to what is shown in Figure 4-26. The second-stage spool diameter is 12.7 mm and the spool valve has a flow force gain of 3.425 N/mm. The spool valve spring rate is 50 N/mm with an assembled preload of 500 N. A 2 mm diameter feedback wire is used between the spool and the flapper in which the dimensions $l = 25.4$ mm and $R = 23$ mm (these dimensions are shown in Figure 4-29). The flapper uses a torsion spring that perfectly cancels the magnetic spring rate of the torque motor. The torque constant for the torque motor is 120 Nm/Amp. Calculate the effective spring rate for the flapper and the initial current that is required to overcome the spring preload on the main-stage spool valve. Is the valve stable? Why or why not?

4.16   For the two-stage electrohydraulic valve in Problem 4.15 calculate the time constant and bandwidth for the valve. What is the steady-state constant of proportionality that exists between the input current and the output displacement of the main-stage spool valve? Assuming that 60 mA of current are applied to the torque motor, how far will the spool valve travel?

5

HYDRAULIC PUMPS

5.1 INTRODUCTION

5.1.1 Overview

Hydraulic pumps are the power source for hydraulic control systems. The task of the hydraulic pump is to covert rotating shaft power into fluid power that may be controllably transmitted downstream to other output devices. Two types of pumps may potentially be used for this task: (1) hydrodynamic, non-positive-displacement pumps, or (2) hydrostatic, positive-displacement pumps. Hydrodynamic pumps are constructed as turbines or fans and are generally inefficient due to the large amount of fluid slippage that occurs at the pumping elements. These pumps are also low-pressure machines and are not well suited for higher performance hydraulic control systems. On the other hand, hydrostatic pumps are constructed using devices that displace a specific volume of fluid for each revolution of the pump independent of the operating pressure of the pump, thus they are called positive-displacement pumps. These pumps are capable of operating at substantial pressures with much higher efficiencies and are therefore well suited for hydraulic control systems. For this reason this chapter is concerned only with hydrostatic, positive-displacement pumps.

By way of introduction this chapter begins with an overview of hydrostatic pump types. In this introduction typical applications are noted for each pump type and well-known advantages and disadvantages of these pumps are mentioned. The efficiency of the hydrostatic pump is then considered with its standard definitions and the measurement uncertainty of the efficiency is discussed. In considering the dynamic behavior of the hydrostatic pump the flow ripple must often be accounted for; therefore, the flow ripple of the pump is discussed as it pertains to gear pumps and swash-plate axial-piston pumps (two of the most common pumps used in hydraulic control systems). Finally, the control of the swash-plate axial-piston pump is discussed for both displacement and pressure control applications.

5.1.2 Hydrostatic Pump Types

Overview.   This subsection presents a brief survey of hydrostatic pumps that are commonly used within hydraulic control systems. As previously mentioned, the pumps discussed in this subsection are of the positive-displacement type, which means that each pump delivers a specific amount of fluid independent of the operating pressures of the machine. As is shown in the following paragraphs, there are many pump designs that may be used to accomplish this task and each one of these designs has its own advantages and disadvantages when compared to other machines of similar function. The list if items that may be compared are cost, noise, operating range, efficiency, and control.

Gear Pumps.   Figure 5-1 shows a cross-sectional view taken through the gear set of a typical gear pump. Like most actual gear pump designs this pump is shown with two identical gears that are used for displacing fluid. The gears are contained within a close tolerance housing that separates the discharge port from the intake port. An external shaft is connected to the driven gear (not shown in Figure 5-1) while the other gear is supported by an internal shaft and bearing. The gear pump shown in Figure 5-1 is more precisely identified as an external gear pump since the gear geometry is external. Internal gear geometry may also be used for a gear pump, however, this configuration is less common.

When considering the operation of a gear pump it is a common mistake to assume that the fluid flow occurs through the center of the pump; that is, through the meshed gear geometry, however, as shown in Figure 5-1 this is not what happens. To produce flow with a gear pump fluid is carried around the outside of each gear within each tooth gap from the intake side of the pump to the discharge side of the pump. As the gear teeth mesh within the gear set fluid is squeezed out of each tooth gap by a mating tooth and is thereby displaced into the discharge line of the pump. On the intake side the gear teeth are coming out of the mesh. In this condition fluid backfills for the volume of the mating teeth that are now evacuating each tooth space. This backfilling draws fluid into the pump through the intake port of the pump housing. This process repeats itself for each revolution of the pump and thereby displaces fluid at a rate proportional to the pump speed.

The gear pump exhibits advantages over other hydrostatic pumps in two main areas: (1) the gear pump is inexpensive to buy, and (2) there are only two moving parts in the

Figure 5-1. Sectional view of an external gear pump.

gear pump, which tends to make this design very reliable. Disadvantages of the gear pump are: (a) the gear pump typically operates at a medium pressure level (up to 21 MPa), (b) efficiencies of gear pumps tend to be low (between 80% and 90%), (c) gear pumps tend be fairly loud (between 70 dB and 80 dB), and (d) gear pumps are fixed-displacement units, which means the only control variable is the shaft speed of the driven gear. Gear pumps are typically used as a power supply for inexpensive and inefficient systems. They are often used as oil pumps for diesel engines, charge pumps for hydrostatic pilot systems, or plastic versions are sometimes used for water supply; for example, windshield washing fluid in an automotive application. Gear pumps and their flow characteristics are discussed in greater detail in Section 5.3 of this chapter.

Gerotor Pumps. Figure 5-2 shows a cross-sectional view of a gerotor pump. As shown in the figure this pump is comprised of an inner and outer element both rotating in the same direction relative to the fixed discharge and intake ports shown with dashed lines. The inner element rotates slightly faster than the outer element so that the inner element is always in sliding contact with the outer element to form sealed pockets of fluid. Fluid enters the intake port as the sealed space between the inner and outer element grows. Fluid exits the discharge port as the sealed space between the inner and outer element decreases. Views (a), (b), and (c) of Figure 5-2 illustrate the function of the gerotor pump by tracking the location of two teeth which define a sealed pocket of fluid. The design of the gerotor pump is indeed clever with an unobvious mechanism for displacing fluid.

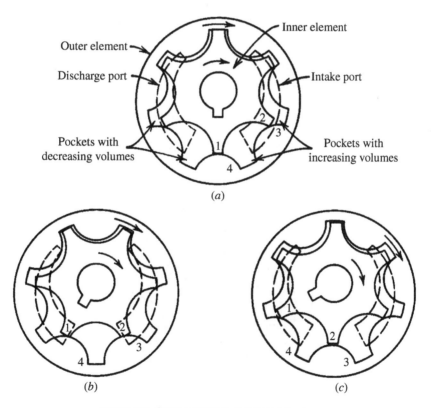

Figure 5-2. Cross-sectional view of gerotor pump.

The advantages of the gerotor design are similar to those of the standard gear pump: (a) it is an inexpensive design, and (b) there are few moving parts, which makes this design very reliable. The disadvantages of the gerotor design are also similar to those of the gear pump: (a) the gerotor pump operating pressures are generally low (c.a., 7 MPa), (b) the gerotor pump operating efficiencies are low (60–80%), and (c) the gerotor pump is a fixed-displacement pump, which means that the discharge flow can only be controlled by adjusting the input speed of the pump shaft. Gerotor pumps are often used as charge pumps for supplying pilot pressures for hydraulic control systems and for providing makeup flow for hydrostatic transmissions.

Vane Pumps.   Figure 5-3 shows a cross-sectional view of a vane pump. As shown in this figure the vain pump consists of an internal rotor, a housing, and spring loaded elements that are called vanes. The vanes are contained within the rotor and the rotor is offset from the center of the housing by the eccentricity distance e. As the rotor turns relative to the fixed housing the vanes are thrown outwardly by their own inertia. The springs are also used to assist in pushing the vanes against the housing and the discharge pressure is sometimes supplied to the vanes as well for accomplishing the same purpose. As the vanes pass over the intake port on the housing the space between the vanes and the rotor and the housing increases, thereby drawing fluid into the pump. As the vanes pass over the discharge port on the housing the space between the vanes and the rotor and the housing decreases, thereby pushing fluid out of the pump. One can see from Figure 5-3 that the pumping action depends on the eccentricity distance e. If this distance were zero there would be no fluid displaced by the pump. The vane pump is sometimes made to be a variable-displacement pump by utilizing a mechanism that varies the eccentricity distance by a controlled amount. Usually, however, the eccentricity distance is a fixed dimension and the vane pump is designed as a fixed-displacement machine.

The vane pump has the following advantages associated with its design: (a) the vane pump is fairly inexpensive, (b) the vane pump has a few number of parts and is therefore very reliable, and (c) the vane pump is self-adjusting for vane wear. The disadvantages of this design are: (a) the vane pump operating pressures are generally low (up to 21 MPa), (b) the vane pump operating efficiencies are low (70–80%), and (c) the vane pump is usually a fixed-displacement pump, which means that the discharge flow can only be controlled by adjusting the input speed of the pump shaft. Vane pumps are typically used in automotive applications for supplying hydraulic power to power steering systems.

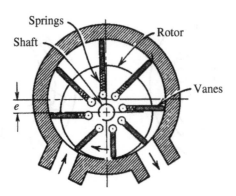

Figure 5-3. Cross-sectional view of a vane pump.

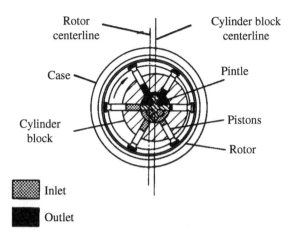

Figure 5-4. Cross-sectional view of a radial piston pump.

Radial Piston Pumps.   Figure 5-4 shows a cross-sectional view of the radial piston pump. This pump consists of pistons that are oriented in a radial fashion within the cylinder block. The intake and discharge ports are located at the center of the cylinder block. This pump has features in common with the vane pump in that it depends on an eccentricity offset shown by the difference in the rotor centerline and the cylinder block centerline. As the rotor turns the pistons follow the profile of the rotor thereby drawing fluid into the intake side of the pump and pushing fluid out of the discharge side of the pump. By varying the eccentricity distance this pump can be made into a variable-displacement unit; however, the radial pump is usually a fixed-displacement design with a fixed eccentricity distance.

Advantages of the radial piston pump are: (a) radial piston pumps typically exhibit high efficiencies (85–98%), (b) radial piston pumps are capable of operating at high pressures (up to 42 MPa), (c) radial piston pumps exhibit a low starting torque, which makes them a good rotary actuation device when operating in the reverse mode as a motor, and (d) radial piston pumps are inherently short and flat, which means that they are well suited for applications in which space limitations are extreme. Disadvantages of the radial piston pump are: (a) it is a fairly expensive pump to purchase ($1000 to $5000 per machine), and (b) it is typically a fixed-displacement device which means that the flow rate must be varied by varying the shaft speed. Due to their high efficiency and low starting torque these pumps are often used in their reverse operating mode as an output motor for hydrostatic transmissions.

Axial-Piston Swash-Plate Pumps.   Figure 5-5 shows a picture of the swash-plate axial-piston pump. As shown in this figure the pump consists of several pistons within a common cylindrical block. The pistons are nested in a circular array within the cylinder block at equal intervals about the shaft centerline. As shown in Figure 5-5 the cylinder block is held tightly against a port plate using the compressed force of the cylinder block spring. A thin film of oil separates the port plate from the cylinder block, which under normal operating conditions forms a hydrodynamic bearing between the cylinder block and the port plate. The cylinder block is connected to the shaft through a set of splines that run parallel to the shaft. A ball and socket joint connects the base of each piston to a slipper. The slippers themselves are kept in reasonable contact with the swash plate by a retainer (not shown in Figure 5-5) where a hydrostatic and hydrodynamic bearing surface separates the slippers from the swash plate. The swash plate angle α may generally

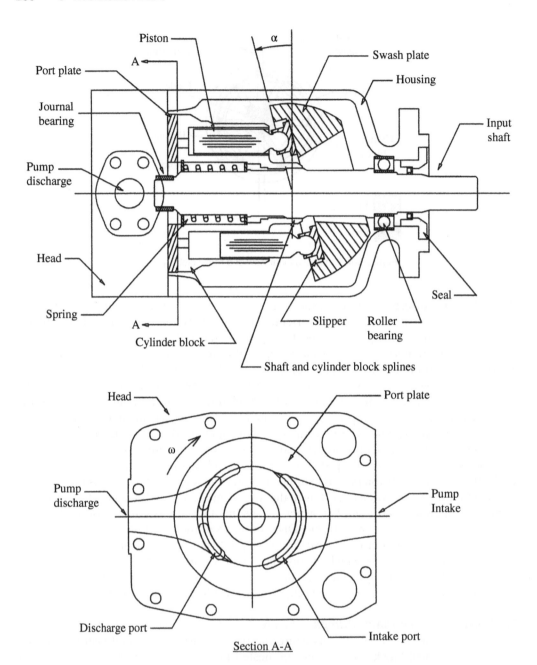

Figure 5-5. Cross-sectional view of an axial-piston swash-plate pump.

vary with time to accomplish a variable displacement of the pump. The entire machine is contained within a housing that is filled with hydraulic fluid.

While the port plate is held in a fixed position against the head the shaft and cylinder block rotate about the shaft centerline at a constant angular speed ω. The reader should recall that the shaft and cylinder block are connected through a set of splines that run parallel to the shaft. During this rotational motion each piston periodically passes over the discharge and intake ports on the port plate. Furthermore, because the slippers are held against the inclined plane of the swash plate the pistons also undergo an oscillatory displacement in and out of the cylinder block. As the pistons pass over the intake port the piston withdraws from the cylinder block and fluid is drawn into the piston chamber. As the pistons pass over the discharge port the piston advances into the cylinder block and fluid is pushed out of the piston chamber. This motion repeats itself for each revolution of the cylinder block and the basic task of displacing fluid is accomplished. The intake and discharge flow of the machine are distributed using the head component, which functions as a manifold to the rest of the hydraulic system.

Swash-plate axial-piston pumps are the most commonly used pumps in hydraulic control systems. Advantages of using these pumps are: (a) swash-plate axial-piston pumps exhibit high operating efficiencies (85%–95%), (b) swash-plate axial-piston pumps are capable of operating at high pressure (up to 42 MPa), and (c) swash-plate axial-piston pumps lend themselves nicely to variable-displacement control, which means that the discharge flow from these pumps may be controlled by adjusting the swash plate angle independent of the pump shaft speed. Disadvantages of using swash-plate axial-piston pumps are: (a) swash-plate pumps are fairly expensive to purchase ($1000–$5000 per machine), and (b) their construction is more complex than the pumps that have been previously mentioned. This higher degree of complexity leads to a lower degree of reliability. The applications in which swash-plate axial-piston pumps are used vary. Typical applications include hydrostatic transmissions, fuel pumps for diesel engines, implement pumps for robots, earthmoving equipment, agricultural equipment, and aircraft flight control systems. The axial-piston swash-plate pump will be considered in greater detail in Section 5.4.

Axial-Piston Bent-Axis Pumps. Figure 5-6 shows a bent-axis axial-piston pump. From this figure it can be seen that the cylinder block is at an angle to the drive shaft and that the pistons are joined to the drive shaft with connecting rods. The cylinder block rotation is synchronized with the drive shaft by a constant velocity universal joint.

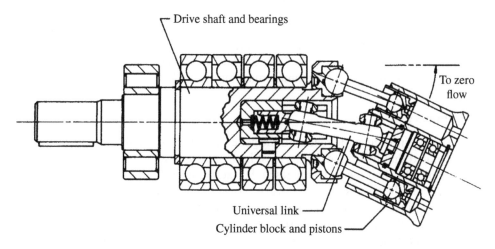

Figure 5-6. Cross-sectional view of an axial-piston bent-axis pump.

The universal joint carries torque only – no side loading of this link is allowed as all thrust loads are carried by the connecting rods themselves. As the pistons move with the shaft toward the top location in the cylinder block fluid is drawn into the piston chamber through the intake port of the pump. As the pistons move toward the bottom location in the cylinder block fluid is pushed out of the piston chamber through the discharge port of the pump. The ports are separated by a port plate much like that of the swash-plate axial-piston pump design (see Figure 5-5). By varying the angle of inclination for the cylinder block the displacement of the pump may be varied as well. For the swash-plate pump shown in Figure 5-5 the maximum swash plate angle that may be achieved is approximately 21°. For the bent-axis pump shown in Figure 5-6 the maximum cylinder block angel that may be achieved is approximately 45°. This maximum angle difference produces a higher degree of displacement variability for the bent-axis design as compared to the swash plate design; however, the penalty for achieving this higher degree of variability is usually observed in higher machine cost and lower reliability due to the complexity of moving parts within the bent-axis axial-piston pump.

Advantages of using the bent-axis axial-piston pump are: (a) the bent-axis design exhibits high efficiencies (85–98%), (b) the bent-axis design is capable of operating at high pressures (up to 42 MPa), (c) the bent-axis design exhibits the highest degree of variability which means that its output flow can be easily adjusted by simply altering the angle of the cylinder block, and (d) the bent-axis design exhibits a very low starting torque requirement, which makes it well suited for a motor operating in the reverse mode of a pump. This characteristic is common with the radial piston pump. Disadvantages of the bent-axis axial-piston pump are: (a) it is the most expensive pump to purchase ($2000–$7000 per machine), and (b) it is the most complex pump design, which translates into a lower reliability as compared to some of its competitors. Though the bent-axis axial-piston pump is not used as widely as the swash-plate axial-piston pump, it is used most commonly as a motor due to its high degree of variability and its low starting torque characteristics. The maximum speed ratio that may be achieved with the bent-axis motor is 5:1 as compared to a maximum speed ratio of 3:1 for the swash plate motor.

Summary.   In this subsection a brief survey of hydrostatic pump technology has been presented. The reader should be impressed by the number of machine variations that exist for accomplishing the positive displacement of fluid; however, it must be mentioned that not all machine types have been presented here. Rather, we have only presented the most common machines that are currently used in the marketplace today. The reader is referred to other references for a discussion of less common machine types [1, 2].

5.1.3 Summary

In this section we introduce the reader to the general topic of hydraulic pumps. In particular it has been noted that hydraulic pumps exist in many different configurations, all of which are accompanied by their own advantages and disadvantages. In the sections that follow the efficiency of hydraulic pumps in general will be considered along with modeling and measuring techniques. Since gear and axial-piston pumps are the most common pump types used in hydraulic control systems today, these pumps are discussed in more detail in the sections that follow. Of particular interest is the control of variable-displacement swash-plate type pumps. Under this heading both pressure-controlled and displacement-controlled pumps are presented.

5.2 PUMP EFFICIENCY

5.2.1 Overview

As previously mentioned, the task of the hydraulic pump is to convert rotating mechanical shaft power into fluid power that may be utilized downstream of the pump. This process is carried out by the machinery described in the previous section; however, none of these machines are 100% efficient. They all lose power in the process of converting power. This power loss results from fluid flow that leaks away from the main path of power transmission and from the friction that exists within machine. In this section pump efficiency is discussed as it is defined, modeled, and measured.

5.2.2 Efficiency Definitions

Figure 5-7 shows a schematic of power that flows in and out of a typical hydraulic pump. In this schematic power is supplied to the pump through the rotating shaft by an external drive device not shown in Figure 5-7. As the shaft rotates, the pump draws fluid into the inlet side and pushes fluid out of the discharge side.

The input power to the shaft is calculated based on the standard formula of shaft torque times rotating speed $T\omega$. Power is also delivered to the pump on the inlet side by any pressure that may exist at the intake port of the pump. This power is described by the standard formula for hydraulic power, which is given as pressure times volumetric flow rate $P_i Q_i$. The discharge power of the pump is given similarly as $P_d Q_d$ where P_d is the discharge pressure and Q_d is the discharge volumetric flow rate. In Figure 5-7 power is also shown to leak away from the pump in the form of internal leakage – this power loss is calculated as $P_l Q_l$ where P_l is the pressure drop across the leak path and Q_l is the leakage volumetric flow rate. Finally, power also leaves the pump in the form of dissipating heat. This heat dissipation is shown in Figure 5-7 by the squiggly line exiting the dashed control volume.

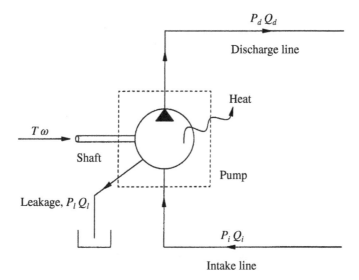

Figure 5-7. Schematic of power flowing in and out of the pump.

The overall pump efficiency is defined as the useful output power divided by the supplied input power. From the flow of power shown in Figure 5-7 this definition may be mathematically expressed as

$$\eta = \frac{P_d Q_d}{T\omega}, \tag{5.1}$$

where P_d is the discharge pressure of the pump, Q_d is the volumetric flow rate measured at the discharge port of the pump, T is the input torque on the pump shaft, and ω is the angular velocity of the pump shaft. Using the volumetric displacement of the pump V_d the overall efficiency may be separated into two components – the volumetric efficiency and the torque efficiency. This separation is given by

$$\eta = \eta_t \eta_v, \tag{5.2}$$

where the torque efficiency is expressed as

$$\eta_t = \frac{V_d P_d}{T}, \tag{5.3}$$

and where the volumetric efficiency is

$$\eta_v = \frac{Q_d}{V_d \omega}. \tag{5.4}$$

In Equations (5.3) and (5.4) the volumetric displacement is given in units of volume per radian. The volumetric displacement per revolution is given by $V_D = 2\pi V_d$. The torque efficiency shown in Equation (5.3) is used for describing power losses that result from fluid shear and internal friction. The volumetric efficiency shown in Equation (5.4) is used for describing power losses that result from internal leakage and fluid compressibility.

5.2.3   Modeling Pump Efficiency

Figure 5-8 shows a typical graph of the pump efficiency plotted against the nondimensional group $\mu\omega/P$, where μ is the fluid viscosity, ω is the angular shaft speed, and P is the pressure drop across the pump.

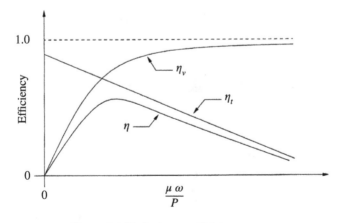

Figure 5-8. Typical pump efficiency curves.

Many attempts have been made to model pump efficiency with some degree of accuracy and a good summary of this work may be found in the literature [3]. At this present time there does not seem to be an accurate way to predict pump efficiency characteristics in an a priori way and therefore experimental coefficients are still required in the modeling process. The pump efficiency terms may be grouped into physical expressions and modeled as follows:

$$\eta_v = 1 - C_l \frac{P}{\mu\omega} - C_t \sqrt{\frac{P}{\mu\omega}}, \tag{5.5}$$

and

$$\eta_t = 1 - C_s - C_c \frac{\mu\omega}{P} - C_h \sqrt{\frac{\mu\omega}{P}}, \tag{5.6}$$

where C_l accounts for fluid compressibility effects and low Reynolds number leakage, C_t accounts for high Reynolds number leakage, C_s accounts for starting torque losses, C_c accounts for Coulomb friction torque losses that are proportional to applied loads within the pump, and C_h accounts for hydrodynamic torque losses that result from fluid shear. Again, all of these coefficients must be determined from experiments and substituted into Equations (5.5) and (5.6) for the purposes of modeling the pump efficiency. The determination of these coefficients is best achieved by using a least squares evaluation of Equations (5.5) and (5.6) with a sufficiently large number of experimental data points that have been used for determining the actual efficiencies corresponding to specific values for the dimensionless group $\mu\omega/P$. Equations (5.5) and (5.6) are a set of well-posed equations for least squares analysis. As just stated, experimental values are required for this work; therefore, measurements of pump efficiency are discussed in the next subsection.

5.2.4 Measuring Pump Efficiency

Overview.   The topic of pump efficiency is more important than it may at first appear. The reader will recall that hydraulic control systems find their niche in transmitting large amounts of power while producing a favorable dynamic response. Since large amounts of power are being transmitted by these systems, small percent changes in efficiency correspond to large amounts of power either being saved or lost. Within a cost-sensitive environment these small percent changes in efficiency translate into a significant economic factor. As a result it is not uncommon within the marketplace to see the sale of a hydraulic pump being either made or broken based on a reported efficiency difference of 2–5%. The question is this: Can the efficiency of a hydraulic pump be regularly measured to this accuracy? This subsection is being written in part to consider this issue.

Uncertainty Analysis.   To consider the uncertainty associated with measuring the efficiency of the pump a definition of uncertainty is required. This definition is given as

$$\varepsilon = \frac{\eta' - \eta}{\eta'}, \tag{5.7}$$

where η' is the measured efficiency value and η is the true efficiency, which is strictly speaking unknown. Throughout this discussion primed notation denotes measured quantities while unprimed variables denote exact or true quantities. If we assume that the uncertainty of our measurements is small then a Taylor series may be written for Equation (5.7) with respect to the measured values of pressure, flow, torque, and speed. This expression is given by

$$\varepsilon = \left(\frac{\partial \varepsilon}{\partial P_d}\right)'(P_d - P'_d) + \left(\frac{\partial \varepsilon}{\partial Q_d}\right)'(Q_d - Q'_d) + \left(\frac{\partial \varepsilon}{\partial T}\right)'(T - T') + \left(\frac{\partial \varepsilon}{\partial \omega}\right)'(\omega - \omega'). \quad (5.8)$$

Using the definition of the overall efficiency given in Equation (5.1), it may be shown that

$$\left(\frac{\partial \varepsilon}{\partial P_d}\right)' = -\frac{1}{P'_d}, \quad \left(\frac{\partial \varepsilon}{\partial Q_d}\right)' = -\frac{1}{Q'_d}, \quad \left(\frac{\partial \varepsilon}{\partial T}\right)' = \frac{1}{T'}, \quad \left(\frac{\partial \varepsilon}{\partial \omega}\right)' = \frac{1}{\omega'}. \quad (5.9)$$

Substituting these results into Equation (5.8) produces the following equation for the uncertainty associated with measuring the overall efficiency of the pump:

$$\varepsilon = -\frac{(P_d - P'_d)}{P'_d} - \frac{(Q_d - Q'_d)}{Q'_d} + \frac{(T - T')}{T'} + \frac{(\omega - \omega')}{\omega'}. \quad (5.10)$$

Similarly, it may be shown that the uncertainty associated with the measurement of the torque efficiency is given by

$$\varepsilon_t = -\frac{(P_d - P'_d)}{P'_d} + \frac{(T - T')}{T'} - \frac{(V_d - V'_d)}{V'_d}, \quad (5.11)$$

and the uncertainty associated with the measurement of the volumetric efficiency is

$$\varepsilon_v = -\frac{(Q_d - Q'_d)}{Q'_d} + \frac{(\omega - \omega')}{\omega'} + \frac{(V_d - V'_d)}{V'_d}. \quad (5.12)$$

Using Equations (5.10) through (5.12) it can be shown that

$$\varepsilon = \varepsilon_t + \varepsilon_v. \quad (5.13)$$

The accuracy of the instrumentation that is used to measure the efficiency of the pump is the deciding factor in determining the uncertainty of the measurement. For almost all measurement devices the accuracy of the measurement may be modeled as a percentage of the full-scale reading of the instrument [4]. In other words, the maximum deviation of the measured value from the true value may be modeled as

$$(P_d - P'_d) = \pm \xi_p P_{\max},$$
$$(Q_d - Q'_d) = \pm \xi_q Q_{\max},$$
$$(T - T') = \pm \xi_t T_{\max}, \quad (5.14)$$
$$(\omega - \omega') = \pm \xi_s \omega_{\max},$$
$$(V_d - V'_d) = \pm \xi_v V_{\max},$$

where the accuracy value of the instrument is given by the symbol ξ and the full-scale reading of the instrument is denoted by the "max" subscript. *Note*: These instrument parameters are provided either by the manufacturer of the instrument or by some other source of a priori knowledge. Substituting Equation (5.14) into Equations (5.10) through (5.12) produces the following uncertainties for the measurements of the overall efficiency and the torque and volumetric efficiency respectively:

$$\varepsilon = \pm \left(\xi_p \frac{P_{\max}}{P'_d} + \xi_q \frac{Q_{\max}}{Q'_d} + \xi_t \frac{T_{\max}}{T'} + \xi_s \frac{\omega_{\max}}{\omega'} \right),$$

$$\varepsilon_t = \pm \left(\xi_p \frac{P_{\max}}{P'_d} + \xi_t \frac{T_{\max}}{T'} + \xi_v \frac{V_{\max}}{V'_d} \right), \tag{5.15}$$

$$\varepsilon_v = \pm \left(\xi_q \frac{Q_{\max}}{Q'_d} + \xi_s \frac{\omega_{\max}}{\omega'} + \xi_v \frac{V_{\max}}{V'_d} \right).$$

Clearly, the uncertainties of the measurements can be minimized by using very accurate instrumentation (small values of ξ) and by taking measurements that are very close to the maximum measurement capability of the instrument. For instance, by using an instrument to measure a quantity that is half of the instrument's maximum capability the uncertainty in that measurement has been doubled when compared to a measurement that has been taken at the full-scale capability. Obviously, the situation gets worse as the measurement value gets smaller.

The uncertainty results of the previous analysis are dependent on the accuracy of the instrumentation that is used to make the efficiency measurement. For pressure measurements it is common to purchase strain-gauged transducers with an accuracy range of $\pm 0.5\%$ of full-scale readings; in other words, for pressure measurements $\xi_p = 0.005$. Volumetric flow measurements must be taken on the high-pressure side of the pump for determining the delivered pump power. Usually, this high-pressure requirement for the sensor dictates the use of a positive-displacement flow meter as turbine meters are generally not capable of operating at pressures above 21 MPa. Positive-displacement flow meters are usually capable of measuring volumetric flow rates with an accuracy range of $\pm 0.5\%$ of full-scale readings. In other words, for volumetric flow measurements, $\xi_q = 0.005$. The accuracy range for torque meters is typically within $\pm 0.25\%$ of full-scale readings, therefore, $\xi_t = 0.0025$. Shaft speed measurements can be made very accurately using pulse counting with a Hall effect sensor. The accuracy of this measurement tends to increase with the number of pulses that are taken per revolution of the shaft. According to Doebelin [4] the Hall effect sensor can measure shaft speeds within $\pm 0.06\%$ of full-scale readings, which is essentially a negligible error. From Doebelin's reported accuracy we conclude that $\xi_s = 0.00006$. As strange as it may sound the true volumetric displacement of the pump is not known with absolute certainty even when mechanical drawings are available for calculating this number. The primary reason for such uncertainty is due to inaccuracies in the manufacturing process and deflection under load. Though the amount of uncertainty varies from pump to pump a reasonable approximation of this uncertainty is $\pm 0.5\%$ of the maximum pump displacement, though this percentage might increase with pressure. For a first calculation, it is reasonable to assume that $\xi_v = 0.005$. For fixed-displacement pumps $V_{\max}/V'_d = 1$. For variable-displacement pumps, this ratio may be greater than unity, which will increase the uncertainty of the measurement.

Using the preceding analysis with the general form of Equation (5.7) the confidence interval of the efficiency measurement may be written as

$$\eta'(1 - \varepsilon) < \eta < \eta'(1 + \varepsilon),$$
$$\eta_t'(1 - \varepsilon_t) < \eta_t < \eta_t'(1 + \varepsilon_t), \qquad (5.16)$$
$$\eta_v'(1 - \varepsilon_v) < \eta_v < \eta_v'(1 + \varepsilon_v),$$

where, once again, the primed notation denotes a measured value and the unprimed notation denotes actual or true values. The uncertainty values for ε, ε_t, ε_v and are determined using Equation (5.15).

Case Study

Within the laboratory an engineer has measured the discharge pressure, discharge flow, shaft speed, and shaft torque of a pump and has concluded from the calculations of Equations (5.3) and (5.4) that both the volumetric and torque efficiencies are 95%. In the measurements taken, all instruments have been used at half of their maximum capacity, which means that $P_{max}/P_d' = T_{max}/T' = Q_{max}/Q_d' = 2$. The accuracy of the instrumentation used in the experiment is given by $\xi_p = 0.005$, $\xi_t = 0.0025$, and $\xi_q = 0.005$. You are to assume that the pump speed and volumetric displacement of the pump are known perfectly; that is, $\xi_s = \xi_v = 0$. Using these numbers calculate the uncertainty of the efficiency measurements. What is the confidence interval in the efficiency measurement?

The uncertainty of the efficiency measurements may be calculated from Equation (5.15) as:

$$\varepsilon = \pm \left(\xi_p \frac{P_{max}}{P_d'} + \xi_q \frac{Q_{max}}{Q_d'} + \xi_t \frac{T_{max}}{T'} + \xi_s \frac{\omega_{max}}{\omega'} \right)$$
$$= \pm(0.005 \times 2 + 0.005 \times 2 + 0.0025 \times 2 + 0) = \pm 2.5\%,$$

$$\varepsilon_t = \pm \left(\xi_p \frac{P_{max}}{P_d'} + \xi_t \frac{T_{max}}{T'} + \xi_v \frac{V_{max}}{V_d'} \right)$$
$$= \pm(0.005 \times 2 + 0.0025 \times 2 + 0) = \pm 1.5\%,$$

$$\varepsilon_v = \pm \left(\xi_q \frac{Q_{max}}{Q_d'} + \xi_s \frac{\omega_{max}}{\omega'} + \xi_v \frac{V_{max}}{V_d'} \right)$$
$$= \pm(0.005 \times 2 + 0 + 0) = \pm 1.0\%.$$

Using these results with Equation (5.16) the confidence interval for the efficiency measurements may be calculated. For the overall efficiency measurement the confidence interval is given by

$$\eta'(1 - \varepsilon) < \eta < \eta'(1 + \varepsilon)$$
$$90.25\% \times (1 - 0.025) < \eta < 90.25\% \times (1 + 0.025)$$
$$87.99\% < \eta < 92.51\%.$$

Similarly, for the torque and volumetric efficiencies the confidence interval is

$$\eta_t'(1 - \varepsilon) < \eta_t < \eta_t'(1 + \varepsilon)$$
$$95\% \times (1 - 0.015) < \eta_t < 95\% \times (1 + 0.015)$$
$$93.58\% < \eta_t < 96.43\%,$$

and

$$\eta_v'(1 - \varepsilon) < \eta_v < \eta_v'(1 + \varepsilon)$$
$$95\% \times (1 - 0.01) < \eta_v < 95\% \times (1 + 0.01)$$
$$94.05\% < \eta_v < 95.95\%.$$

The confidence interval for the overall efficiency of this pump measurement is about 4.5%, which illustrates how crude many of our efficiency measurements are.

Summary. As can be shown by the previous case study, typical pump efficiency measurements are not within the confidence range that most end users believe they are. Therefore, it would be a recommended practice to report the measured efficiency with the confidence interval as defined in Equation (5.16).

5.2.5 Summary

This section discusses the topic of pump efficiency. As noted, in this section the issue of pump efficiency is an important one since pumps are typically used to transmit large amounts of power. Since efficiencies vary over a wide range of operating conditions for the pump it is useful to model pump efficiency using forms that have been presented in Equations (5.5) and (5.6). These forms lend themselves nicely to the method of least square for evaluating the experimental coefficients. Furthermore, it has been shown that pump efficiency measurements carry with them levels of uncertainty that depend on the accuracy of the instrumentation used for making the measurements and the range of the measurements relative to the full-scale reading of the instrument. Last, it should be noted that the efficiency discussion in this section does not depend on a particular pump construction and can be applied to all types of pumps, including gear pumps, vane pump, piston pumps.

5.3 GEAR PUMPS

5.3.1 Overview

The basic construction and operating principle of a gear pump was discussed in Section 5.1. In this introductory section it was noted that gear pumps are popularly used as an inexpensive auxiliary power supply for larger hydraulic systems. In this capacity gear pumps tend to be one of the more widely used pump constructions; and, therefore, a section of this chapter is devoted to the topic of gear pumps. In this section we primarily intend to discuss the gear pump flow characteristics.

Figure 5-9 shows a cross-sectional view of the gear pump. This figure shows two identical gears in mesh with one another, both gears having 15 involute-shaped teeth. The number of gear teeth is an important design parameter and is generally represented in the analysis that

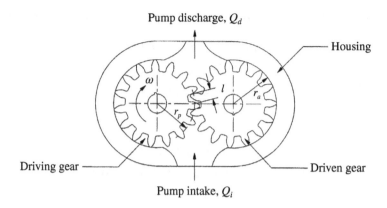

Figure 5-9. Gear pump geometry.

follows by the symbol N. The left gear in Figure 5-9 is the driving gear and the right gear is the driven gear. Both gears rotate in opposite directions at an angular velocity given by ω. As the gears rotate fluid is carried around the outside of the gears within the tooth spaces that are defined by the housing geometry and the gears themselves. This function draws fluid into the pump on the intake side at a volumetric flow rate given by Q_i and expels fluid out of the pump on the discharge side at a volumetric flow rate given by Q_d.

In Figure 5-9 the outside radius of the gear, known as the addendum radius, is given by r_a while the pitch radius is shown by the symbol r_p. In this figure the dimension l describes the instantaneous mesh length of the gear, which is used and discussed further in the following subsection.

5.3.2 Pump Flow Characteristics

As simple as the gear pump of Figure 5-9 may appear, the analysis of its discharge flow is far from trivial. Indeed, the basic equation for describing the instantaneous flow of the gear pump depends on a close evaluation of the meshing geometry of the gear teeth [5]. In the end this analysis results in the following equation for the instantaneous flow rate of the pump:

$$Q_d = w(r_a^2 - r_p^2 - l^2)\omega, \tag{5.17}$$

where w is the face width of the gear teeth into the paper, r_a is the addendum radius, r_p is the pitch radius, l is the instantaneous meshing length shown in Figure 5-9, and ω is the angular velocity of the pump shaft. As the reader may see from this equation and Figure 5-9 the instantaneous meshing length l is the varying parameter that results in a repeating flow ripple for the pump. As it turns out this parameter is closely approximated by

$$l = l_s - \theta r_p \cos(\alpha) \qquad \text{for} \qquad 0 < \theta < 2\pi/N, \tag{5.18}$$

where the starting mesh length is given by

$$l_s = \sqrt{r_a^2 - r_p^2 \cos^2(\alpha)} - r_p \sin(\alpha). \tag{5.19}$$

The starting mesh length l_s is identified when two teeth first make contact within the gear mesh. In Equation (5.18) θ is the angular rotation of the gear pump, α is the pressure angle of the gear set, which is usually between 12° and 24°, and N is the total number of teeth on a single gear. By letting θ equal $2\pi/N$ Equation (5.18) may be used to show that the final mesh length before another set of teeth come in contact with each other is given by

$$l_f = l_s - \frac{2\pi r_p \cos(\alpha)}{N}, \qquad (5.20)$$

where the starting mesh length l_s is given in Equation (5.19). For most practical gear pump designs, the final mesh length l_f is a negative number. To calculate the average discharge flow for the gear pump an integral average of Equation (5.17) will be taken. This result is given by

$$\overline{Q} = \frac{1}{(l_s - l_f)} \int_{l_f}^{l_s} w(r_a^2 - r_p^2 - l^2)\, \omega\, dl = V_d \omega, \qquad (5.21)$$

where V_d is the volumetric displacement of the pump. By evaluating the integral in Equation (5.21) it may be shown that the volumetric displacement of the gear pump is given by

$$V_d = w(r_a^2 - r_p^2 - k^2), \qquad (5.22)$$

where

$$k^2 = \frac{1}{3}\left(\frac{\pi r_p \cos(\alpha)}{N}\right)^2 + \left(\frac{\pi r_p \cos(\alpha)}{N} - l_s\right)^2. \qquad (5.23)$$

Again, for this equation the starting mesh length l_s is given in Equation (5.19). If we assume that the final mesh length l_f is less than zero then it may be shown from Equation (5.17) that the maximum instantaneous pump flow occurs when the mesh length l is zero. Furthermore, if we assume that the magnitude of the starting mesh length is greater than the magnitude of the final mesh length then the minimum instantaneous pump flow occurs at the starting position of the mesh when two pump teeth are just making contact. Using these two conditions the magnitude of the flow ripple amplitude is given by

$$\Delta Q = w l_s^2 \omega. \qquad (5.24)$$

Clearly, the flow ripple amplitude of the gear pump may be reduced by reducing the magnitude of the starting mesh length l_s.

The flow ripple amplitude described in Equation (5.24) is a measure of the gear pump flow variability. Generally, large values of the flow ripple amplitude correspond to vibration and noise problems that are undesirable; therefore, it is advantageous to minimize this quantity by minimizing the starting mesh length l_s. From Equation (5.19) it is not exactly clear how to minimize the starting length without impacting the average volumetric displacement of the pump shown in Equation (5.22). If the recommendations of the American Gear Manufacturing Association (AGMA) are followed for the pump design of Figure 5-9, then the radius of the addendum will be related to the pitch radius according to the following formula:

$$r_a = \frac{(2 + N)}{N} r_p, \qquad (5.25)$$

where N is the total number of teeth on a single gear.[1] Using this result with Equations (5.21) and (5.24), the ratio of the flow ripple amplitude ΔQ to the average pump flow $\overline{Q}$ may be

---

[1] It must be mentioned that this recommendation is for gear transmissions and that gear pump designers are not constrained by such recommendations.

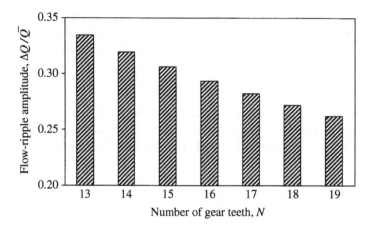

Figure 5-10. Gear pump flow ripple amplitude for $\alpha = 21°$.

calculated for a given number of gear teeth N and for a specified pressure angle α. This calculation is presented graphically in Figure 5-10 for a pressure angle of 21°. From this figure one may ascertain two things: (1) that the flow ripple amplitude is a significant percentage of the average flow rate, and (2) that the flow ripple amplitude decreases as the number of gear teeth increases. Although it is not shown in Figure 5-10, larger pressure angles also tend to decrease the flow ripple amplitude of the pump.

Case Study

A 13-tooth gear pump is manufactured with a pitch radius of 16 mm, an addendum radius of 19 mm and a pressure angle of 21°. The tooth thickness into the paper is 30 mm. Calculate the volumetric displacement of the gear pump. Also, calculate the average flow rate and the flow ripple amplitude of the pump for an input speed of 1800 rpm.

The starting mesh length is an integral part of both calculations that have been requested in this problem. Using Equation (5.19) this calculation is given by

$$l_s = \sqrt{r_a^2 - r_p^2\cos^2(\alpha)} - r_p \sin(\alpha)$$

$$= \sqrt{(19 \text{ mm})^2 - [16 \text{ mm} \times \cos(21°)]^2} - 16 \text{ mm} \times \sin(21°) = 6 \text{ mm}.$$

Using this result with Equation (5.23) it may be shown that $k^2 = 10 \text{ mm}^2$. From Equation (5.22), the volumetric displacement of the pump may be calculated as

$$V_d = w(r_a^2 - r_p^2 - k^2) = 30 \text{ mm} \times [(19 \text{ mm})^2 - (16 \text{ mm})^2 - 10 \text{ mm}^2]$$

$$= 2.85 \frac{\text{cm}^3}{\text{rad}} = 17.91 \frac{\text{cm}^3}{\text{rev}}.$$

The average flow rate of this pump is given by Equation (5.21). With an input speed of 1800 rpm this result is given by

$$\overline{Q} = V_d\omega = 17.91 \quad \text{cm}^3/\text{rev} \times 1800 \text{ rpm} = 32.24 \text{ lpm}.$$

Equation (5.24) may be used to calculate the flow ripple amplitude as

$$\Delta Q = wl_s^2\omega = 30 \text{ mm} \times (6 \text{ mm})^2 \times 1800 \text{ rpm} = 12.21 \text{ lpm}.$$

From this result, it may be seen that the flow ripple amplitude for this pump is nearly 38% of the average flow rate. *Note*: This pump was not designed in accordance with the AGMA recommendation of Equation (5.25) and so the amplitude of the flow ripple is slightly higher than expected.

5.3.3 Pump Control

As shown in Equation (5.21), the average output flow of the gear pump is given by the average volumetric displacement, times the angular input speed of the pump shaft. This result is written as

$$\overline{Q} = V_d\,\omega, \tag{5.26}$$

where the volumetric displacement V_d is given in Equation (5.22). Since the volumetric displacement of the pump is a constant value determined by the geometry of the meshing gears, this pump is called a fixed-displacement pump that displaces a fixed amount of fluid for every revolution of the pump shaft. As such, the only flow control option available for this machine is the adjustment of the pump input speed ω. If a variable speed prime mover is used to drive the pump this control feature is not difficult to implement. On the other hand, if the gear pump is used as an auxiliary power supply to a larger hydraulic system it is usually driven at a constant speed and a relief valve is used to expel any excess fluid flow that is not immediately needed by the application. In other words, the gear pump may be used in both controlled and uncontrolled modes depending on the application.

5.3.4 Summary

This section of the chapter is used to describe the ideal flow characteristics of an external gear pump and to briefly mention the control options that are available for this machine. As shown in Figure 5-10 gear pumps demonstrate large flow variations due to the geometry of the meshing gears; therefore, these pumps typically exhibit larger noise and vibration problems than other positive-displacement pumps of similar size. Since the gear pump is a fixed-displacement machine the output flow is proportional to the input speed. Therefore, the only control option available for this machine is to control the input speed to the pump. Pump speed controls are occasionally used with a fixed-displacement gear pump to control the output of a hydraulic actuator, and this arrangement is discussed in Chapter 8.

5.4 AXIAL-PISTON SWASH-PLATE PUMPS

5.4.1 Overview

The basic construction and operating principle of an axial-piston swash-plate type pump was discussed in Section 5.1. In this introductory section it was noted that these pumps are among the most frequently used machines for providing the main power supply to a hydraulic control system. One reason for the popular use of the swash-plate type pump is that this machine is commonly manufactured in a variable-displacement configuration in which the swash plate is readily adjusted by a control mechanism. Due to the popularity of this machine an entire section of this chapter is devoted to the topic of axial-piston swash-plate type pumps. In this section both the pump flow characteristics and various control designs are discussed. Figure 5-11 shows a cross-sectional view of the axial-piston

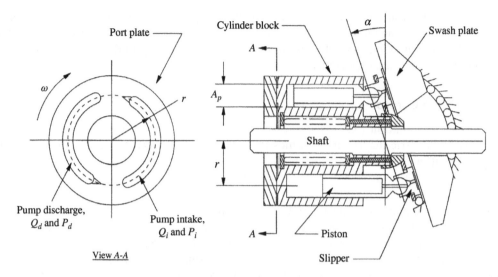

Figure 5-11. Axial-piston swash-plate pump geometry.

swash-plate type pump. This figure shows a sectional view of the pumping elements of the machine as well as a plan view of the port plate noted in Figure 5-11 as View A-A. Relevant features of the pump are identified in Figure 5-11 by noting the cross-sectional area of a single piston A_p, the piston pitch radius r, the swash plate angle α, and the angular velocity of the pump ω. The port plate view identifies the pump discharge port and the intake port. The reader will recall from Section 5.1 that as the pistons pass over the discharge port they advance into the cylinder block and push fluid out of the pump. As the pistons pass over the intake port they withdraw from the cylinder block thereby drawing fluid into the pump. By varying the swash plate angle α the magnitude of the piston stroke is altered, thereby creating a variable fluid displacement for the pump. The total number of pistons within the pump is not explicitly shown in Figure 5-11; however, this integer quantity is noted in the analysis that follows by the symbol N. For most pumps the number of pistons is usually nine; however, seven pistons are occasionally used as well. The reason for using an odd number of pistons within the pump design become apparent in the subsection that follows.

5.4.2 Pump Flow Characteristics

The discharge flow characteristics of an axial-piston pump are similar to those of a gear pump in the fact that both pumps exhibit an oscillatory flow ripple that may be roughly described by the flow ripple amplitude. The flow ripple characteristics of an axial-piston pump have been described in previous literature [6], where it has been noted that the discharge flow rate of a single piston, say the nth piston, is given by

$$Q_n = A_p r \, \tan(\alpha)\omega \, \sin(\theta_n), \qquad (5.27)$$

where A_p is the cross-sectional area of a single piston, r is the piston pitch radius, α is the pump swash plate angle, ω is the angular velocity of the pump shaft, and θ_n is the angular location of the nth piston about the shaft centerline measured relative to the bottom dead center where the piston is pulled out of the cylinder block as far as it can go. Since the pistons are spaced evenly in a circular array about the centerline of the pump shaft the angular position of the nth piston may be written as

$$\theta_n = \theta_1 + \frac{2\pi}{N}(n-1), \tag{5.28}$$

where θ_1 is the position of the reference piston and N is the total number of pistons within the machine. The total instantaneous discharge flow from the pump is then given by

$$Q_d = A_p r \, \tan(\alpha)\omega \sum_{n=1}^{n'} \sin(\theta_n), \tag{5.29}$$

where n' is the total number of pistons that are instantaneously located over the discharge port of the pump shown in Figure 5-11. Using Equation (5.28) and (5.29) with a number of trigonometric identities it may be shown that

$$Q_d = \begin{cases} A_p r \, \tan(\alpha)\omega\frac{1}{2}\csc\left(\frac{\pi}{2N}\right)\cos\left(\theta_1 - \frac{\pi}{2N}\right) & 0 < \theta_1 < \frac{\pi}{N} \quad \text{odd pistons} \\[2mm] A_p r \, \tan(\alpha)\omega \csc\left(\frac{\pi}{N}\right)\cos\left(\theta_1 - \frac{\pi}{N}\right) & 0 < \theta_1 < \frac{2\pi}{N} \quad \text{even pistons} \end{cases}. \tag{5.30}$$

Using this result the integral average of the instantaneous flow rate may be taken to develop an expression for the average flow rate of the pump. Using the even piston expression in Equation (5.30) the average flow rate is given by

$$\overline{Q} = \frac{N}{2\pi} \int_0^{2\pi/N} A_p r \, \tan(\alpha)\omega \csc\left(\frac{\pi}{N}\right)\cos\left(\theta_1 - \frac{\pi}{N}\right) d\theta_1 = V_d\omega, \tag{5.31}$$

where V_d is the volumetric displacement of the pump. By evaluating the integral in Equation (5.31) the volumetric displacement may be expressed as

$$V_d = \frac{N A_p r \, \tan(\alpha)}{\pi}. \tag{5.32}$$

Note: This same result may have been generated by taking the integral average of the odd piston result in Equation (5.30) as well. It is important to notice that the volumetric displacement of the axial-piston pump is dependent on the swash plate angle α. By adjusting this parameter online the volumetric displacement of the pump is made variable, which is one of the most attractive features of using this pump construction. Equation (5.32) describes the volumetric displacement of the pump per radian. To calculate the volumetric displacement per revolution this equation should be multiplied by 2π. Using Equation (5.30), it may be shown that the minimum instantaneous flow rate of the pump is calculated when $\theta_1 = 0$ and the maximum instantaneous flow rate is calculated when θ_1 equals half of the permissible range for each calculation. Using these two conditions the magnitude of the flow ripple amplitude for axial-piston swash-plate type pumps, using an odd or an even number of pistons, is given by

$$\Delta Q = \begin{cases} A_p r \, \tan(\alpha)\omega\frac{1}{2}\tan\left(\frac{\pi}{4N}\right) & \text{odd pistons} \\[2mm] A_p r \, \tan(\alpha)\omega \tan\left(\frac{\pi}{2N}\right) & \text{even pistons} \end{cases}. \tag{5.33}$$

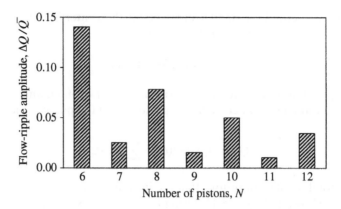

Figure 5-12. Axial-piston pump flow ripple amplitude.

The flow ripple amplitude described in Equation (5.33) is a measure of the axial-piston pump flow variability. Generally, large values of the flow ripple amplitude correspond to vibration and noise problems that are undesirable; therefore, it is advantageous to minimize this quantity as much as possible. By inspection, Equation (5.33) may be used to show that the flow ripple amplitude is generally greater for an axial-piston pump with an even number of pistons as compared to one of similar size with an odd number of pistons. For this reason pumps with an odd number of pistons are used almost exclusively within the axial-piston pump industry. Figure 5-12 shows a bar graph of the piston pump flow ripple amplitude divided by the average volumetric flow rate of the pump. This figure illustrates the points just mentioned. Although it has not been explicitly shown or discussed in this subsection the flow ripple frequency for a piston pump utilizing an odd number of pistons is two times the piston pass frequency, whereas the flow ripple frequency for a pump utilizing an even number of pistons is equal to the piston pass frequency. A comparison of Figures 5-10 and 5-12 shows that the flow ripple amplitude for an axial-piston machine is significantly less than that of gear pump.

5.4.3 Pressure-Controlled Pumps

Overview.   In this subsection of our chapter we consider the very important topic of pressure-controlled pumps. This topic is important for hydraulic control systems since pressure-controlled pumps provide an improvement in efficiency that is needed in order for hydraulic control systems to remain competitive with other control technologies. Since pressure-controlled pumps are usually of the axial-piston swash-plate type design, this discussion is included within this section of our chapter.

Figure 5-13 shows a schematic of the typical pressure-controlled pump that is used to supply power for high-performance hydraulic control systems. As shown on the right-hand side of this figure the pump to be controlled is an axial-piston swash-plate type machine.

The swash plate itself is contained by a cradle bearing system that allows a single degree of rotational freedom about the swash plate pivot with an axis that is normal to the paper. The angle of rotation for the swash plate is shown by the symbol α. The control system for the swash plate is shown in Figure 5-13 by two actuators: a bias actuator on the bottom and a control actuator on the top. The bias actuator is open to the discharge pressure of the pump P_d and is arranged in parallel with a bias spring that is used to force the pump into stroke. This actuator is used to overcome the destroking pressure effects of the pumping torque shown generally in Figure 5-13 by the symbol T. The control actuator near the top of the pump is larger than the bias actuator and is used to destroke the pump with a force that is proportional to the control pressure P_c. As shown on the left-hand side of Figure 5-13 the control pressure P_c is regulated by an open-centered three-way hydromechanical valve similar to the three-way valve shown in Figure 4-12 of Chapter 4. Since the pump is a pressure-controlled pump the hydraulic signal to the control valve contains the pump discharge pressure P_d as shown on the bottom side of the valve. To actuate the valve the discharge pressure acts on the cross-sectional area of the valve A_v to force the valve in the positive x-direction. When this force overcomes the preload of the valve spring the three-way valve is moved away from the null position and the control pressure P_c increases to destroke the pump. By destroking the pump the discharge pressure P_d is reduced, assuming that the load resistance downstream of the pump has not changed. When this happens the discharge pressure reaches its desired value and the valve/pump system seeks a steady operating point, thus the control objective of the pump is satisfied. In the paragraphs that follow this system is analyzed and discussed in more detail.

Steady Analysis and Pump Design. Before a dynamic analysis of the pump shown in Figure 5-13 may be carried out it is important to design the basic actuation system that will be used to control the pump. As with most design problems this task is performed by considering the steady-state operating condition of the swash plate. By summing moments

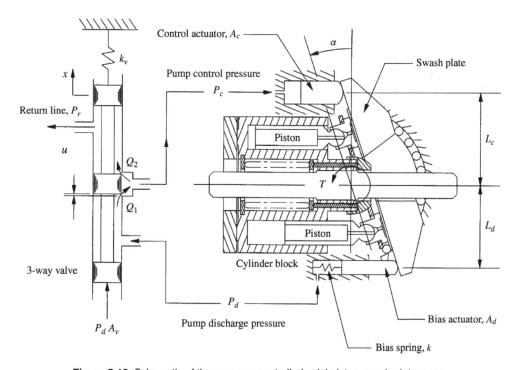

Figure 5-13. Schematic of the pressure-controlled axial-piston swash-plate pump.

about the swash plate pivot point in Figure 5-13 and setting them equal to zero it may be shown that

$$0 = T + F_d L_d - F_c L_c, \tag{5.34}$$

where T is the net torque induced on the swash plate by the pumping elements of the pump, F_d is the force exerted on the swash plate by the bias actuator, F_c is the force exerted on the swash plate by the control actuator, and L_d and L_c are the moment arms for each actuator shown in Figure 5-13.

The swash plate torque T has been the topic of considerable research within the past 20 years [7–9]. This quantity is indeed complicated to evaluate; however, it has been shown that a very good approximation for this torque is given by

$$T = C_i \alpha - C_p (P_d - P_i), \tag{5.35}$$

where

$$C_i = \frac{N M_p r^2 \omega^2}{2} \quad \text{and} \quad C_p = \frac{N A_p r \gamma}{2\pi}. \tag{5.36}$$

In this expression N is the total number of pistons within the pump, M_p is the mass of a single piston/slipper assembly, r is the piston pitch radius shown in Figure 5-11, ω is the angular velocity of the pump shaft, α is the pump swash plate angle, A_p is the cross-sectional area of a single piston shown in Figure 5-11, P_d and P_i are the discharge and intake pressures of the pump respectively, and γ is the pressure carry over angle on the port plate, which is discussed shortly. The first term in Equation (5.35) describes the reciprocating inertia of each piston/slipper assembly that has a net effect of forcing the swash plate into stroke. For high operating pressures of the machine this term is typically much smaller than the second term in Equation (5.35), which describes the swash plate torque resulting from the skewed pressure profile on the port plate. The skewed pressure profile is quantified by the pressure carry over angle γ, which is shown in the schematic of Figure 5-14.

In Figure 5-14 the pressure profile shown is used to illustrate the fluid pressure that is felt by a single piston as it passes over the ports on the port plate. From this schematic

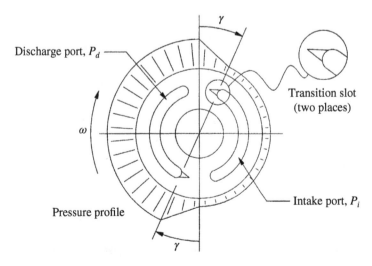

Figure 5-14. Port plate pressure profile.

it may be seen that each piston sees a constant pressure as it passes directly over either the intake or the discharge port and that it undergoes a transition in pressure as it passes over the transition slots on the port plate. This transition occurs through an average angular distance, which is noted in Figure 5-14 by the symbol γ, and which is commonly referred to as the pressure carry over angle on the port plate. As it turns out the pressure carry over angle is dependent on the operating pressures of the machine $P_d - P_i$ and the rotational speed of the pump ω and tends to grow as each of these quantities increases. The pressure carry over angle is also heavily dependent on the port plate geometry and may be particularly shaped by altering the characteristics of the port plate transition slots [10]. For initial design purposes it is useful to consider the pressure carry over angle as a constant that describes the dominant operating condition of the pump. For practical pump designs it has been observed that a reasonable range for the pressure carry over angle is given by

$$10° < \gamma < 24°,\tag{5.37}$$

depending on the port plate design. For Equation (5.36) γ must be specified in radians.

Now that the swash plate torque T has been quantified it is necessary to consider the forces exerted on the swash plate by the bias actuator and the control actuator of the pump. By inspection of Figure 5-13 it may be seen that the linearized force exerted on the swash plate by the bias actuator is given by

$$F_d = A_d P_d + \underbrace{F_o - kL_d\alpha}_{\text{Bias spring force}},\tag{5.38}$$

where A_d is the cross-sectional area of the bias actuator shown in Figure 5-13, P_d is the discharge pressure of the pump, F_o is the preload on the bias spring when $\alpha = 0$, and k is the rate of the bias spring. Similarly, the force exerted on the swash plate by the control actuator is

$$F_c = A_c P_c,\tag{5.39}$$

where A_c is the cross-sectional area of the control actuator shown in Figure 5-13 and P_c is the control pressure provided by the three-way control valve.

At this point all of the quantities of effort shown in the steady result of Equation (5.34) have been identified. Substituting Equations (5.35), (5.38), and (5.39) into Equation (5.34) the following equation for the moments acting on the swash plate may be written:

$$A_c L_c P_c - (A_d L_d - C_p)P_d = (C_i - kL_d^2)\alpha + F_o L_d,\tag{5.40}$$

where the intake pressure of the pump P_i has been assumed to be zero, which is a common inlet condition for a pressure-controlled pump. As it turns out, for reasonable operating pressures of the pump the left-hand side of Equation (5.40) is much, much greater than the right-hand side. This means that the forces from the bias spring and the reciprocating inertia of the pistons are negligible when compared to the pressure terms of the actuator and the stroke decreasing effects of the swash plate torque T. If it is assumed that the pump is operating at a sufficiently high pressure the right-hand side of Equation (5.40) may be neglected and the swash plate moment equation may be further reduced to the following result:

$$P_c = \frac{(A_d L_d - C_p)}{A_c L_c}P_d.\tag{5.41}$$

This result tells us what the control pressure needs to be for a given discharge pressure of the pump. Since the control pressure is provided by an open-centered three-way valve, and since the supply pressure to the valve is the discharge pressure of the pump, it is helpful from a valve design point of view to design the control pressure so that it equals half of the discharge pressure. This design condition may be summarized as:

$$\frac{1}{2} = \frac{(A_d L_d - C_p)}{A_c L_c}, \qquad P_c = \frac{1}{2} P_d. \tag{5.42}$$

Again, since C_p is a function of the pressure carry over angle γ this design condition needs to be satisfied at the dominant operating condition of the pump. In the dynamic analysis that follows we assume that Equation (5.42) is satisfied near the nominal operating conditions of the pressure-controlled pump shown in Figure 5-13.

Dynamic Analysis and Pump Response.   To consider the dynamic operation of the pressure-controlled pump shown in Figure 5-13 we begin by considering the parameter that is to be controlled; namely, the discharge pressure of the pump P_d. The governing equation for the pump discharge pressure may be derived by using the standard pressure-rise-rate-equation presented in Subsection 2.6.3 of Chapter 2. This equation assumes that the fluid in the discharge line is slightly compressible and that the flow in and out of the discharge line may be instantaneously different. For the pump application of Figure 5-13 the discharge pressure may be modeled as:

$$\frac{V}{\beta} \dot{P}_d + K P_d = Q - Q_{sys}, \tag{5.43}$$

where V is the volume of fluid in the discharge line, β is the effective bulk modulus of the discharge line, K is the leakage coefficient of the pump, Q is the volumetric flow rate from the pump into the discharge line, and Q_{sys} is the volumetric flow rate being drawn out of the discharge line by the downstream hydraulic system. For steady-state operating conditions of the pump Equation (5.43) may be used to show

$$Q_{sys} = Q_o - K P_{d_o}, \tag{5.44}$$

where Q_o is the nominal flow rate of the pump and P_{d_o} is the nominal discharge pressure. Substituting this result into Equation (5.43) produces the governing equation for the discharge pressure of the pump:

$$\frac{V}{\beta} \dot{P}_d + K(P_d - P_{d_o}) = Q - Q_o. \tag{5.45}$$

Since pressure-controlled pumps generally operate using a fixed rotational speed of the input shaft, it may be shown that

$$Q - Q_o = G_p(\alpha - \alpha_o), \tag{5.46}$$

where G_p is the pump gain given by

$$G_p = \frac{N A_p r \omega}{\pi}. \tag{5.47}$$

The parameters of this equation have been previously defined. It should be noted that Equation (5.46) has been linearized for small values of the swash plate angle α and that α_o is the nominal swash plate angle of the pump. Using Equations (5.45) and (5.46) the governing equation for the pump discharge pressure may be finally written as

$$\frac{V}{\beta}\dot{P}_d + K(P_d - P_{d_o}) = G_p(\alpha - \alpha_o). \tag{5.48}$$

From this equation it is clear that adjustments in the swash plate angle α may be used to adjust the discharge pressure P_d. In the following paragraph the dynamics of the swash plate angle are considered.

Since the swash plate's time rate of change of angular momentum is extremely small compared to the moments acting on the swash plate, the second-order dynamics of the swash plate may be safely neglected without loss of accuracy in our modeling process. Furthermore, since the volume of fluid in the control actuator of Figure 5-13 is very small the transient effects of this control pressure may also be safely neglected. Therefore, from Figure 5-13 it may be shown using the conservation of mass for an incompressible fluid that

$$0 = Q_1 - Q_2 + A_c L_c \dot{\alpha}, \tag{5.49}$$

where Q_1 and Q_2 are the volumetric flow rates of the three-way valve and where the other parameters are shown in Figure 5-13. From the open-centered three-way valve analysis of Chapter 4 it may be shown that

$$Q_1 - Q_2 = 2K_q x, \tag{5.50}$$

where K_q is the valve flow gain and x is the valve displacement shown in Figure 5-13. *Note*: This result has assumed that the design conditions of Equation (5.42) have been satisfied and that the control pressure P_c is approximately half of the discharge pressure P_d. From Equations (5.49) and (5.50), the governing equation for the dynamics of the swash plate may be written as

$$\dot{\alpha} = -\frac{2K_q}{A_c L_c} x. \tag{5.51}$$

From this result it may be seen that positive displacements of the control valve x may be used to decrease the swash plate angle α. In the next paragraph an expression for the valve displacement is developed.

From Figure 5-13 it may be seen that the valve is forced in the positive x-direction by the discharge pressure of the pump while the valve is forced in the negative x-direction by the valve spring. The reader will also recall from Chapter 4 that the three-way valve experiences both transient and steady flow forces. While the steady flow forces tend to close the valve the transient flow forces may operate in either direction depending on their sign. Since the linear momentum and viscous drag of the valve is small compared to these forces the governing equation of motion for the three-way spool valve may be written as

$$k_v x = A_v P_d + F_x - F_{v_o}, \tag{5.52}$$

where k_v is the rate of the valve spring, A_v is the cross-sectional area of the spool valve, F_x is the flow force acting on the valve, and F_{v_o} is the valve spring preload acting on the valve when $x = 0$. From Equation (4.72) of Chapter 4 the steady flow force exerted on the open-centered three-way valve may be written as

$$F_x = -2K_{f_q} x, \tag{5.53}$$

where once again the design conditions of Equation (5.42) are assumed. In this equation K_{f_q} is the flow force gain of the valve. At the nominal operating conditions of the valve Equation (5.52) may be used to show that the valve spring preload is given by

$$F_{v_o} = A_v P_{d_o},$$
(5.54)

where P_{d_o} is the nominal or desired operating pressure of the pump. Substituting Equations (5.53) and (5.42) into Equation (5.52) produces the following equation of motion for the three-way spool valve:

$$x = \frac{A_v}{(k_v + 2K_{f_q})}(P_d - P_{d_o}).$$
(5.55)

This equation shows that adjustments to the spool valve position are made by a sensed error in the discharge pressure.

Using Equations (5.48), (5.51), and (5.55) the dynamic equations for the pressure-controlled pump of Figure 5-13 may be summarized as:

$$\dot{P}_d = -\frac{\beta K}{V}(P_d - P_{d_o}) + \frac{\beta G_p}{V}(\alpha - \alpha_o),$$

$$\dot{\alpha} = -\frac{2K_q A_v}{A_c L_c(k_v + 2K_{f_q})}(P_d - P_{d_o}).$$
(5.56)

From these equations the characteristic equation for the dynamic system may be written using either state space methods or the Laplace transform. This result is given by

$$a_o s^2 + a_1 s + a_2 = 0,$$
(5.57)

where the coefficients are

$$a_o = \frac{V}{\beta}, \quad a_1 = K, \quad a_2 = \frac{2K_q A_v G_p}{A_c L_c(k_v + 2K_{f_q})}.$$
(5.58)

From the characteristic equation the undamped natural frequency and the damping ratio of the pressure-controlled pump may be written as

$$\omega_n = \sqrt{\frac{2K_q A_v G_p \beta}{A_c L_c(k_v + 2K_{f_q})V}} \quad \text{and} \quad \zeta = \frac{K}{2}\sqrt{\frac{A_c L_c(k_v + 2K_{f_q})\beta}{2K_q A_v G_p V}}.$$
(5.59)

These results may be used to calculate the rise time, settling time, and maximum percent overshoot to be expected from the pressure-controlled pump design. Figure 5-15 shows a block diagram for this system. In this diagram the desired swash plate angle determines that amount of flow needed from the pump and sudden changes in this parameter may be viewed as a disturbance to the pressure controlled system.

Case Study

A pressure-controlled pump similar to Figure 5-13 is designed for a maximum displacement of 45 cm³/rev. The desired discharge pressure of the pump is 20 MPa and the pump shaft speed is 1800 rpm. From this information the pump gain may be calculated as $G_p = 4.15 \times 10^{-3}$ m³/s. The cross-sectional area of the control actuator is 3.76 cm² and the moment arm for this actuator is 6.75 cm. The leakage coefficient for the pump is

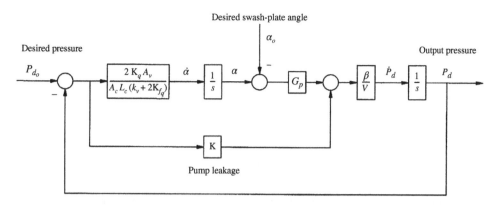

Figure 5-15. Block diagram of the pressure-controlled pump in Figure 5-13.

3.38×10^{-12} m³/(Pa s), the effective fluid bulk modulus is 0.8 GPa, and the discharge volume is 1 liter. The three-way control valve has a flow gain equal to 0.19 m²/s and a flow force gain equal to 5.76 N/mm. The spring rate for the valve is 56.55 N/mm and the cross-sectional area of the valve is 28.27 mm². Calculate the undamped natural frequency and the damping ratio for this system. Do you expect this system to exhibit overshoot and oscillation during a transient response? Why or why not? What is the bandwidth frequency of this system? What is the system rise time, settling time, and maximum percent overshoot?

The natural frequency and damping ratio for the system may be determined using Equation (5.59). For this problem these parameters are calculated as

$$\omega_n = \sqrt{\frac{2K_q A_v G_p \beta}{A_c L_c (k_v + 2K_{f_q}) V}}$$

$$= \sqrt{\frac{2 \times 0.19 \text{ m}^2/\text{s} \times 28.27 \text{ mm}^2 \times 4.15 \times 10^{-3} \text{m}^3/\text{s} \times 0.8 \text{ GPa}}{3.76 \text{ cm}^2 \times 6.75 \text{ cm} \times (56.55 + 2 \times 5.76) \text{ N/mm} \times 1 \text{ liter}}}$$

$$= 142.87 \frac{\text{rad}}{\text{s}} = 22.74 \text{ Hz}$$

$$\zeta = \frac{K}{2} \sqrt{\frac{A_c L_c (k_v + 2K_{f_q}) \beta}{2K_q A_v G_p V}} = \frac{3.38 \times 10^{-12} \text{ m}^3/(\text{Pa s})}{2}$$

$$\times \sqrt{\frac{3.76 \text{ cm}^2 \times 6.75 \text{ cm} \times (56.55 + 2 \times 5.76) \text{N/mm} \times 0.8 \text{ GPa}}{2 \times 0.19 \text{ m}^2/\text{s} \times 28.27 \text{ mm}^2 \times 4.15 \times 10^{-3} \text{m}^3/\text{s} \times 1 \text{ liter}}}$$

$$= 0.01.$$

Since the damping ratio is less than unity, the pressure-controlled pump will exhibit oscillation and overshoot during a transient response. From Chapter 3, the bandwidth frequency of the pump response may be calculated as

$$\omega_b = \omega_n \frac{2 + \sqrt{2 - 8\zeta^2}}{2 + 8\zeta^2} = 22.74 \text{ Hz} \times \frac{2 + \sqrt{2 - 8 \times 0.01^2}}{2 + 8 \times 0.01^2} = 38.8 \text{ Hz}.$$

Also from Chapter 3 the rise time, settling time, and maximum percent overshoot may be calculated as

$$t_r = \frac{\pi - \arctan\left(\sqrt{1-\zeta^2}/\zeta\right)}{\omega_n\sqrt{1-\zeta^2}} = \frac{\pi - \arctan\left(\sqrt{1-0.01^2}/0.01\right)}{142.87 \text{ rad}/s \times \sqrt{1-0.01^2}} = 11 \text{ ms}$$

$$t_s = \frac{4}{\zeta\omega_n} = \frac{4}{0.01 \times 142.87 \text{ rad}/s} = 3 \text{ s}$$

$$M_p = \text{Exp}\left(\frac{-\pi\zeta}{\sqrt{1-\zeta^2}}\right) \times 100\% = \text{Exp}\left(\frac{-\pi \times 0.01}{\sqrt{1-0.01^2}}\right) \times 100\% = 97\%.$$

These numbers describe a typical pressure-controlled pump response. To reduce the settling time and maximum percent overshoot, the damping ratio may be increased by increasing pump leakage.

Summary.   In this subsection the design and control of the pressure-controlled pump of Figure 5-13 has been considered. A guideline for designing the actuation system for this machine has been presented in Equation (5.42) where the control pressure is designed to be exactly half of the pump discharge pressure. This result assumes that the pump is operating at a sufficiently large pressure so as to make the bias spring and reciprocating inertia effects negligible in comparison to the pressure effects. In order to satisfy Equation (5.42) the port plate needs to be designed to provide a pressure carry over angle that is acceptable for the dominant operating point of the pump. Port plate design is beyond the scope of this text; however, the reader may refer to previous work [10] for a broader consideration of this topic. By considering the dynamics of the control system shown in Figure 5-13, it is clear that this system is dominantly characterized as an underdamped second-order system with a natural frequency and damping ratio given in Equation (5.59). As shown in this equation the damping ratio is proportional to the pump leakage coefficient K, which tends to reduce the maximum percent overshoot and the settling time of the dynamic response. In many applications leakage is added to the system of the pressure-controlled pump for the purposes of improving the dynamic response characteristics. Though this leakage tends to make the pump response more favorable it also reduces the overall operating efficiency of the machine. Therefore, a reasonable balance between efficiency and dynamic response must be sought in order to provide an acceptable pressure-controlled pump for a given application.

5.4.4   Displacement-Controlled Pumps

Overview.   In this subsection of our chapter we wish to consider the important topic of displacement-controlled pumps. This topic is important for hydraulic control systems since displacement-controlled pumps provide a method for the direct control of a hydraulic actuator without the use of an intermediate valve. Since intermediate valves are always an energy drain for hydraulic control systems, displacement-controlled pumps provide a means for improved efficiency. Systems that are directly controlled by a displacement-controlled pump are discussed at length in Chapter 8. Since displacement-controlled pumps are usually of the axial-piston swash-plate type design, a discussion of these pumps is included within this section of our chapter.

Figure 5-16 shows a schematic of the typical displacement-controlled pump that is used to supply power for high-performance hydraulic control systems. As shown on the right-hand

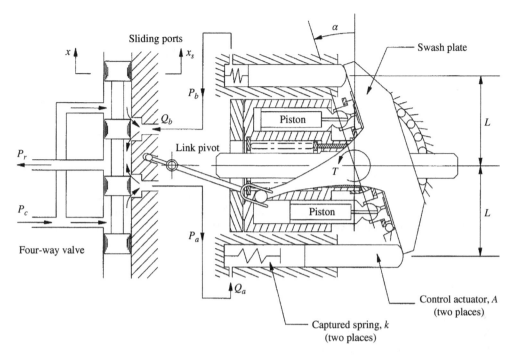

Figure 5-16. Schematic of the displacement-controlled axial-piston swash-plate pump.

side of this figure, the pump to be controlled is an axial-piston swash-plate type machine. The swash plate itself is contained by a cradle bearing system that allows for a single degree of rotational freedom about the swash plate pivot with an axis that is normal to the paper. The angle of rotation for the swash plate is shown by the symbol α and due to the symmetry of this pump design the swash plate angle may be positive or negative. A positive swash plate angle pumps fluid out of one pump port, while a negative swash plate angle pumps fluid out of the other pump port. This ability to pump fluid both ways is an important feature of the displacement-controlled pump. The control system for the swash plate is shown in Figure 5-16 by two identical actuators on the top and bottom of the figure. Each actuator is equipped with a captured spring that only exerts a force on the spring-compressing actuator.

As shown in Figure 5-16 positive swash plate angles tend to compress the top actuator spring while the bottom spring remains captured and inactive. Negative swash plate angles would tend to compress the bottom spring while leaving the top spring inactive. The fluid pressure within each actuator is controlled by the four-way spool valve shown on the left-hand side of Figure 5-16. This spool valve is supplied by an auxiliary pressure source that is also commonly used to charge the inlet side of the pump. This supply pressure and pump inlet pressure is known as the charge pressure and is shown in Figure 5-16 by the symbol P_c. The return pressure from the valve P_r is typically a reservoir pressure with a gauge value of zero. The valve ports that supply flow from the four-way valve to the actuation system are shown in Figure 5-16 to slide up and down by the dimension x_s based on the swash plate angle feedback that occurs through the linkages shown. These linkages are typical four-bar linkages with members including the swash plate, the link, the sliding ports, and the ground. The linkage feedback is used to generate a swash plate angle error signal to the valve; this signal is discussed later. By moving the four-way valve in the positive x-direction fluid is allowed to flow into the bottom actuator and out of the top actuator in Figure 5-16.

This function may be used to force the pump into a positive stroke, and thereby create a positive flow rate from the pump. By moving the four-way valve in the negative x-direction fluid is allowed to flow into the top actuator and out of the bottom actuator. This function may be used to force the pump into a negative stroke, thereby creating a negative flow rate from the pump. Once the four-way valve moves in either direction the pump reacts through the feedback linkage to move the sliding ports into a position that matches the spool valve location. When this occurs the error signal to the pump vanishes and a steady operating condition of the pump is achieved. In the paragraphs that follow the displacement-controlled pump is analyzed and discussed in more detail.

Steady Analysis and Pump Design.   Before a dynamic analysis of the pump shown in Figure 5-16 is carried out it is important to design the basic actuation system that is used to control the pump. As with most design problems, this task is carried out by considering the steady-state operating condition of the swash plate. By summing moments about the swash plate pivot point in Figure 5-16 and setting them equal to zero it may be shown that

$$0 = T + F_a L - F_b L, \tag{5.60}$$

where T is the net torque induced on the swash plate by the pumping elements of the pump, F_a is the force exerted on the swash plate by the bottom actuator in Figure 5-16, F_b is the force exerted on the swash plate by the top actuator in Figure 5-16, and L is the moment arm for each actuator shown in this figure.

The swash plate torque T was presented and discussed in Equations (5.35) through (5.37). These results may be used and substituted in Equation (5.60) for the purposes of analyzing the displacement-controlled pump. By inspection of Figure 5-16 it may be seen that the force exerted on the swash plate by the bottom actuator is given by

$$F_a = AP_a, \tag{5.61}$$

where A is the cross-sectional area of the actuator and P_a is the fluid pressure within the bottom actuator provided by the four-way control valve. Similarly, the linearized force exerted on the swash plate by the top actuator in Figure 5-16 is given by

$$F_b = AP_b + \underbrace{F_o + kL\alpha}_{\text{Spring Force}}, \tag{5.62}$$

where once again A is the cross-sectional area of the actuator, P_b is the fluid pressure within the top actuator provided by the four-way control valve, F_o is the assembled preload on the captured spring when $\alpha = 0$, and k is the rate of the captured spring.

At this point all of the quantities of effort shown in the steady result of Equation (5.60) have been identified. Substituting Equations (5.35), (5.61), and (5.62) into Equation (5.60), the following equation for the moments acting on the swash plate may be written:

$$C_p(P_d - P_c) - AL(P_a - P_b) = (C_i - kL^2)\alpha - F_o L, \tag{5.63}$$

where the intake pressure of the pump P_i has been set equal to the charge pressure of the system P_c. As it turns out, for reasonable operating pressures of the pump the left-hand side of Equation (5.63) is much, much greater than the right-hand side. This means that the forces from the captured spring and the reciprocating inertia of the pistons are negligible when compared to the pressure terms of the actuator and the stroke decreasing effects of the swash plate torque T. If it is assumed that the pump is operating at a sufficiently high pressure the right-hand side of Equation (5.63) may be neglected and the swash plate moment equation for the displacement-controlled pump may be further reduced to the following result:

$$(P_a - P_b) = \frac{C_p}{AL}(P_d - P_c). \tag{5.64}$$

This result tells us what the difference in actuator pressures needs to be for a given discharge pressure of the pump. This equation must be satisfied for all high-pressure operating conditions of the pump and may be used to appropriately size the actuator system shown by the product $A\,L$.

Dynamic Analysis and Pump Response. To analyze the dynamic response of the displacement-controlled pump we begin by considering the flow rates in and out of the control actuators. These flow rates are shown in Figure 5-16 by the symbols Q_a and Q_b. Since the volume of fluid between the four-way control valve and the pump actuation system is small, the pressure transients in each actuator may be neglected and it may be shown that

$$Q_a = Q_b = AL\dot{\alpha}, \tag{5.65}$$

where the positive sign convention for these flow rates is shown by the directional arrows in Figure 5-16. The four-way valve in this figure is shown to be an open-centered valve with a supply pressure given by the charge pressure P_c. If it is assumed that nominal pressure in each actuator is half of the charge pressure, the linearized flow analysis of the four-way valve may be used to show that the volumetric flow rates in and out of the actuators are given by

$$\begin{aligned}
Q_a &= 2K_q(x - x_s) - 2K_c(P_a - P_c/2) \\
Q_b &= 2K_q(x - x_s) + 2K_c(P_b - P_c/2),
\end{aligned} \tag{5.66}$$

where K_q is the flow gain, K_c is the pressure flow-coefficient, x is the absolute displacement of the spool, and x_s is the absolute displacement of the sliding ports shown in Figure 5-16. Using the geometry of the feedback links shown in Figure 5-16 it may be shown that

$$x_s = G_c\alpha, \tag{5.67}$$

where G_c is the control gain and small angles of the swash plate angle have been assumed. Using the results of Equations (5.64) through (5.67) the dynamic equation of motion for the swash plate may be written as

$$\frac{AL}{2K_q}\dot{\alpha} + G_c\alpha = x - \frac{C_p}{ALK_p}(P_d - P_c), \tag{5.68}$$

where K_p is the pressure sensitivity of the four-way control valve. As it turns out, due to the large pressure sensitivity of the four-way valve, the pressure term on the right-hand side of Equation (5.68) is very small compared to other terms. By neglecting this term the equation of motion for the swash plate may be written as

$$\tau\dot{\alpha} + \alpha = \psi x, \tag{5.69}$$

where the time constant and the constant of proportionality between the swash plate angle and the spool position are given respectively as

$$\tau = \frac{AL}{2K_q G_c} \quad \text{and} \quad \psi = \frac{1}{G_c}. \tag{5.70}$$

For a quick response and a high-frequency bandwidth it is desirable to make the time constant τ as small as possible. From Equation (5.70) it may be shown that small time constants may be achieved by using small actuators with a high flow gain valve and a high control gain G_c. From a steady perspective the constant of proportionality or the control gain must be designed to match the maximum swash plate angle with the maximum spool displacement.

Case Study

A displacement-controlled pump similar to the one shown in Figure 5-16 is designed with a maximum volumetric displacement of 75 cm³/rev. The actuator area is given by 18.38 cm² and the actuator moment arm is 8 cm. The feedback linkage provides a control gain for the sliding port geometry of 0.167 mm/deg. The flow gain for the four-way spool valve is 0.0616 m²/s. In order to assess the pump's dynamic response calculate the system time constant, the settling time, and the bandwidth frequency.

The time constant for the displacement-controlled pump is given in Equation (5.70) and may be calculated from the problem statement as

$$\tau = \frac{AL}{2K_q G_c} = \frac{18.38 \text{ cm}^2 \times 8 \text{ cm}}{2 \times 0.0616 \text{ m}^2/\text{s} \times 0.167 \text{ mm/deg}} = 125 \text{ ms.}$$

From this calculation it may be determined that the settling time for the first-order response is given by 0.5 seconds and the bandwidth frequency is 1.27 Hz. These are typical response characteristics for displacement-controlled pumps similar to the one shown in Figure 5-16.

Summary.   In this analysis the time-rate-of-change of angular momentum and the viscous damping for the swash plate has been neglected. This is due to the fact that these terms are negligible when compared to the actuator forces that are exerted on the swash plate. Furthermore, the displacement-controlled pump has not required an analysis of the discharge pressure transients as the discharge pressure has a negligible impact on the transient response of the displacement-controlled pump. In a word, the response of the displacement-controlled pump is very nice as it exhibits first-order behavior without overshoot or oscillation. Furthermore, the steady-state error for the displacement-controlled pump vanishes as time approaches infinity. Last, it should be noted that the position of the spool valve in Figure 5-16 is typically controlled using a two-stage electrohydraulic valve similar to what is shown in Figure 4-26 of Chapter 4. This allows for the displacement of the pump to be controlled electronically, thus facilitating a flexible control structure for this pump when it is applied to advanced applications.

5.4.5   Summary

This section of the chapter discusses the flow and control characteristics of axial-piston swash plate type pumps. In this section it is shown that axial-piston pumps exhibit a lower amplitude of flow pulsation when compared to gear pumps and that axial-piston pumps with an odd number of pistons exhibit a lower amplitude of flow pulsation than pumps with an even number of pistons. For this reason axial-piston pumps are usually designed with an odd number of pistons. In this section the control of a swash-plate type pump is also discussed. In particular it has been shown that pressure-controlled pumps exhibit second-order characteristics in which they produce overshoot and oscillation before reaching a steady-state condition. On the other hand, it is shown in this part of the chapter that displacement-controlled pumps are generally first-order systems, which exhibit no overshoot and oscillation, but, rather, these pumps asymptotically approach the desired operating point at a rate that is governed by the system time constant.

5.5 CONCLUSION

In conclusion, this chapter is written to acquaint the reader with hydraulic pump technologies that are available for use within hydraulic control systems. As shown within the introductory remarks of this chapter there are several machine types that can be used for supplying hydraulic power to a hydraulic control system. As mentioned in subsequent sections of this chapter the two pump types that tend to dominate this field are gear pumps and swash-plate type axial-piston pumps. Gear pumps are generally used for lower performance applications in which a great deal of pump variability is not needed and where the operating pressures are reasonably low (below 20 MPa). On the other hand, axial-piston swash-plate pumps are used in high-performance applications in which either pressure or displacement control may be desired and/or where operating pressures are much higher. In this chapter the efficiencies of these machines are discussed and the flow ripple characteristics for both gear pumps and axial-piston pumps are presented. Typical control issues for these machines are also visited. These aspects of pump control are more thoroughly utilized in Chapters 7 and 8 where pumps are used to supply hydraulic power for both valve-controlled and pump-controlled hydraulic systems.

5.6 REFERENCES

[1] Lambeck R. P. 1983. *Hydraulic Pumps and Motors: Selection and Application for Hydraulic Power Control Systems*. Marcel Dekker, New York.

[2] Merrit H. E. 1967. *Hydraulic Control Systems*. John Wiley & Sons, New York.

[3] Shi X., and N. D. Manring. 2001. A torque efficiency model for an axial-piston swash-plate type, hydrostatic pump. *Power Transmission and Motion Control* (PTMC 2001), C. R. Burrows and K. A. Edge (Eds.). Antony Rowe Ltd., Great Britain, pp. 3–20.

[4] Doebelin, E. O. 1990. *Measurement Systems: Application and Design*, 4th ed. McGraw-Hill, New York.

[5] Manring, N. D., and S. B. Kasaragadda. 2003. The theoretical flow ripple of an external gear pump. ASME *Journal of Dynamic Systems, Measurement, and Control* 125: 396–404.

[6] Manring, N. D. 2000. The discharge flow ripple of an axial-piston swash-plate type hydrostatic pump. ASME *Journal of Dynamic Systems, Measurement, and Control* 122: 263–268.

[7] Zeiger, G., and A. Akers. 1985. Torque on the swashplate of an axial-piston pump. ASME *Journal of Dynamic Systems, Measurement, and Control* 107: 220–226.

[8] Manring, N. D., and R. E. Johnson. 1996. Modeling and designing a variable displacement open-loop pump. ASME *Journal of Dynamic Systems, Measurement, and Control* 118: 267–271.

[9] Manring, N. D. 1999. The control and containment forces on the swash plate of an axial-piston pump. ASME *Journal of Dynamic Systems, Measurement, and Control* 121: 599–605.

[10] Manring, N. D. 2003. Valve-plate design for an axial-piston pump operating at low displacements. ASME *Journal of Mechanical Design*. Technical Brief. 125: 200–205.

5.7 HOMEWORK PROBLEMS

5.7.1 Pump Efficiency

5.1 A gear pump has been designed to displace 19 cm³/rev. When the pump is operating at 14 MPa the torque efficiency is measured to be 92%. What is the input torque to the pump?

5.2 An axial-piston pump is designed to displace 100 cm³/rev. The following volumetric efficiencies have been measured while this pump is operating at 1800 rpm:

Pressure	η_v
2 MPa	0.99
20 MPa	0.95
40 MPa	0.91

From this data calculate the average coefficient of leakage for the pump.

5.3 The overall efficiency of an axial-piston pump is calculated to be 0.89 from taking measurements of torque, speed, pressure, and volumetric flow. The measurement instruments are all being used at half of their full-scale reading. The torque and pressure measurements are accurate within ±1.5% of their full-scale reading while the flow measurements are accurate within ±3.0% of its full-scale reading. The speed reading is known with near perfect accuracy. What is the overall uncertainty within the measurement? What is the confidence interval for the pump efficiency number?

5.4 You have just taken the following pump measurements in the lab: The discharge pressure is 3000 psi, the flow rate is 20.36 gpm, the input speed is 1800 rpm, and the torque on the shaft is 1382 in-lbf. The volumetric displacement of the pump is 2.75 in³/rev. What is the overall, the volumetric, and the torque efficiency of the pump?

5.5 You are given the following information from the instrument manufacturer regarding the accuracy of each instrument.

Instrument description	Full-scale measurement capability	Accuracy (as a percentage of full-scale readings)
Pressure measurement	6000 psi	1.5%
Flow measurement	30 gpm	2.5%
Torque measurement	5000 in-lbf	1.5%
Speed measurement	10000 rpm	0.0%

Using the data of the previous problem what is the range of uncertainty associated with the overall efficiency measurement? What is the efficiency confidence interval?

5.7.2 Gear Pumps

5.6 A gear pump is designed with 17 teeth and a pressure angle of 20 degrees. The pitch radius is 1.6 cm and the addendum radius is 1.86 cm. Calculate the necessary width of the gear teeth in order to create a pump with an average volumetric displacement of 15 cm^3/rev. What is the average volumetric flow rate of this machine for an input speed of 1800 rpm. What is the magnitude of the flow ripple for this operating condition?

5.7 For the results of Problem 5.6 change the number of teeth to 13 keeping everything else the same. What has happened to the volumetric displacement of the pump? Has it gone up or gone down? What has happened to the flow ripple amplitude?

5.7.3 Axial-Piston Swash-Plate Pumps

5.8 An axial-piston pump utilizes seven pistons and a maximum swash-plate angle of 18 degrees. The pitch radius is 2.55 cm and the diameter of each piston is 1.66 cm. Calculate the volumetric displacement of this pump per revolution. What is the average volumetric flow rate and the amplitude of the flow ripple for this pump assuming that the pump speed is 2000 rpm? What is the pump gain G_p at this operating speed?

5.9 An axial-piston pump utilizes nine pistons and a maximum swash-plate angle of 18 degrees. The pitch radius is 2.78 cm and the diameter of each piston is 1.40 cm. Calculate the volumetric displacement of this pump per revolution. What is the average volumetric flow rate and the amplitude of the flow ripple for this pump assuming that the pump speed is 2000 rpm? How does this problem compare to the previous problem?

5.10 A pressure-controlled pump utilizes a discharge hose that is 2 m long with a 25.4 mm inside diameter. The effective fluid bulk modulus is 0.82 GPa. Calculate the necessary leakage coefficient for the pump to achieve a settling time of 0.5 s.

5.11 A pressure compensated swash-plate type, axial-piston pump is designed to displace a maximum of 215 cm^3/rev. The pump is normally operated at 1800 rpm with an operating pressure of 20 MPa. The maximum swash plate angle is 18°. The discharge hose for the pump is 2 m long with a 25.4 mm inside diameter and the fluid bulk modulus is 0.8 GPa. The control actuator for this pump is 35 mm in diameter and is located 75 mm away from the pivot point of the swash plate. The three-way control valve for this pump is 6 mm in diameter and the valve spring rate is 56.55 N/mm. The flow gain for the valve is 0.19 m^2/s and the flow force gain is 5.76 N/mm. Calculate the necessary coefficient of leakage for the pump to achieve a damping ratio of 0.05. What is the rise time, settling time, maximum percent overshoot, and bandwidth frequency for this system?

5.12 A displacement-controlled swash-plate type, axial-piston pump is designed to displace a maximum of 100 cm^3/rev. The control actuator is 5.3 cm in diameter and is located 8.8 cm away from the pivot point of the swash plate. A four-way valve with 3 mm × 3 mm square porting is used to control the pump. A supply pressure of 2 MPa is provided to the valve by an auxiliary pump. Calculate the require control gain for the pump that will produce a settling time of 0.5 s. Based on this gain, what is the maximum displacement of the valve spool assuming that the maximum swash plate angle is 18 degrees?

6

HYDRAULIC ACTUATORS

6.1 INTRODUCTION

In this chapter the output device of the hydraulic control system is considered. This device, which we call the actuator, is the mechanism that is responsible for delivering force and motion to the external load system of a given application. In this chapter two actuator types are considered: linear actuators and rotary actuators. Linear actuators are often called hydraulic cylinders or rams while a rotary actuator is commonly called a hydraulic motor. As the names suggest, the linear actuator is used to create translational motion while the rotary actuator is used to create rotational motion – both functions are widely used in hydraulic control systems. In this chapter the basic construction of each actuator type is described with a consideration of machine performance. The reader will recall that the primary advantage of using a hydraulic control system over other control systems is the high effort to inertia ratio that is exhibited by hydraulics. Since the inertial effects of hydraulic actuators are small compared to the forces generated by these devices, the dynamics of the hydraulic actuator are not considered in this chapter. Rather, the output dynamics for the hydraulic control system are considered when an entire system is assembled and when the load dynamics become substantial. These dynamic considerations are presented in Chapters 7 and 8. In this chapter it is sufficient to discuss the output effort and displacement characteristics for both linear and rotary actuators from a quasi-steady point of view. The outcome of this chapter is useful for building systems of hydraulic components and for designing adequate hydraulic control systems that are used in everyday practice.

6.2 ACTUATOR TYPES

6.2.1 Linear Actuators

Figure 6-1 shows a cross-sectional view of a linear actuator that is typically used to generate translational output motion for hydraulic control systems. The basic construction of this actuator is described by a simple piston/cylinder arrangement; however, there are necessary machine elements that are used in practice to guarantee satisfactory performance of

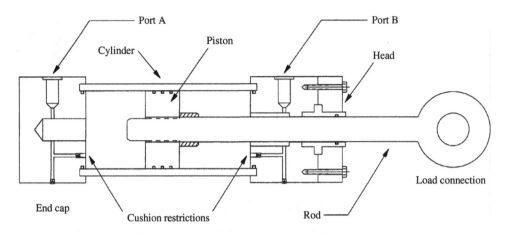

Figure 6-1. Linear actuator geometry.

this machine. These machine elements include seals, bolts, welded joints, plugs, bearings, threaded rods, and specially machined end caps and housings. Some of these elements are shown in Figure 6-1. In this figure, the two end caps of the linear actuator are connected to each other by a sealed cylindrical tube. These end caps are machined with hydraulic ports that are labeled Port A and Port B in Figure 6-1. Within each end cap there is machined geometry that is used for creating a fluid cushion at both ends of the piston stroke. This geometry involves a mechanism for plugging the main flow passage to the port and providing an alternate route for fluid to pass through the cushion restrictions. The piston within the linear actuator is attached to a displacement rod, which is supported by a bearing/seal arrangement and contained by the bolted head. The load device for the hydraulic control system is connected to the rod at the load connection point.

The operation of the linear actuator is accomplished by forcing fluid into Port A, which then causes the piston/rod assembly to move to the right. As the piston moves to the right fluid is forced out of Port B into the return side of the hydraulic system. As the rod approaches its maximum displacement to the right the cross-hatched cushion geometry, which is attached to the piston, enters into the annular geometry of the left end cap and plugs the main flow passage to Port B. The fluid exiting the cylinder must then be routed through the cushion restrictor shown in Figure 6-1, which then slows down the travel velocity of the piston near the end of the piston stroke. A reverse operation of the linear actuator is achieved by forcing fluid into Port B, which then causes the piston/rod assembly to move to the left. Fluid then exits Port A and the travel velocity near the end of the stroke is retarded in a fashion previously described for motion toward the right. *Note:* In the case of leftward motion the cross-hatched geometry does not provide the plug for the main port passage; rather, the rod extension through to piston provides this function for the cushioning effect.

Figure 6-1 depicts a linear actuator that is designed in a double-acting, single-rod configuration. The design in Figure 6-1 is considered to be double acting because it can be forced in either the right or left direction depending on the direction of fluid flow through Ports A or B. Figure 6-1 is considered to be a single-rod design because it only has one rod and head end of the actuator. Having said this, it is important to note that multiple configurations of the linear actuator exist. Figure 6-2 shows schematic configurations of four common designs: the double-acting single rod, the double-acting double rod, the

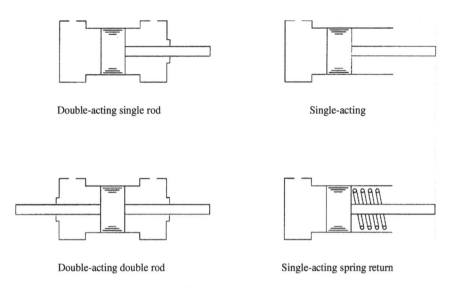

Double-acting single rod

Single-acting

Double-acting double rod

Single-acting spring return

Figure 6-2. Linear actuator configurations.

single-acting, and the single-acting spring return. The schematic in Figure 6-2 sufficiently describes the differences between these four designs. It is important to note that practical texts exist in which various issues of linear actuator design are considered. These texts are listed in the References section of this chapter [1–3].

6.2.2 Rotary Actuators

Rotary actuators are used to generate rotational output for hydraulic control systems. As such, these devices may be considered as hydraulic pumps that operate in a reverse mode. Whereas pumps convert rotating shaft power into fluid power, rotary actuators do just the opposite – they convert fluid power into rotating shaft power. Rotary actuators are commonly called hydraulic motors and are constructed using the same basic configurations that have been shown for hydraulic pumps in Chapter 5. Rather than representing these configurations here, the reader is directed to Figures 5-1 through 5-6. The functional difference between pumps and motors may be understood by considering the mechanical shaft of these machines. When the mechanical shaft is being driven by an engine, electric motor, or even a load source, the machine generates fluid flow and is called a pump. When the mechanical shaft is driving the hydraulic load, fluid flow is provided to the machine and it is called a motor or a rotary actuator. In certain applications rotary actuators may be used to carry out both pumping and motoring functions on an intermittent basis; however, these machines are generally applied in either a dominant pumping mode or a dominant actuator mode for a given control system.

Each hydraulic pump that is shown in Figures 5-1 through 5-6 of Chapter 5 is capable of working as a rotary actuator; however, there are preferred choices for the use of some of these machines in particular applications. For instance, Figures 5-1 through 5-4 show a gear motor, a gerotor motor, a vane motor, and a radial piston motor, respectively. These devices are usually designed to deliver or receive a fixed volume of fluid for each rotation of the shaft. This means that the output speed of the fixed displacement motor cannot be adjusted without a larger or smaller volume of fluid being supplied to the device. On the other hand, Figures 5-5 and 5-6 show a swash plate type motor and a bent axis type motor. These devices are commonly designed using an adjustable configuration for determining the amount of shaft rotation for a given fluid input. In other words, the output speed of the variable displacement motor can be adjusted without a larger or smaller volume of fluid being supplied to the device. The ability to vary the output speed of the motor without changing the motor's volumetric flow rate is a feature that is very important to many hydraulic control applications.

Hydraulic rotary actuators exhibit the capability of maintaining a high shaft torque during zero speed conditions. In applications where this feature is valued the torque required to start shaft motion (i.e., the starting torque) is often a significant performance characteristic of the machine. Typically, the highest starting torque is observed when either the radial piston motor of Figure 5-4 or the bent axis piston motor of Figure 5-6 is used; therefore, these motors are preferred when the starting torque characteristics are critical. In general, the piston type motor tends to exhibit a higher operating efficiency over that of either a gear type or a vane type motor. This efficiency advantage is observable over the entire operating range of the machine.

6.3   LINEAR ACTUATORS

6.3.1   Overview

In this section of the chapter we discuss the performance characteristics of linear actuators. These characteristics include a consideration of the actuator efficiency and the quasi-steady operation and function of the actuator. Since the double-acting, single-rod actuator design shown in Figure 6-1 may be generalized to describe other linear actuators, this design is the basis of our consideration for this section. Using the information presented here, the actuation of a larger hydraulic control system is carried out in subsequent chapters of this text.

6.3.2   Efficiency

Figure 6-3 shows a schematic of the power distribution for the linear actuator. This figure shows that hydraulic power is delivered to the system through Port A as a product of pressure P_A and volumetric flow rate Q_A. As the piston/rod assembly moves to the right power is expelled from the actuator system through Port B as a product of pressure P_B and volumetric flow rate Q_B. Figure 6-3 shows that heat is also lost to the atmosphere as a result of viscous shear and Coulomb friction within the system. The useful output power of the linear actuator is shown in Figure 6-3 as the product of the actuator force F and the piston/rod velocity v.

The overall efficiency of the linear actuator is defined as the ratio of the useful output power to the supplied input power. Figure 6-3 shows that the supplied input power is the hydraulic power delivered to the actuator while the useful output power is the power delivered by the translating piston and rod. In this case the overall actuator efficiency is given by

$$\eta = \frac{Fv}{P_A Q_A}, \tag{6.1}$$

where F is the actuator force, v is the actuator velocity, P_A is the fluid pressure in Port A, and Q_A is the volumetric flow rate into Port A. Using the pressured area of the actuator on

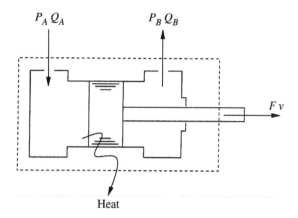

Figure 6-3. Schematic of power flow for the linear actuator.

the side A of the piston A_A, the overall efficiency of the actuator may be separated into two components: the volumetric efficiency and the force efficiency. This result is given by

$$\eta = \eta_v \eta_f, \tag{6.2}$$

where the volumetric efficiency is

$$\eta_v = \frac{A_A v}{Q_A}, \tag{6.3}$$

and the force efficiency is given by

$$\eta_f = \frac{F}{P_A A_A}. \tag{6.4}$$

In general, the volumetric efficiency of the actuator will be less than unity due to fluid compression and leakage past the piston. The force efficiency will be less than unity due to Coulomb friction and viscous shear. If the actuator of Figure 6-3 is operated in the reverse direction the subscript A in Equations (6.3) and (6.4) must be changed to B to denote the input power being supplied to side B of the actuator. Also, the force and velocity of these equations must be considered in an absolute sense.

The volumetric efficiency of the linear actuator generally increases with speed and decreases with pressure, while the force efficiency decreases with speed and increases with pressure. Figure 6-4 shows a schematic of the actuator efficiency as a function of the nondimensional group $\mu v/(PL)$ where μ is the fluid viscosity, v is the actuator velocity, P is the fluid pressure in the actuator, and L is the length of the actuator stroke. *Note:* This schematic is very similar to the pump efficiency schematic shown in Figure 5-8 and may therefore be modeled using equations that are similar to Equations (5.5) and (5.6).

6.3.3 Actuator Function

As previously mentioned, the function of the linear actuator is to convert hydraulic power into linear mechanical power. Since the momentum effects of the moving actuator parts are small compared to the forces that are generated by the actuator, and since the pressure transients in the actuator occur much faster than the dynamics of the actuator load, the power conversion of the actuator may be considered as a quasi-steady process in which most of the time varying quantities may be neglected.

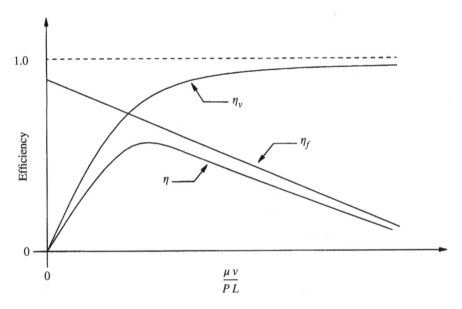

Figure 6-4. Typical linear actuator efficiency curves.

Figure 6-5 is presented to describe a positive output motion for the single-rod linear actuator. In this figure the displacement and velocity of the actuator are shown by the symbols y and v, respectively. An applied load to the actuator is shown in this figure by the symbol F and sides A and B are noted in the drawing. In Figure 6-5 the fluid pressure within side A is given by P_A while the fluid pressure on side B is given by P_B. The pressurized area on side A is shown in this figure by the symbol A_A while the pressurized area on side B is given by A_B. *Note:* A_A and A_B are different only because the actuator is a single-rod design. If the actuator were a double-rod design, using rods of the same size, the actuator areas would be identical. The volumetric flow rates into sides A and B are shown in Figure 6-5 by the symbols Q_A and Q_B, respectively. These flow rates are shown in a positive sense as they produce a positive displacement and velocity for the actuator.

Using quasi-steady assumptions with Figure 6-5 in which the momentum effects of the actuator have been neglected, it may be shown that the applied force to the actuator is given by

$$F = -(A_A P_A - A_B P_B)\eta_f, \tag{6.5}$$

where η_f is the force efficiency of the actuator. *Note:* This equation is signdependent and may be used to describe the actuator pressures for an applied load that is either positive or negative. Similarly, the velocity of the actuator shown in Figure 6-5 may be calculated as

$$v = \frac{Q_A}{A_A}\eta_v = \frac{Q_B}{A_B}\eta_v, \tag{6.6}$$

where η_v is the volumetric efficiency of the actuator. This equation, too, has sign dependency where a negative sign on the flow rate terms must be applied when the flow arrows in Figure 6-5 are pointing in the opposite directions. In order to determine the

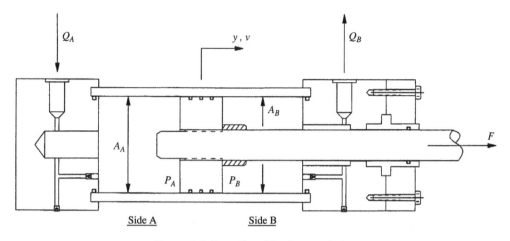

Figure 6-5. Operation of the linear actuator.

instantaneous displacement y of the linear actuator the integral of Equation (6.6) must be taken with respect to time. When designing hydraulic control systems the linear actuator and its power supply must be designed to produce the output requirements of the system. Usually these output requirements are given in terms of effort and power, which means that the load capacity of the actuator and the velocity at which this load is applied must fit the application. The load requirements and actuator size may be adequately matched using Equation (6.5) while the required flow rate of the actuator power supply may be determined using Equation (6.6). The following case study is used to illustrate this point.

Case Study

Four hydraulic actuators are used to operate a car lift that is used within an auto mechanic garage. Each actuator must be designed to lift 2000 lbf at a velocity of 12 feet per minute. The maximum available pressure to each actuator is 3000 psi; the force and volumetric efficiencies are given by 0.95 and 0.97, respectively. Assuming that the actuators are of a double-rod design, specify the minimum required pressurized area of each actuator and the associated volumetric flow rate that is required from the hydraulic power supply. What is the maximum hydraulic power that will be consumed by this application?

The minimum pressurized area for each linear actuator may be calculated by rearranging Equation (6.5) and setting $A_A = A_B = A$ for the double-rod design. This result is given by

$$A = \frac{F}{(P_A - P_B)\eta_f} = \frac{2000 \text{ lbf}}{3000 \text{ psi} \times 0.95} = 0.70 \text{ in}^2.$$

If we were to assume that the rod diameter was 75% of the cylinder diameter, it may be shown that the rod and cylinder diameter of the linear actuator are given by 1.07 in and 1.43 in, respectively. Equation (6.6) may be used to show that the volumetric flow rate required by the linear actuator is

$$Q = \frac{Av}{\eta_v} = \frac{0.70 \text{ in.}^2 \times 12 \text{ ft/min}}{0.97} = 0.45 \frac{\text{gal}}{\text{min}}.$$

The maximum hydraulic power that is required by this application is given by

$$\Pi = 4 \times P_{\max} \times Q = 4 \times 3000 \text{ psi} \times 0.45 \text{ gpm} = 3.1 \text{ hp},$$

where the multiplication of 4 is due to the existence of four actuators for this system. *Note:* This is the amount of power required at the actuator. If a control valve is used to direct this flow the actual power source for the entire system may need to be larger.

6.3.4 Summary

In summary, this section has been written to describe the performance characteristics of a linear actuator. As shown in this section the actuator performance depends on the application requirement and the efficiency characteristics of the actuator design. From this information the actuator size and power supply to the actuator may be designed or specified using quasi-steady analysis that is justified based on the insignificant contribution of linear momentum and the quick transient response of the pressure characteristics within the actuator. This information is used more extensively in the following chapters.

6.4 ROTARY ACTUATORS

6.4.1 Overview

In this section of the chapter we discuss the performance characteristics of rotary actuators. These characteristics include a consideration of the actuator efficiency and the quasi-steady operation and function of the actuator. As previously mentioned, the rotary actuator can be designed using one of several design configurations that are also common with pump technologies. These configurations are shown in the introductory remarks of Chapter 5. In the discussion that follows the rotary actuator is considered generally as being any one of these machines and the presentation applies equally for all machine types. Using the information presented here the rotary actuation of a larger hydraulic control system is carried out in subsequent chapters of this text.

6.4.2 Efficiency

In Section 5.2 of Chapter 5 the efficiency for a hydraulic pump was discussed at length. As it turns out, the efficiency equations for the rotary actuator (hydraulic motor) are similar to those of the hydraulic pump and the discussion of pump efficiency measurement is also applicable to the efficiency measurement of the hydraulic motor. Due to these similarities only the very basic efficiency equations for the hydraulic motor are presented here. For a broader discussion of this topic, the reader is referred to Section 5.2 of Chapter 5. Figure 6-6 shows the schematic of power flowing in and out of a typical hydraulic motor. In this schematic power is supplied to the motor through the intake line $P_i Q_i$ while the useful output power of the motor is shown by the shaft power $T\omega$. The discharge line returns power back to the hydraulic circuit while the internal leakage of the motor creates a power loss. The motor shown in Figure 6-6 is different from the pump of Figure 5-7 in that the intake pressure is higher than the discharge pressure and power is delivered to the shaft as opposed to being delivered by the shaft.

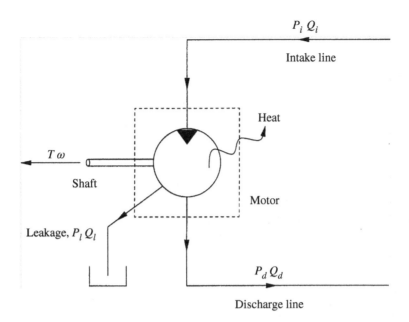

Figure 6-6. Schematic of power flow for the rotary actuator.

The overall motor efficiency is defined as the useful output power divided by the supplied input power. From the flow of power shown in Figure 6-6 this definition may be mathematically expressed as

$$\eta = \frac{T\omega}{P_i Q_i}, \tag{6.7}$$

where T is the output torque on the motor shaft, ω is the angular velocity of the shaft, P_i is the input pressure to the motor, and Q_i is the volumetric flow rate into the motor. Using the volumetric displacement of the motor V_d the overall efficiency may be separated into two components: the volumetric efficiency and the torque efficiency. This separation is given by

$$\eta = \eta_t \eta_v, \tag{6.8}$$

where the torque efficiency is expressed as

$$\eta_t = \frac{T}{V_d P_i}, \tag{6.9}$$

and where the volumetric efficiency is

$$\eta_v = \frac{V_d \omega}{Q_i}. \tag{6.10}$$

In Equations (6.9) and (6.10) the volumetric displacement is given in units of volume per radian. The volumetric displacement per revolution is given by $V_D = 2\pi V_d$. The torque efficiency of Equation (6.9) is useful for describing the power that is lost due to friction while the volumetric efficiency is used for describing power losses that result from leakage. *Note:* The efficiency equations for the hydraulic pump shown in Equations (5.3) and (5.4) are essentially the reciprocal of the efficiency equations for the motor shown in Equations (6.9) and (6.10).

The efficiency characteristics of rotary actuators tend to behave similarly to those of hydraulic pumps and may be graphically depicted using the general form of Figure 5-8 where the efficiency is plotted against the dimensionless group $\mu\omega/P$. In this dimensionless group μ is the fluid viscosity, ω is the angular velocity of the output shaft, and P is the pressure drop across the rotary actuator. Since this dimensionless group is common with the pump efficiency curve the efficiency of the rotary actuator may be modeled using Equations (5.5) and (5.6) of Chapter 5; however, a word of caution is in order. Even though the general efficiency characteristics of the rotary actuary are similar to those of a pump the dimensionless coefficients in Equations (5.5) and (5.6) may be different for the same machine operating as a motor versus a pump. This is primarily due to the fact that internal parts of the rotary actuator are loaded differently when the machine is running in a pumping mode as opposed to an actuation mode. Therefore, it is recommended that efficiency measurements for a machine operating as a motor be taken while the machine is functioning as an actuator, and that pumping efficiencies should not be used as equivalent actuation efficiencies for the same machine.

6.4.3 Actuator Function

Similar to the linear actuator the momentum of the rotating parts within the rotary actuator is negligible when compared to the torque that is generated by this machine. Therefore, the actuator function may be considered using quasi-steady assumptions in which the momentum effects are ignored. These assumptions may be used to show that the output torque on the rotary actuator shaft is given by

$$T = -V_d(P_A - P_B)\eta_t, \tag{6.11}$$

where V_d is the volumetric displacement of the actuator, $P_A - P_B$ is the pressure drop across the actuator, and η_t is the torque efficiency of the machine. This equation is sign dependent and may be used to describe the actuator pressures for an applied load that is either positive or negative. Similarly, the angular velocity of the rotary actuator may be calculated using the following equation:

$$\omega = \frac{Q}{V_d}\eta_v, \tag{6.12}$$

where Q is the volumetric flow rate into the actuator and η_v is the volumetric efficiency of the actuator. In order to determine the instantaneous displacement of the rotary actuator the integral of Equation (6.12) must be taken with respect to time. When designing hydraulic control systems the rotary actuator and its power supply must be designed to produce the output requirements of the system. Usually these output requirements are given in terms of effort and power, which means that the load capacity of the actuator and the velocity at which this load is applied must fit the application. The load requirements and the size of the rotary actuator may be adequately matched using Equation (6.11) while the required flow rate of the actuator power supply may be determined using Equation (6.12).

Case Study

A rotary actuator is used to operate a factory hoist in which a 45 kg payload is lifted using a pulley with a radius of 1 meter. The velocity of the payload is 1 m/s and the available pressure to operate the actuator is 42 MPa. Calculate the required volumetric displacement of the rotary actuator and the volumetric flow rate that must be supplied to the actuator assuming that the volumetric and torque efficiencies are both given by 0.95. How much power must be supplied to the actuator to accomplish the output objective of this system?

From the information given in this problem the load torque exerted on the rotary actuator may be calculated as $T = 45\,\text{kg} \times 9.81\,\text{m/s}^2 \times 1\,\text{m} = 441.45\,\text{N}\,\text{m}$. Using this calculation with Equation (6.11) the required volumetric displacement of the rotary actuator is given by

$$V_d = \frac{T}{(P_A - P_B)\eta_t} = \frac{441.45\,\text{N}\,\text{m}}{42\,\text{MPa} \times 0.95} = 11\frac{\text{cm}^3}{\text{rad}} = 69.5\frac{\text{cm}^3}{\text{rev}}.$$

Since the payload must travel at a velocity of 1 m/s it can be shown that the angular velocity of the actuator must be given by 1 rad/s, which equals about 10 rpm. Using Equation (6.12) the required volumetric flow rate for the actuator may be calculated as

$$Q = \frac{V_d \omega}{\eta_v} = \frac{69.5\,\text{cm}^3/\text{rev} \times 10\,\text{rev/min}}{0.95} = 0.73\frac{\text{liters}}{\text{min}}.$$

The amount of power that must be delivered to the actuator is given by

$$\Pi = 42\,\text{MPa} \times 0.73\ \text{lpm} = 512\ \text{W},$$

which is the equivalent of operating about seven light bulbs.

6.4.4 Summary

In summary, this section is written to describe the performance characteristics of a rotary actuator. Similar to the linear actuator, the rotary actuator performance depends on the application requirement and the efficiency characteristics of the actuator design. From this information the rotary actuator size and power supply to the actuator may be designed or specified using quasi-steady analysis that is justified based on the insignificant contribution of linear momentum and the quick transient response of the pressure characteristics of the fluid supplied to the actuator. It should be mentioned that the rotary actuator differs from the single-rod linear actuator in that it exhibits symmetry in both directions of actuation. In this regard the rotary actuator demonstrates similarities with the double-rod linear actuator design. Again, this information is used more extensively in the following chapters.

6.5 CONCLUSION

In conclusion, this chapter describes and models the performance characteristics of two types of hydraulic actuators: linear actuators and rotary actuators. In both cases the operating efficiency of each device is described as the useful output power divided by the supplied input power. In this analysis energy dissipative effects are separated into losses that occur due to volumetric reductions resulting from leakage and compressibility effects, and effort reductions resulting from coulomb friction and viscous shear effects. These reductions, with the exception of compressibility effects, are noted to produce heat that must be dissipated using natural or forced convection methods. Beyond considering the efficiency characteristics of the actuator the performance characteristics of effort and velocity are presented for both actuator types. These characteristics are based on the actuator size and the hydraulic power that is supplied to the actuator. In this chapter calculations for determining the required actuator size and the power supply are presented. These methods of design are used more extensively in the chapters that follow.

6.6 REFERENCES

[1] Yeaple, F. D. 1996. *Fluid Power Design Handbook*, 3rd ed. Marcel Dekker, New York.

[2] Wolansky, W. D., J. Nagohosian, and R. W. Henke. 1986. *Fundamentals of Fluid Power*. Waveland Press, Prospect Heights, IL.

[3] Kokernak, R. P. 1999. *Fluid Power Technology*, 2nd ed. Prentice-Hall, Upper Saddle River, NJ.

6.7 HOMEWORK PROBLEMS

6.7.1 Linear Actuators

6.1 Laboratory measurements have been taken for a single-rod, double-acting linear actuator. The input pressure and flow to the actuator are given by 3000 psi and 38 gpm, respectively. The output force is measured at 4000 lbf and the output velocity is 100 in/s. Calculate the overall efficiency of the linear actuator.

6.2 If the pressurized area of the linear actuator in Problem 6.1 is given by 1.43 in^2, what is the volumetric and torque efficiency of the actuator?

6.3 A double-rod, double-acting linear actuator is used to exert 4500 N on a load. When generating this load the fluid pressures on each side of the actuator are given by 21 MPa and 7 MPa, respectively. The force efficiency of the actuator is 90%. Calculate the required pressurized area of the actuator to generate the specified load. If the rod diameter is 80% of the piston diameter calculate the rod and piston diameter for this actuator.

6.4 If the actuator in Problem 6.3 has a volumetric efficiency of 98%, what is the required volumetric flow rate supplied to the actuator for a linear velocity of 1.5 m/s?

6.7.2 Rotary Actuators

6.5 Laboratory measurements have been taken for a rotary actuator operating at 40 MPa and 1800 rpm. The torque on the output shaft is measured to be 350 N m and the input volumetric flow rate is 112 lpm. Calculate the overall efficiency of the actuator.

6.6 Assuming that the volumetric displacement of the rotary actuator in Problem 6.5 is 60 cm^3/rev, calculate the volumetric and torque efficiency of the machine.

6.7 A rotary actuator is used to generate 3000 in-lbf of torque using a 3000 psi pressure source. The torque efficiency of the actuator is 92%. Calculate the required volumetric displacement of the actuator.

6.8 If the actuator in Problem 6.7 is operating at 1800 rpm calculate the required supply flow to the actuator assuming that the actuator has a volumetric efficiency of 96%.

7

AUXILIARY COMPONENTS

7.1 INTRODUCTION

The previous chapters of this book discussed the primary components of a hydraulic system, which are used to generate, control, and deliver fluid power. These primary components include pumps, valves, and actuators. In this chapter, we seek to present the auxiliary components that must also be considered for building a practical hydraulic control system. These auxiliary components include accumulators, conduits, reservoirs, coolers, and filters. As the reader will see, the auxiliary components are not optional and are the very things that allow the system to operate smoothly and for a long duration. Proper attention must be given to the design and specification of these auxiliary components – this chapter aims to describe the salient features that must be considered for each device.

7.2 ACCUMULATORS

7.2.1 Function of the Accumulator

The reader will recall from the introduction of this book, that the primary advantage of using a hydraulic control system over, say, an electrical control system, is the stiff dynamic response that results from using low-mass actuators that deliver extremely high effort. Hydraulic control systems are fast and typically exhibit less overshoot and oscillation than their electrical counterparts. While this is the primary advantage of using a hydraulic system, this advantage is sometimes "too much" and so the design engineer may be interested in softening the system by adding compliance. To add compliance to the system, a hydraulic accumulator is used.

Another reason for using a hydraulic accumulator is to capture kinetic or potential energy that would ordinarily be dissipated in the form of heat. Once this energy is captured, it may be redelivered to the hydraulic system during an appropriate time in order to displace energy that would normally be provided by the primary power source. While there are limits concerning the speed at which a hydraulic accumulate may be charged and discharged [1, 2], this energy storage and delivery process can normally occur at a very fast rate, unlike an electric battery, which generally takes more time and space to perform the same function.

This fast-acting response of a hydraulic accumulator has often encouraged engineers to say that the device stores both energy and power – a statement that is strictly not accurate since power cannot be stored.

Figure 7-1 presents an example of a hydraulic system that uses an accumulator to either alter the stiffness characteristics of the design or to store energy. In this system the pump is used to deliver flow Q to a hydraulic conduit, which is connected to a load valve. The flow across the load valve is constant and given by Q_L. A low-Reynolds number leakage coefficient is shown in the figure by the symbol K and the accumulator itself has a compliance shown by the symbol C.

Using a form of the pressure-rise rate equation given in Equation (2.122) of Chapter 2, the fluid pressure in the hydraulic conduit may be modeled as

$$C \frac{dP}{dt} + KP = Q - Q_L, \tag{7.1}$$

where t is time. The reader will recognize this as a first-order dynamical system, which exhibits an exponential decay to a step input. The time constant associated with the exponential decay is given by

$$\tau = \frac{C}{K}. \tag{7.2}$$

If the time constant is large, the decay occurs slowly. If the time constant is small, the decay occurs quickly.

For a step response, the settling time is equal to four time constants, which means that added compliance extends the settling time and slows the system down. This is true also for underdamped second-order systems where perhaps a mass is being moved, and overshoot and oscillation result from a sudden input. In the case of a second-order system, added compliance will not only increase the settling time but it will also reduce the damping ratio and increase the maximum percent overshoot. Generally speaking, added compliance does not benefit a step response, and tends to work against the primary advantage of using a hydraulic control system, which is to produce stiff dynamic behavior.

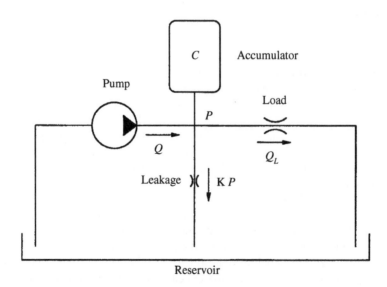

Figure 7-1. Example of a hydraulic system that uses an accumulator.

Perhaps the primary benefit of using a hydraulic accumulator may be illustrated using Figure 7-1 to consider the frequency response of the fluid pressure to a sinusoidal flow input. Such a sinusoidal input may result from the flow characteristics of the pump, similar to the discussion presented in Chapter 5 concerning the flow ripple of gear pumps and axial-piston pumps. A stiff hydraulic system responds to these high-frequency flow variations while a softer system is more forgiving.

As shown in Equation (3.37) of Chapter 3, the bandwidth frequency of a first-order system is given by the inverse of the time constant τ. This means that increased compliance reduces the bandwidth frequency of the system and tends to filter out high-frequency content in the pressure line. Thus, the arrangement in Figure 7-1 is sometimes referred to as a low-pass hydraulic filter. It may also be shown that by adding compliance the phase angle of the system increases creating greater lag, and the peak-to-peak amplitude of the pressure pulsation is reduced. This reduction in the pressure amplitude is usually the reason for adding an accumulator to the system, as noise generation is associated with large pressure pulsations and accumulators tend to make the system quieter.

7.2.2 Design of the Accumulator

Accumulator Types. Figure 7-2 shows a schematic of three different accumulator designs that are considered in this subsection. The first design is simply a fixed-volume container that holds a mass of liquid shown by the symbol M. As shown in the figure this liquid is pressurized to the system pressure P. Compliance and energy storage is achieved by compressing the liquid. As is discussed later, this design for an accumulator is not tremendously helpful as the fluid is nearly incompressible and can alter the compliance and energy storage only a small amount.

The second design shown in Figure 7-2 is a spring/piston arrangement where the system pressure P is applied to the underside of the piston which has a cross-sectional area given by A. The top side of the piston remains exposed to atmospheric pressure. As the pressure increases in the accumulator, the piston moves up and compresses the spring, which is shown to have a spring constant k. Obviously, compliance and energy storage is achieved by compressing the spring. This design requires a well-designed spring that is able to withstand an infinite number of cycles without breaking. The design may also suffer from wear due to the piston motion, causing increased leakage in the system that may not be desirable. If a high-frequency response is needed from the accumulator, the spring/piston design may have limitations associated with the mass of the spring and piston.

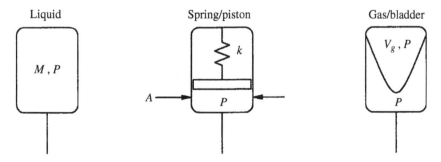

Figure 7-2. A schematic of three accumulator designs.

The third design shown in Figure 7-2 is a gas/bladder arrangement where the system pressure P is the same for both the gas-side and the liquid-side of the bladder. The volume of gas inside the bladder is shown by the symbol V_g, which decreases as the system pressure increases. Compliance and energy storage is achieved by compressing the gas within the bladder. For a typical gas accumulator, nitrogen is used as the gaseous medium in order to prevent inadvertent combustion. Since the bladder is made of an elastic, lightweight material, a gas accumulator is usually considered to be more responsive than a spring accumulator.

Compliance.   As previously noted, one of the main uses of an accumulator is to add compliance to the system, usually for purposes of reducing the amplitude of a pressure pulsation. By definition, the compliance of each accumulator shown in Figure 7-2 is given by

$$C = \frac{dV}{dP},\tag{7.3}$$

where V is the volume of the fluid within the pressurized system and P is the fluid pressure.
    For the liquid accumulator shown in Figure 7-2, the mass of liquid is in the pressurized system is given by

$$M = \rho V,\tag{7.4}$$

where ρ is the liquid density. Taking the derivative of this equation with respect to the fluid pressure P and rearranging terms it may be shown that

$$\frac{dV}{dP} = \frac{V}{\rho}\frac{d\rho}{dP}.\tag{7.5}$$

Using the definition of the fluid bulk modulus of elasticity presented in Equation (1.3) of Chapter 1, it may be shown from Equation (7.5) that the compliance of the liquid accumulator shown in Figure 7-2 is

$$C = \frac{dV}{dP} = \frac{V}{\beta},\tag{7.6}$$

where β is the fluid bulk modulus. Since liquid is nearly incompressible, and β is a very large number on the order of 1 GPa, it is evident that a liquid accumulator does not add much compliance to the system and maintains the stiff dynamical behavior that is expected from an ordinary hydraulic system.
    For the spring/piston accumulator shown in Figure 7-2, the volume of fluid in the pressurized system is given by

$$V = Ay,\tag{7.7}$$

where y is the vertical displacement of the piston, which is not shown explicitly in the figure. Summing forces on the piston produces the following force balance for static equilibrium:

$$0 = AP - ky,\tag{7.8}$$

where A is the cross-sectional area of the piston and k is the spring rate. Solving Equation (7.8) for y and substituting the result into Equation (7.7) produces an expression for the fluid volume in the pressurized system given by

$$V = \frac{A^2}{k}P.\tag{7.9}$$

Taking the derivative of this expression with respect to P produces the following result for the compliance of the spring/piston accumulator shown in Figure 7-2:

$$C = \frac{dV}{dP} = \frac{A^2}{k}. \tag{7.10}$$

This equation shows that the compliance is inversely proportional to the spring rate, which is rather expected – stiff springs should produce less compliance. The less obvious aspect of this result is that the compliance is proportional to the square of the cross-sectional area of the piston, which suggests that the piston area is the most significant design parameter to adjust when seeking to alter the compliance of a spring accumulator.

For the gas/bladder accumulator shown in Figure 7-2, the volume of pressurized fluid in the accumulator is given by

$$V = V_c - V_g, \tag{7.11}$$

where V_c is the charged volume of the accumulator when there is no pressurized fluid inside the device, and V_g is the compressed volume of the gas. For an isothermal process, the ideal gas law may be used to show that the volume of gas in the accumulator is given by

$$V_g = V_c \frac{P_c}{P}, \tag{7.12}$$

where P_c is the initial charged pressure for the gas. Substituting this result into Equation (7.11) produces the following expression for the instantaneous volume of fluid within the gas/bladder accumulator shown in Figure 7-2:

$$V = V_c - V_c \frac{P_c}{P}. \tag{7.13}$$

Taking the derivative of this expression with respect to P, the following result for the compliance of the gas/bladder accumulator may be written:

$$C = \frac{dV}{dP} = \frac{P_c V_c}{P^2}. \tag{7.14}$$

Unlike the liquid and spring/piston accumulator, this result shows that the compliance of the gas/bladder accumulator is not a constant and that the instantaneous compliance is inversely proportional to the square of the fluid pressure. Notice also that a large charge pressure and volume for the accumulator tends to increase the compliance.

Energy Storage. It was previously mentioned that accumulators are also used to capture kinetic or potential energy that would ordinarily be dissipated in the form of heat. The purpose for capturing this energy is to redeliver the energy to the hydraulic system during an opportune time during the working cycle. By definition, the energy that is able to be captured by the accumulator is

$$E = \int P \, dV = \int P C \, dP, \tag{7.15}$$

where P is the fluid pressure, V is the fluid volume, and C is the compliance of the pressurized system. As shown in Equation (7.15), the energy storage capacity for the accumulator increases with increased compliance.

For the liquid accumulator shown in Figure 7-2, the energy storage capacity of the accumulator may be calculated by substituting the liquid compliance of Equation (7.6) into Equation (7.15) and integrating the expression. This result is given by

$$E = \frac{1}{2} \frac{V}{\beta} P_{max}^2, \tag{7.16}$$

where P_{max} is the maximum fluid pressure that is expected in the system. Not surprisingly, the capacity to store energy in this accumulator is significantly hampered by the fact that the fluid is nearly incompressible and that the fluid bulk modulus β is so large. In general, liquid accumulators do not make good energy-storage devices.

For the spring accumulator shown in Figure 7-2, a similar substitution may be made using Equations (7.10) and (7.15). In this case, the energy storage capacity for the spring accumulator is given by

$$E = \frac{1}{2} \frac{A^2}{k} (P_{max}^2 - P_c^2), \tag{7.17}$$

where P_c is the fluid pressure required to overcome the assembled preload on the spring and to begin the motion of the piston. With a greater compliance, this device is much better at storing energy than the liquid accumulator.

For the gas accumulator the energy storage capacity may be calculated by substituting Equation (7.14) into Equation (7.15) and integrating with respect to pressure. This result is given by

$$E = P_c V_c \ln \left(\frac{P_{max}}{P_c} \right). \tag{7.18}$$

As shown by this expression, the ability for the gas accumulator to store energy is increased by increase the charge volume of the system, V_c.

Case Study

A spring/piston accumulator has the capacity to hold 0.25 gallons of fluid while operating between 250 and 1500 psi. The piston diameter is 3 inches, which means the cross-sectional area of the piston is 7 in². Calculate the required spring rate and assembled preload for the spring. What is the compliance for this accumulator, and how much energy can it store?

Using a form of Equation (7.9), it can be seen that the displaced volume of the accumulator is given by

$$V = \frac{A^2}{k} (P_{max} - P_{min}).$$

Rearranging this equation produces the following calculation for the required spring rate:

$$k = \frac{A^2}{V} (P_{max} - P_{min})$$

$$= \frac{(7 \text{ in}^2)^2}{0.25 \text{ gal}} \times (1500 - 250) \text{ psi} = 1,081 \frac{\text{lbf}}{\text{in}}.$$

The force balance on the piston requires the assembled preload on the spring to be

$$F = A P_{min} = 7 \text{ in}^2 \times 250 \text{ psi} = 1,750 \text{ lbf}.$$

Dividing this result by the spring rate shows that the spring is compressed 1.6 inches in its assembled state. The compliance for the spring/piston accumulator is calculated as

$$C = \frac{A^2}{k} = \frac{(7 \text{ in}^2)^2}{1081 \text{ lbf/in}} = 0.0462 \frac{\text{in}^3}{\text{psi}}.$$

The amount of energy that can be stored in the accumulator and redelivered to the system is given by

$$E = \frac{1}{2} C (P_{\max}^2 - P_c^2)$$

$$= \frac{1}{2} \times 0.0462 \frac{\text{in}^3}{\text{psi}} \times \left[(1500 \text{ psi})^2 - (250 \text{ psi})^2 \right] = 4{,}211 \text{ ft-lbf}.$$

This would be comparable to the potential energy of a 200 pound individual, standing on a 21-foot high dive at the swimming pool.

7.3 HYDRAULIC CONDUITS

7.3.1 Function of Hydraulic Conduits

Hydraulic conduits are the elements that connect major hydraulic-components together within a hydraulic control system. As such, hydraulic conduits are mainly comprised of flexible hoses, or rigid tubing that carries hydraulic fluid from one component to the other. For instance, a system that uses a four-way valve to control a linear actuator must connect the pump to the valve, and the valve to the actuator. The connecting lines or tubes are called hydraulic conduits. Hydraulic conduits, however, are not simply hoses and tubes. They are also the fittings, elbows, and joints that are used to make critical and terminal connections for all tubing that is used within the circuit. As such, all hydraulic conduits must perform the following three tasks:

1. Facilitate flow without inducing unnecessary turbulence and pressure drop,
2. Avoid burst and collapse from inside and outside pressure, and
3. Prevent leakage.

The following subsection discusses topics and references that are relevant to each of these functions that are performed by hydraulic conduits.

7.3.2 Specification of Hydraulic Conduits

Facilitate Flow. The facilitation of fluid flow from one component to the other is perhaps the most obvious function to be performed by a hydraulic conduit. As discussed in Chapter 2, a pressure drop through a hydraulic conduit is referred to as a major or minor pressure loss and is shown to be proportional to the velocity of the fluid, squared. See Equations (2.91) and (2.95). Major and minor pressure loss creates heat and energy dissipation, and, therefore, these losses should be minimized by keeping the fluid velocity as low as possible. Similarly, high-velocity turbulent flow is harsh on the fluid and tubing and so it is recommended that the Reynolds number within hydraulic conduits remain below 2100 – the Reynolds number associated with the onset of turbulence in straight, round pipes. Using the definition of the Reynolds number presented in Chapter 2, it can be shown that the inside diameter of the hydraulic conduit should be sized so that

$$d > \frac{4\rho}{\mu\pi\text{Re}}Q, \tag{7.19}$$

where d is the inside diameter of the conduit, ρ is the fluid density, μ is the absolute fluid viscosity, Re is the Reynolds number (recommended not to exceed 2100), and Q is the volumetric flow rate through the conduit. The important thing to notice from this equation is that the recommended inside diameter of the conduit increases proportionally with the volumetric flow rate.

The reader should note that the inlet-pipe diameter for a naturally aspirated pump requires special design consideration. The primary concern for designing this hydraulic conduit is to guarantee that the fluid does not cavitate or vaporize due to high fluid velocities that are not properly matched to the inlet pressure conditions. For a naturally aspirated pump, these pressure conditions may change depending on the location of the reservoir with respect the pump. For instance, if the pump is placed above the reservoir a larger inlet-pipe diameter will be required as compared to an arrangement where the pump is placed below the reservoir. See Reference [3] for a mathematical consideration of this principle.

Avoid Burst and Collapse.   In order to prevent a stress failure of the hydraulic conduit, the wall thickness and strength of the conduit material must be sufficient to avoid either burst or collapse of the conduit wall. Hydraulic conduits may be made of rigid metal tubing or flexible hose, sometimes reinforced with stainless steel braids. Manufacturers will general design and test their hydraulic conduits to satisfy a rated burst pressure, and to withstand at least 1 million cycles of oscillating pressure conditions. An example standard that is often referenced is the International Standard Organization (ISO) Standard 18752 for rubber hoses and hose assemblies.

Prevent Leakage.   One of the greatest disadvantages of using a hydraulic control system as opposed to, say, an electric control system is the fact that hydraulic systems leak. Hydraulic leakage wastes hydraulic fluid. It creates a dangerous work environment by exposing flammable substances to unpredictable conditions, and produces slippery surfaces that can cause the fall and injury of workers. The leakage of hydraulic fluid can also produce environmental problems, contaminating soils and killing natural foliage. Manufacturers of hydraulic systems often cite leakage as their number one maintenance problem.

Since hydraulic conduits are used to connect hydraulic components within the hydraulic circuit, the joints and points of connection are liable for developing undesired leaks. There are numerous sealing technologies that have been developed over the years to prevent hydraulic leaks for both static and dynamic/sliding connection points [4]. For hydraulic conduits, the seals at connection points are typically static in nature, with most designs being sealed using elastomer O-rings within rectangular grooves, compressed against a flat surface. The pressurization of these sealed joints is usually arranged in a self-energizing way so that increased pressure increases the effectiveness of the seal. Reasons for seal failure typically include poor geometry design, improper material selection, incorrect seal installation, and unexpected environmental conditions.

7.4 RESERVOIRS

7.4.1 Functions of the Reservoir

The hydraulic reservoir is the basin of fluid from which fluid is drawn to supply flow for the hydraulic system. Obviously, the reservoir must also be replenished by return flow from the hydraulic system in order to provide a semi-closed circuit for fluid circulation. While the reservoir itself is a passive component, it nevertheless plays a significant role in maintaining the condition of the hydraulic fluid. The following list describes four important functions of the reservoir:

1. Storing hydraulic fluid
2. Separating contaminants of water and solid particles from the hydraulic fluid
3. Separating entrained air from the liquid
4. Cooling the hydraulic fluid

Figure 7-1 shows a schematic for a typical reservoir used in a hydraulic system for storing fluid. The components of the reservoir include a pump line through which fluid is drawn into the hydraulic system by the pump, and a return line that routes all displaced fluid in the system back to the reservoir. These two sides of the reservoir are general separated by a baffle, which is intended to mix the fluid in the reservoir and to keep the pump from recycling the return fluid directly. In other words, the baffle allows fluid to stay in the reservoir long enough to benefit from the intended functions that have been listed earlier.

Figure 7-3 shows the pump line connected to a strainer, which is used to prevent large particles and flakes of material from entering the pump. Similarly, on the return side of the reservoir a mesh screen is shown in the right-hand view inclined at a 45-degree angle for removing entrained air from the fluid. *Note*: The mesh screen has been omitted from the other views for clarity. The angular cut on the bottom of the return-line tube is used to direct fluid flow toward the reservoir walls where convective and conductive heat transfer may occur most efficiently as the moving fluid passes over the cool walls of the reservoir. In the next subsection, considerations for the design of the reservoir are presented as they pertain to these functions.

7.4.2 Design of the Reservoir

Storing Hydraulic Fluid. Perhaps one of the most common rules of thumb for designing hydraulic control systems is to size the reservoir so that it stores a volume of fluid that is equal to three times the volumetric flow rate of the pump when measured in units of volume per minute [1–3]. In other words, if the pump is capable of drawing 100 liters per minute from the reservoir, the nominal volume of fluid held in the reservoir should be at least 300 liters. This rule of thumb allows the hydraulic fluid to "dwell" in the reservoir for at least three minutes before it is pumped back into the hydraulic system for use. This dwell time allows for the separation of contaminants and entrained air prior to reusing the fluid.

If the reservoir is used to service a hydraulic control system that uses double-acting, single-rod linear actuators and/or accumulators (to be discussed later), then the fluid level in the reservoir will increase and diminish during the operation of the system. The fluid level in the reservoir is shown in Figure 7-3 by the symbol h. This change in fluid level is

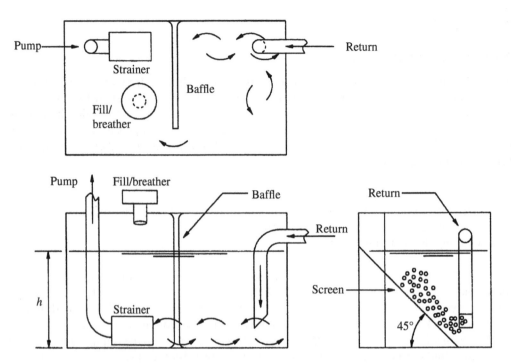

Figure 7-3. Schematic for a typical reservoir with the mesh screen only shown in one view.

due to the fact that single-rod linear actuators are nonsymmetric and are capable of storing hydraulic fluid within the hydraulic system as the actuators extend. Hydraulic accumulators are also capable of storing hydraulic fluid. *Note*: Double-rod actuators and rotary actuators are symmetric and do not create an alteration in the fluid level within the reservoir. In order to ensure that the reservoir is large enough to store the exchanged hydraulic fluid within the system, the maximum volume of the reservoir should be designed such that

$$V_{\max} = V_{\min} + \sum_{n=1}^{N} A_n L_n + V_{acc}, \tag{7.20}$$

where $V_{\min}$ is the minimum volume of fluid required to keep the pump line and the return line submerged within the reservoir when the hydraulic systems is full of fluid and all actuators have been fully extended, N is the total number of double-acting, single-rod linear actuators being serviced by the reservoir, A_n is the cross-sectional area of the rod (not the piston) for the nth actuator, L_n is the maximum stroke length for the nth actuator, and V_{acc} is the total volume of all accumulators in the system. It should be noted that the reservoir is equipped with a breather for maintaining an atmospheric pressure within the reservoir while the fluid level goes up and down. This breather is shown in Figure 7-3 as an integral part of the fill cap.

Separating Contaminants. Another important function of the reservoir is to facilitate the natural separation of contaminants that are in the hydraulic fluid. These contaminants may consist of water or solid particles that have been generated by condensation and wear within the system; or, they may have found their way into the hydraulic system through an unintended communication with the outside environment. Water content in hydraulic fluid degrades the fluid properties quickly and any fluid that exceeds 0.2% water should be discarded [5]. Fluid that appears milky is most likely contaminated with water and should be replaced. Solid particles are known to create interference and abrasion within hydraulic machinery. According to the International Standard Organization, Standard 4406, solid particles within hydraulic systems that are 15 microns in size should be kept to less than 50 particles per milliliter by designing adequate filtration systems [6]. Larger particles should be naturally separated from the fluid in the reservoir.

The primarily mechanism for separating contamination from the hydraulic fluid within the reservoir is gravity. Given enough dwell time within the reservoir (a minimum of three minutes being recommended) heavy particles of water and solid material should fall to the bottom of the reservoir, thus being separated from the less-dense hydraulic fluid. In order to keep the pump from scavenging contamination from the bottom of the reservoir, the pump line should be equipped with a strainer as shown in Figure 7-3, and should be kept a minimum of 1 inch from the bottom of the reservoir [1].

Separating Entrained Air from the Liquid. Since the reservoir is vented to the atmosphere through the breather shown in Figure 7-3, the hydraulic fluid is occasionally exposed to air that may become entrained in the hydraulic fluid. The potential for this is greatest when turbulent flow occurs near the fluid-level surface; thus, conditions of this kind should be minimized as much as possible. As previously mentioned in Chapter 1, hydraulic fluid with 1% entrained air by volume has the potential to reduce the fluid bulk modulus by as much as 60%, thus creating significant alterations in the dynamic characteristics of the hydraulic control system. Given enough dwell time, large air bubbles will separate from the hydraulic fluid by rising rapidly to the fluid surface within the reservoir [2]. Increased temperatures and lower fluid viscosities will enhance air separation; however, when entrained air bubbles are small, they are very difficult to remove from the fluid and may take hours of dwell time to naturally separate. Sometimes tiny air bubbles will cause the fluid to appear milky and will be mistaken for high levels of water content.

A mechanical method for encouraging air separation within the reservoir is shown in the right-hand view of Figure 7-3 by the mesh screen that is inclined at 45 degrees relative to the fluid surface. *Note*: The screen has been omitted from the other views in Figure 7-3 for clarity. The screen causes air bubbles to combine on the screen surface as fluid passes through the screen. As the bubbles coalesce they gain buoyancy and rise more quickly to the fluid surface within the reservoir. A 100-mesh screen should be used for large bubbles, and a 400-mesh screen should be used for systems that generate small bubbles [2].

Cooling the Hydraulic Fluid. The efficiency for a hydraulic system using a high-efficiency pump and directional control-valves with linear or rotary actuators may range from 20% to 60%, depending on the operating conditions. This means that a significant amount of input power to the pump will be dissipated in the form of heat. It is desirable to dissipate as much of this heat through the reservoir as possible using mechanisms of convection, conduction, and radiation. Obviously, increasing the surface area of the reservoir will enhance these natural mechanisms of heat dissipation; and sometimes fins are added to the reservoir to do this.

In order to take full advantage of the surface area within the reservoir, the return line should be positioned in such a way as to direct the fluid against the inside wall of the reservoir for facilitating convective and conductive heat transfer. The baffles within the reservoir are also useful for facilitating the motion of fluid and enhancing heat transfer. Figure 7-3 shows the return line with an angular cut pipe for directing flow toward the wall of the reservoir. The return line is also placed near the corner of the reservoir to take advantage of two walls to force bi-directional circulation.

7.5   COOLERS

7.5.1   Function of the Cooler

Elevated temperatures within the hydraulic control system can reduce the service life of the design. High temperatures tend to reduce the fluid viscosity, which in turn degrades the ability of the hydraulic fluid to lubricate sliding machine-parts within the system thus increasing the potential for metal-to-metal contact and wear. In addition to reducing lubrication properties, high temperatures degrade the life and performance of elastomer materials, which are used to make hydraulic conduits and seals. Elevated temperatures enhance the ability of hydraulic fluid to permeate into the elastomer material, while simultaneously softening the elastomer and reducing its strength properties [4]. Typically, steady operating temperatures within the fluid reservoir should be kept below 60°C, with intermittent temperatures being kept below 100°C. If temperatures exceed these recommendations a hydraulic cooler must be added to the low-pressure side of the hydraulic control system to maintain acceptable temperature levels.

7.5.2   Design of the Cooler

One of the most basic coolers to be used for a hydraulic system is the fan-cooled radiator. The radiator itself is comprised of a series of fluid-carrying lines that are attached to "fins" in order to increase the surface area for convective heat transfer to occur. The convection is induced by blowing air from a fan across the finned radiator lines, thus dissipating heat into the surrounding atmosphere.

For a more aggressive removal of heat, a shell-and-tube heat exchanger may be used. This design, shown in Figure 7-4, is made of an outer shell/container with a tube running through the shell. In general, a heat exchanger of this type may have multiple tubes. Around

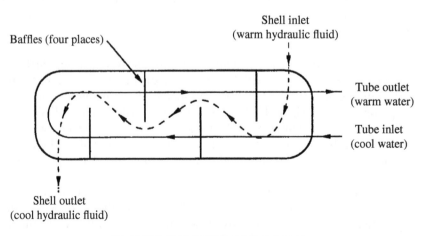

Figure 7-4. Shell-and-tube heat exchanger.

the tubes, and within the shell, hydraulic fluid is caused to pass. The tubes have cold water passing through them, which are designed to carry away heat using a combination of convection and conduction. The use of cold water instead of air as the medium for heat transfer tends to enhance the conduction properties of the cooling process. The design or selection of the shell-and-tube heat exchanger involves the specification of the desired temperature drop for the hydraulic fluid, passing through the tube, and the amount of heat that is to be removed. The temperature drop will be specified to keep the reservoir temperature below 60°C and the amount of heat to be removed will be a small percentage of the average power that is being consumed by the hydraulic system. Heat exchangers are always located in a low-pressure return line for the hydraulic reservoir; however, manufacturers should be consulted for specific design recommendations.

7.6 FILTERS

7.6.1 Function of the Filter

It is sometimes claimed that 75% of hydraulic control system failures occur do to fluid contamination [7]. Fluid contaminants are solid or liquid particles that should not be in the system. Such contaminants tend to create adverse wear or corrosion within the hydraulic machinery, they can cause seizure of machine parts, and they sometimes clog flow passages that are expected to be open during normal operation. These unwanted contaminates may be present due to manufacturing debris, or they may find their way into the system through an outside passageway. Outside passageways may exist through the breather on the reservoir (see Figure 7-3), or through a dynamic seal on the rod of a linear actuator, or through another faulty seal in the system. Sometimes these contaminates may be self-generated within the system by component wear, which is especially common during early hour break-in, or by the decomposing of hydraulic fluid that has not been changed. It is usually recommended that hydraulic fluid be changed every 5000 hours of system operation.

In addition to changing hydraulic fluid on a regular basis, the most common treatment for contamination control within the hydraulic system is to regularly filter the hydraulic fluid. Ordinarily, hydraulic fluid is cleaned using a filter with at least a 25 micron filtration rating, which means that the filter will capture particles that are larger than 25 microns. Most clearances between cylindrical parts within the hydraulic system (spool valves, pistons, etc.) are on the order of 20 microns, so many hydraulic systems will be designed with filtration systems that are capable of capturing particles that are smaller than 20 microns. Some systems will utilize a filter that is rated as low as 5 microns. The lower the filtration rating, the larger the pressure drop will be across the filter. This pressure drop needs to be considered when deciding where to place the filter within the design of the hydraulic circuit.

7.6.2 Placement of the Filter

As it turns out, there are various locations within the hydraulic system where a filter may be placed, and occasionally filters are placed in multiple locations within the same system. Figure 7-5 shows four different filter locations for a hydraulic system that utilizes a fixed-displacement pump and a four-way valve to control a linear actuator. This system is discussed at greater length in Chapter 8. For now, it is important to note from Figure 7-5 that a filter may be placed on the pump inlet, the pump discharge, the four-way valve return

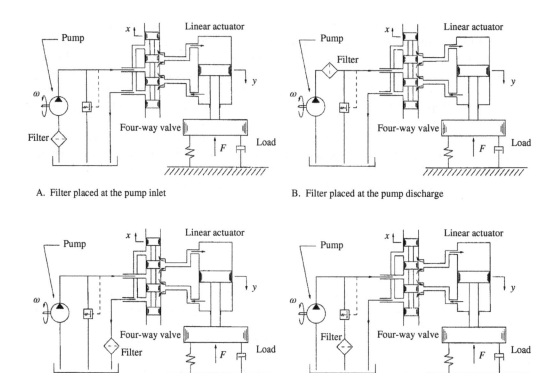

Figure 7-5. A schematic of four different filter locations within a hydraulic control system.

line, or even the relief-valve return line. Each of these locations is illustrated in the figure and are discussed in the following paragraphs.

Schematic A in Figure 7-5 illustrates a hydraulic system where the filter is located at the pump inlet. As shown previously in Figure 7-3, the reservoir is frequently equipped with a strainer in this location, and the difference between a strainer and a filter is not precise. Often, a strainer is simply viewed as a course filter. One advantage of putting a filter at the pump inlet is that the pump is one of the most expensive components in the system and prefiltering the fluid that enters the pump will reduce the chances of pump failure. It is also true that if a pump begins to experience adverse wear due to contamination, by its very nature the pump will distribute these newly created wear particles throughout the rest of the system thus enhancing the chances of downstream contamination. Another advantage of putting a filter at the pump inlet is that the fluid pressure at this location is low, which means that a low-pressure filter may be used thus reducing the cost of the filter and the chances of a pressure failure in the system. A disadvantage of locating the filter at the pump inlet is that the filter will create a pressure drop, which may induce cavitation or vaporization of fluid that enters the pump. Such fluid conditions are generally undesirable and the likelihood of them occurring will be enhanced when the filter is dirty and when the pump is generating a high flow rate. Schematic B in Figure 7-5 illustrates a hydraulic system where the filter is located at the pump discharge. A filter in this location is designed to withstand high pressure and is used primarily to protect downstream components from debris that either passes through the pump, or are generated by pump wear. Placing a filter in a high-pressure line has the advantage of being able to use a finer filter compared to a filter that is used in a low-pressure line, since the high pressure is available to move the fluid through the finer filter element. The disadvantage of using high-pressure filters is the cost and reliability of the filter itself. Such filters require special design considerations and are less frequently available than low-pressure filters.

Schematic C in Figure 7-5 illustrates a hydraulic system where the filter is located within the four-way valve return line. A filter located in this position is used to remove contaminants that are generated in the working elements of the circuit, before the contaminants reach the reservoir. Such contaminants may include dirt that enters into the system from the dynamic seal on the rod of a linear actuator. One advantage of placing a filter in this location is to avoid using a high-pressure filter with its previously noted drawbacks. Another advantage is that the filter keeps clean fluid in the reservoir, thus avoiding the problem of maintaining a reservoir that is difficult to access. Filters that are located in return lines may need to be rated at a higher flow rate compared to filters that are located at the pump inlet or discharge. This is due to the fact the under overrunning conditions the return flow from a linear actuator may be larger than the supply flow from the pump.

Schematic D in Figure 7-5 illustrates a hydraulic system where the filter is located within the return line of the relief valve. In this location, the full amount of pump flow is filtered during standby conditions, since the pump is a constant flow source, while only a portion of the pump flow is filtered during normal operation. The advantage of placing the filter in this location is to avoid using a high-pressure filter. The disadvantage of this arrangement is that filtration of all the pumped fluid is not guaranteed for a given cycle of the system.

There are other arrangements for filtration that have not been presented here, due to the many combinations that may exist for the system shown in Figure 7-5. Perhaps the most common filtration arrangement for the system shown in this figure is represented by Schematic C, where the filter is located at the four-way valve return line. In addition to putting a filter in this location, it is common to place a strainer on the inlet of the pump as a matter of precaution for screening larger chips of debris that may have found their way into the reservoir. For a variable-displacement pump-controlled system (also called a hydrostatic transmission), which is described in Chapter 9, the system filtration is usually accomplished by putting a filter at the inlet of the charge pump. By placing a filter in this location, only about a third of the main pump flow is filtered during a single cycle of the transmission, which illustrates the difficulty in filtering a closed-loop hydraulic control system.

7.7 CONCLUSION

In this chapter we have presented an overview of auxiliary components that are used to build hydraulic control systems. The components include: accumulators, conduits, reservoirs, coolers, and filters. These auxiliary components are passive, in that they do not contribute directly to the dynamic characteristics of the system. However, without giving thoughtful consideration to the design and specification of these components, the hydraulic system may experience large inefficiencies or premature failure. In this chapter, the salient features of these auxiliary components are presented, which should provide direction to the system engineer for completing the overall design of a hydraulic control system.

7.8 REFERENCES

[1] Daines, J. R., 2009. *Fluid Power. Hydraulics and Pneumatics*. Goodheart-Willcox Company, Inc., Tinley Park, IL.

[2] Parker Training. 2001. *Design Engineers Handbook: Volume 1. Hydraulics*. Parker Hannifin Corporation, Motion Control Training Department, Cleveland, OH.

[3] Manring, N. D. 2013. *Fluid Power Pumps and Motors: Analysis, Design, and Control*. McGraw-Hill, New York.

[4] Flitney, R. 2007. *Seals and Sealing Handbook*, 5th ed. Butterworth-Heinemann, Elsevier Ltd., Boston.

[5] Radhakrishnan, M. 2003. *Hydraulic Fluids: A Guide to Selection, Test Methods, and Use*. American Society of Mechanical Engineers, New York.

[6] International Standard Organization. 1999. "Hydraulic fluid power – Fluids – Method for coding the level of contamination by solid particles." Geneva, Switzerland, Standard No. ISO 4406:1999(E).

[7] Henke, R. W. 1983. *Fluid Power Systems & Circuits*, 3rd ed. Hydraulics & Pneumatics magazine, Cleveland, OH.

7.9 HOMEWORK PROBLEMS

7.9.1 Accumulators

7.1   True or False. Accumulators tend to reduce the stiffness of a hydraulic control system, thus, detracting from the inherit advantage that hydraulic systems have over electrical systems.

7.2   What are the reasons that an engineer might consider for using an accumulator within a hydraulic control system? Select all that apply.
(a) To add compliance to the hydraulic system,
(b) To store and redeliver energy that might otherwise be wasted in the form of heat, or
(c) To reduce hydraulic system noise.

7.3   True or False. The ability for an accumulator to store energy is proportional to the compliance of the accumulator.

7.4   Which of the following accumulators demonstrates the least ability to store energy:
(a) A liquid accumulator,
(b) A spring/piston accumulator, or
(c) A gas/bladder accumulator.

7.9.2 Hydraulic Conduits

7.5   In order to avoid turbulent flow within hydraulic conduits, what is the Reynolds number that should be avoided as a rule of thumb?
(a) Reynolds numbers should always be less than 1,
(b) Reynolds numbers should always be less than 2100, or
(c) Reynolds numbers should always be less than 4000.

7.6 Where did the rule of thumb in Problem 7.5 come from?
(a) Experimental observations that high Reynolds numbers tend to burst conduits,
(b) Osborne Reynolds' experiments for turbulence in round, straight pipes, or
(c) The Moody diagram shown in Figure 2-11 of Chapter 2.

7.7 True or False. As the designed volumetric flow rate increases for a conduit, the inside radius of the conduit should increase proportionally.

7.8 When a supplier of hydraulic conduits rates a conduit for a specific burst pressure and fatigue life, how many pressure cycles are the conduits designed to withstand in a safe manner according to ISO Standard 18752?
(a) 1,000,
(b) 100,000, or
(c) 1,000,000.

7.9 True or False. Fluid leakage from conduits and connectors are considered to be one of the greatest disadvantages of using a hydraulic control system.

7.9.3 Reservoirs

7.10 A hydraulic control systems uses a 25 cm^3/rev displacement pump, operating 1800 rpm. Using the rule of thumb presented in this chapter, what should the nominal fluid volume in the reservoir be?
(a) 25 liters,
(b) 45 liters, or
(c) 135 liters.

7.11 Water content in hydraulic fluid degrades the fluid properties quickly. What is the percentage of water content is allowable for the hydraulic fluid contained in the reservoir?
(a) 0%,
(b) less than 0.2%, or
(c) less than 1%.

7.12 What are the most effective things that can be done to the reservoir design to prevent air from becoming entrained in the hydraulic fluid? Select all that apply.
(a) Remove the air breather from the tank,
(b) Provide adequate dwell time for the fluid to rest within the reservoir,
(c) Heat up the fluid to reduce fluid viscosity, or
(d) Add an inclined mesh screen to catch air bubbles as the fluid passes through.

7.9.4 Coolers

7.13 What is the desired steady-state operating temperature for fluid within the reservoir of a hydraulic control system?
(a) Less than 60°C,
(b) Less than 100°C, or
(c) Less than 120°C.

7.14 Why are baffles used in a shell-and-tube heat exchanger, for cooling hydraulic fluid?
(a) To provide structural support for the tubes,
(b) To circulate hydraulic fluid more effectively around the tubes, or
(c) To create a pressure drop across the heat exchanger.

7.9.5 Filters

7.15   What percentage of hydraulic control system failures are attributed to fluid contamination?
(a) 25%,
(b) 50%, or
(c) 75%.

7.16   What are some of the well-known sources of fluid contaminates? Select all that apply.
(a) Sabotage of competitors,
(b) Debris from the manufacturing and assembly process,
(c) Faulty seals,
(d) Internal wear, or
(e) Decomposition of fluid that has not been changed frequently enough.

7.17   What is the minimum filtration rating, for filters that are used within a hydraulic control systems?
(a) 5 microns,
(b) 15 microns, or
(c) 25 microns.

7.18   What is the most common location for a filter within a hydraulic control system?
(a) At the pump inlet,
(b) At the pump discharge,
(c) At the system return line, or
(d) At a relief-valve return line.

HYDRAULIC
CONTROL SYSTEMS

8

VALVE-CONTROLLED
HYDRAULIC SYSTEMS

8.1 INTRODUCTION

In this chapter we consider the analysis, design, and control of valve-controlled hydraulic systems. These systems utilize a control valve for directing available hydraulic power to and from an actuator that is used to generate useful output. It is well known that valve-controlled hydraulic systems exhibit certain advantages over pump-controlled hydraulics systems. Three advantages are noted:

1. The response characteristics of valve-controlled systems can often be improved by eliminating the large amount of compressible fluid that exists between the actuator and the valve. This is done by locating the valve as close to the actuator as possible.
2. If the response characteristics are not crucial for a given application valve-controlled systems can be arranged so that the system is less bulky near the output actuator. This advantage is obtained by locating the hydraulic power supply and the control valve far away from the actuator and by connecting the actuator to the valve using hydraulic transmission lines. This arrangement is typically found on hydraulic excavators where the control valve is located in the driver's cab and the actuators are located far away on the boom, stick, and bucket.
3. Another advantage of using a valve-controlled actuator is that multiple actuation systems can be powered by a single hydraulic power supply. In other words multiple control valves can tap into the same pressure source that is supplied by a single hydraulic pump. This arrangement is very common for mobile and industrial hydraulic systems where several implements are operated using a single high-pressure rail as the power source for each implement.

In view of these advantages over the pump-controlled systems, it must also be noted that valve-controlled hydraulic systems exhibit several disadvantages. Three disadvantages are noted:

1. Valve-controlled systems are generally less efficient than pump-controlled systems. The inefficiency of a valve-controlled system results from two sources: (a) at standby the power source is continually operating to maintain some level of available pressure, and (b) the required pressure drop across the control valve reduces the overall efficiency of the system.
2. The inefficiency of the valve-controlled system is accompanied with increased temperature difficulties, which require sufficient heat exchangers for dissipating the wasted energy.
3. Finally, the very nature of the valve-controlled system requires additional hardware; that is, the valve itself for operating the system. This additional hardware can increase the overall cost of the system depending upon the sophistication of the valve and the hydraulic power supply that is used.

All of these advantages and disadvantages must be weighed against each other when one chooses to use a valve-controlled hydraulic system as opposed to a pump-controlled system. In the discussion that follows it is assumed that the valve-controlled system has been selected based on one or more of the previously mentioned advantages.

The subject of valve-controlled hydraulic systems is most easily divided between the control of linear actuators (hydraulic pistons and cylinders) and rotary actuators (hydraulic motors). The linear actuator system is commonly used to implement devices that utilize translational motion and provide a linear force output. Linear actuator systems can be designed and built using both four-way and three-way valves, as well as using double-rod actuators and single-rod actuators. Example applications for these systems include industrial robots, mobile vehicles that utilize implement systems (e.g., agricultural equipment), and flight surface controls for the aerospace industry. The rotary actuation system is commonly used for driving a rotating shaft, which then produces a rotary torque output. Example applications for the rotary actuation system may include the drum rotation of a cement mixer, the operation of a machining lathe in a manufacturing environment, or the rotary position control of a gun turret on a battleship. These applications illustrate the wide variation of use for hydraulic control systems, which find their niche in high-power applications that require large effort to inertia ratios for achieving a stiff dynamic response.

In the following discussion, three valve-controlled hydraulic systems are presented: the four-way valve-controlled linear actuator, the three-way valve-controlled linear actuator, and the four-way valve-controlled rotary actuator. Each of these hydraulic control systems is analyzed to determine the equation of motion that governs the dynamic behavior of the output load. Once the governing equation has been established the design of the hydraulic system is evaluated using steady-state forms of the governing equation. Finally, each section of this chapter concludes with a presentation of the various control schemes that may be used for each hydraulic system. These control schemes are classical PID controllers that are implemented for achieving position, velocity, and force or torque control of the valve-controlled system.

8.2 FOUR-WAY VALVE CONTROL OF A LINEAR ACTUATOR

8.2.1 Description

In this section of our chapter we introduce the four-way valve-controlled, single rod linear actuator. This system is perhaps one of the most basic hydraulic control systems known to the hydraulic control industry and therefore this system provides an excellent starting point for our study. As shown in Figure 8-1 this system utilizes a standard four-way valve to control the output characteristics of a single rod linear actuator. In this figure the load to be moved by the actuator is shown as a single mass-spring-damper system with a load disturbance force given by F. The mass, spring rate, and viscous drag coefficient for the load are shown in Figure 8-1 by the symbols m, c, and k, respectively. The linear actuator is shown to be constructed with a single rod that is connected to both the load and the actuator piston. Due to the single rod design the pressurized area on the top of the piston A_A is greater than the pressurized area on the bottom of the piston A_B; however, a double rod design may be considered in the analysis that follows by setting these two areas equal to each other. The volumetric flow of hydraulic fluid into and out of the actuator is controlled by the four-way control valve shown to the left of the linear actuator. This valve is shown to be constructed as a symmetric, open center design with an underlapped dimension on the spool given by the dimension u. The displacement of the spool valve x is shown to be positive in the upward direction. As the valve is moved in the positive x-direction flow is directed into side A of the actuator which then requires flow to exit the actuator from side B. These two volumetric flow rates are shown in Figure 8-1 by the symbols Q_A and Q_B, respectively.

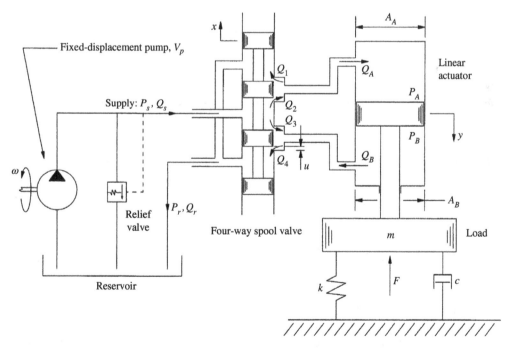

Figure 8-1. Schematic of the four-way valve-controlled linear actuator.

In order to provide an adequate hydraulic power source for the valve-controlled system Figure 8-1 shows a fixed-displacement pump that provides a constant volumetric flow rate into the supply line leading to the four-way control valve. This pump is sized according to its volumetric displacement V_p and is driven by an external power source (not shown in Figure 8-1) at an angular velocity ω. The pressure in the discharge line of the pump is controlled using a high-pressure relief valve that is set at the desired supply pressure P_s. As noted in the introduction of this chapter this standby pressure source results in a disadvantage for valve-controlled systems in that it creates a continual energy drain on the overall system even when the system is doing no useful work.

8.2.2 Analysis

Load Analysis.   The analysis of the system shown in Figure 8-1 rightfully begins with the load. After all it is the load that defines the working objective of the system and therefore its dynamics are the observable output. When we develop the equations of motion for the load we will neglect the inertia of the actuator itself since it is much smaller than the actual forces that are generated by the actuator. The reader will recall that this high force to inertia ratio is one of the principle advantages of using a hydraulic system as opposed to an electric system. By lumping the viscous drag effects of both the actuator and the load into the dashpot shown in Figure 8-1 the equation of motion for the load may be written as

$$m\ddot{y} + c\dot{y} + ky = \eta_{a_f}(A_A P_A - A_B P_B) - F - F_o, \tag{8.1}$$

where P_A and P_B are the fluid pressures on the A and B sides of the actuator respectively, η_{a_f} is the force efficiency of the actuator, and where F_o is the nominal spring or bias load that is applied to the actuator when y equals zero. At the nominal steady-state operating conditions of the system the following design conditions apply:

$$y = 0, \qquad P_A = P_B = P_s/2, \qquad \text{and} \qquad F = 0. \tag{8.2}$$

Under these conditions it can be shown from the steady form of Equation (8.1) that the nominal force exerted on the actuator by the load is given as

$$F_o = \eta_{a_f}(A_A - A_B) P_s/2. \tag{8.3}$$

Substituting this result back into Equation (8.1) produces the following equation for describing the output dynamics of the load:

$$m\ddot{y} + c\dot{y} + ky = \eta_{a_f}[A_A (P_A - P_s/2) - A_B (P_B - P_s/2)] - F. \tag{8.4}$$

From this result it can be seen that the actuator pressures P_A and P_B are the required inputs for adjusting the position of the load. These pressures result from the changing flow and volume conditions within the actuator itself and will be considered in our next paragraph.

Pressure Analysis.   In order to analyze the fluid pressures on sides A and B of the actuator it is assumed that the pressure transients that result from fluid compressibility are negligible. This assumption is especially valid for a system design in which the transmission lines between the valve and the actuator are very short; that is, small volumes of fluid exist on either side of the actuator. The omission of pressure transient effects is also valid for systems in which the load dynamics are much slower than the pressure dynamics themselves.

Since the load dynamics typically occur over a range of seconds and the pressure dynamics typically occur over a range of milliseconds it is usually safe to neglect the time variation of the pressure in favor of the time variation of the overall system. For design and operating conditions in which long transmissions lines between the valve and the actuator are used, or in which a large amount of entrained air is captured within the fluid causing the fluid bulk-modulus to be reduced, it may be necessary to conduct a transient analysis of the pressure conditions on both sides of the actuator. This may also be the case if the actuator dynamics are very fast. For now we assume that this is not the case and that the more typical system conditions describe our problem. Using these assumptions we can use the volumetric efficiency results of Chapter 6 to describe the volumetric flow rates in and out of the actuator shown in Figure 8-1. These results are given by

$$Q_A = \frac{A_A}{\eta_{a_v}} \dot{y} \quad \text{and} \quad Q_B = \frac{A_B}{\eta_{a_v}} \dot{y}, \tag{8.5}$$

where Q_A and Q_B are the volumetric flow rates into and out of the actuator respectively and η_{a_v} is the volumetric efficiency of the actuator. *Note:* In this result it is assumed that $A_A \approx A_B$, which is a simplification that proves useful for the analysis that follows. From the flow rates that are shown at the four-way spool valve in Figure 8-1 it may be seen that

$$Q_A = Q_2 - Q_1 \quad \text{and} \quad Q_B = Q_4 - Q_3. \tag{8.6}$$

From our valve analysis of Chapter 4 the linearized flow equations for fluid passing across the metering lands of the four-way spool valve are given by

$$\begin{aligned}
Q_1 &= \mathrm{K}_c P_s/2 - \mathrm{K}_q x + \mathrm{K}_c (P_A - P_r), \\
Q_2 &= \mathrm{K}_c P_s/2 + \mathrm{K}_q x + \mathrm{K}_c (P_s - P_A), \\
Q_3 &= \mathrm{K}_c P_s/2 - \mathrm{K}_q x + \mathrm{K}_c (P_s - P_B), \\
Q_4 &= \mathrm{K}_c P_s/2 + \mathrm{K}_q x + \mathrm{K}_c (P_B - P_r),
\end{aligned} \tag{8.7}$$

where $\mathrm{K}_c P_s$ is the nominal flow rate across each metering land at the steady operating conditions of the system and where K_q and K_c are the flow gain and the pressure-flow coefficient for the valve, respectively. It is important to recall that the magnitudes of these valve coefficients are the same for each metering land only because the nominal pressure drop across each metering land is equal to half the supply pressure $P_s/2$. If another pressure drop is assumed for the linearization of the valve flow equations the valve coefficients will be different for each flow expression in Equation (8.7) and the analysis that follows will become much more complex. See Chapter 4 for a more thorough discussion of these flow coefficients. By substituting Equation (8.7) into Equation (8.6) the volumetric flow rates into and out of the actuator may be expressed respectively as

$$\begin{aligned}
Q_A &= 2 \, \mathrm{K}_q \, x - 2 \, \mathrm{K}_c \, (P_A - P_s/2), \\
Q_B &= 2 \, \mathrm{K}_q \, x + 2 \, \mathrm{K}_c \, (P_B - P_s/2),
\end{aligned} \tag{8.8}$$

where the return pressure $P_r = 0$. Using this result with Equation (8.5) the following equations for the operating pressures on both sides of the linear actuator may be written as

$$P_A - P_s/2 = \mathrm{K}_p \, x - \frac{A_A}{2 \, \mathrm{K}_c \, \eta_{a_v}} \dot{y},$$

$$P_B - P_s/2 = -\mathrm{K}_p \, x + \frac{A_B}{2 \, \mathrm{K}_c \, \eta_{a_v}} \dot{y}, \tag{8.9}$$

where K_p is the valve's pressure sensitivity given by the ratio of the flow gain to the pressure-flow coefficient K_q/K_c. From Equation (8.9) it can be seen that the fluid pressure on side A of the linear actuator is increased by moving the spool valve in the positive x-direction and that the fluid pressure on side B of the actuator is decreased by the same spool valve motion. As is shown in the summary of the next paragraph this adjustment in fluid pressure by the motion of the spool valve is the mechanism for adjusting the output motion of the load shown in Figure 8-1. *Note:* Equation (8.9) shows a linear velocity dependence for the fluid pressure as well. As is shown subsequently this feature results in favorable damping characteristics for the system.

Analysis Summary.   To summarize the analysis of the linear actuation system shown in Figure 8-1 we substitute Equation (8.9) into Equation (8.4) to produce the following equation of motion for the system:

$$m\,\ddot{y} + \left(c + \frac{A_A^2 + A_B^2}{2\,K_c} \right) \dot{y} + ky = \eta_{a_f}(A_A + A_B)K_p\,x - F. \tag{8.10}$$

In this result it has been assumed that $\eta_{a_f}/\eta_{a_v} \approx 1$. From Equation (8.10) it can be seen that the mechanical design of the linear actuator and the four-way spool valve have a decisive impact on the overall dynamics of the hydraulic control system. In particular it may be seen from Equation (8.10) that these design parameters help to shape the effective damping of the system and to provide an adequate gain relationship between the input motion of the spool valve and the output motion of the load. The following subsection of this chapter is used to consider the mechanical design parameters that must be specified in order to achieve a satisfactory output of the system from a steady-state perspective.

8.2.3 Design

Actuator Design.   When considering the mechanical design of the hydraulic control system that is shown in Figure 8-1 it is always best to start by sizing the linear actuator in accordance with the expected load requirements. Usually for a given application the working load force is known and can be used to determine the required pressurized areas that are needed within the actuator to develop this load for a specified working pressure. If we consider the right-hand side of Equation (8.4) it is clear that the steady working force generated by the single rod actuator is given by

$$F = \eta_{a_f}\left[A_A(P_A - P_s/2) - A_B(P_B - P_s/2)\right], \tag{8.11}$$

where A_A and A_B are the designed pressurized areas on sides A and B of the actuator, η_{a_f} is the force efficiency of the actuator, and P_A and P_B are the fluid pressures on the each side of the actuator. At the working force that is specified for the application it is common to use fluid pressures in the linear actuator that are less than the full supply pressure so as to provide a margin of excessive force capability for the system. Typical design specifications at the working force conditions are given by

$$F = F_w, \qquad P_A = \frac{3}{4}\,P_s, \qquad \text{and} \qquad P_B = \frac{1}{4}\,P_s, \tag{8.12}$$

where F_w is the working force of the hydraulic control system and P_s is the supply pressure to the hydraulic control valve. Substituting these conditions into Equation (8.11) produces the following equation for sizing the pressurized areas of the linear actuator that is shown in Figure 8-1:

$$A_A + A_B = 4 \frac{F_w}{\eta_{a_f} P_s}. \tag{8.13}$$

Similarly, from the Equation (8.3) it can be shown that

$$A_A - A_B = 2 \frac{F_o}{\eta_{a_f} P_s}, \tag{8.14}$$

where F_o is the nominal bias force when $y = 0$ and $P_A = P_B = P_s/2$. A simultaneous solution for Equations (8.13) and (8.14) may be used to show that

$$A_A = \frac{2 F_w + F_o}{\eta_{a_f} P_s} \quad \text{and} \quad A_B = \frac{2 F_w - F_o}{\eta_{a_f} P_s}. \tag{8.15}$$

Obviously, if a double rod linear actuator is used for the design

$$A_A = A_B = \frac{2 F_w}{\eta_{a_f} P_s}. \tag{8.16}$$

These results may be used to design or select a linear actuator that provides a sufficient working force for a given supply pressure. Other design considerations must also be taken into account when designing the linear actuator. For example, the stroke of the actuator or distance of piston travel must be sufficient for the application. Other machine design aspects must also be considered regarding acceptable pressure vessel stresses that will exist within the actuator itself and the sealing mechanisms that must be used to minimize both internal and external leakage. Unfortunately, these machine design topics are beyond the scope of this current text and the reader is referred to other texts that can provide a more thorough address of these subjects [1, 2].

Valve Design. Once the actuator has been designed to generate the necessary working force for the control application the next step is to design a control valve that provides sufficient flow and pressure characteristics for the linear actuator. These characteristics are designed by specifying the appropriate flow gain K_q and pressure-flow coefficient K_c for the valve. The reader will recall that the pressure sensitivity K_p is simply the ratio of these two quantities. The first equation in Equation (8.9) is useful for specifying the needed valve coefficients for our design. This equation is given explicitly in terms of the flow gain and the pressure-flow coefficient as

$$K_c(P_A - P_s/2) = K_q x - \frac{A_A}{2 \eta_{a_v}} \dot{y}, \tag{8.17}$$

where $\dot{y}$ is the velocity of the load. For specifying the appropriate valve coefficients there are usually two operating conditions that are considered: (1) the steady working force condition, and (2) the steady no-load velocity of the actuator. For both of these conditions the displacement of the open center valve is usually taken to be 1/4 of the underlapped valve dimension u that is shown in Figure 8-1. A specified working valve displacement of this magnitude is helpful since this value keeps the valve operating near the vicinity of the null position about which the valve flow equations have been linearized and it provides a margin of excessive capability for the system design. By using the following parameters for the steady working conditions of the hydraulic control system:

$$x = \frac{1}{4}u, \qquad P_A = \frac{3}{4}P_s, \qquad \text{and} \qquad \dot{y} = 0, \tag{8.18}$$

Equation (8.17) may be used to write the following equation that will satisfy the working specifications of the design:

$$K_c P_s = K_q u. \tag{8.19}$$

Similarly, by using the following parameters to identify the no-load velocity conditions of the hydraulic control system:

$$x = \frac{1}{4}u, \qquad P_A = \frac{1}{2}P_s, \qquad \text{and} \qquad \dot{y} = v_o, \tag{8.20}$$

where v_o is the no-load velocity that is specified when the load disturbance force F is zero, Equation (8.17) may be used to write the following equation that will satisfy the no-load velocity requirements of the design:

$$0 = K_q u - \frac{2A_A}{\eta_{a_v}} v_o. \tag{8.21}$$

A simultaneous solution for Equations (8.19) and (8.21) produces the following specifications for the valve flow gain, the pressure-flow coefficient, and the pressure sensitivity, respectively:

$$K_q = \frac{2A_A v_o}{u \eta_{a_v}}, \quad K_c = \frac{2A_A v_o}{P_s \eta_{a_v}}, \qquad \text{and} \qquad K_p = \frac{K_q}{K_c} = \frac{P_s}{u}. \tag{8.22}$$

In Chapter 4 it was shown that the valve coefficients are heavily dependent on the shape of the flow passages that are used to design the valve. If rectangular flow passages are used in the design, it can be shown that the flow gain and pressure-flow coefficient are given by the following expressions:

$$K_q = h\,C_d \sqrt{\frac{P_s}{\rho}} \qquad \text{and} \qquad K_c = \frac{uh\,C_d}{\sqrt{\rho P_s}}, \tag{8.23}$$

where h is the height of the rectangular flow passage, C_d is the discharge coefficient for the flow passage, and ρ is the fluid density. By equating the valve coefficients in Equation (8.23) with those given in Equation (8.22) it may be shown that the nominal open area of the valve must be given by

$$uh = \frac{2A_A v_o}{C_d \sqrt{P_s/\rho}\,\eta_{a_v}}. \tag{8.24}$$

If it is assumed that the underlapped dimension of the open center valve is 1/4 of the flow passage height the following design characteristics of the valve may be written:

$$u = \sqrt{\frac{A_A v_o}{2\,C_d \sqrt{P_s/\rho}\,\eta_{a_v}}} \qquad \text{and} \qquad h = 4\,u. \tag{8.25}$$

From these results it can be seen that as the actuator area A_A and the no-load velocity requirement v_o becomes large the valve size must also grow accordingly. Thus, the valve must be matched with the load requirements in order to satisfy the overall design objectives of the hydraulic control system.

So far in our consideration of the valve design we have only addressed issues regarding the standard flow coefficients and the nominal flow passage size that is required for adequate performance of the actuator. Issues regarding the physical design of the valve and the means of actuation for the valve have not been considered in this discussion. For consideration of these design aspects the reader is referred to the more in-depth valve analysis that is presented in Chapter 4.

Pump Design. Now that the linear actuator and the control valve have been designed it is time to specify the hydraulic pump that is used to power the hydraulic control system. As shown in Figure 8-1 this pump is a fixed-displacement pump that produces a volumetric flow rate that is proportional to the angular input speed of the pump shaft ω. Within the pump itself there is internal leakage that results in a reduction of the pump volumetric efficiency η_{p_v}. If we assume for now that the relief valve shown in Figure 8-1 is closed we may see that the supply flow to the four-way control valve is given by

$$Q_s = \eta_{p_v} V_p \omega, \tag{8.26}$$

where V_p is the volumetric displacement of the pump per unit of rotation and η_{p_v} is the volumetric efficiency of the pump. From Figure 8-1 it can be seen that the supply flow must equal the sum of the volumetric flow rates that are crossing metering lands 2 and 3 on the four-way spool valve. Using Equation (8.7) it may be shown that this supply flow is given by

$$Q_s = Q_2 + Q_3 = 3K_c P_s - K_c(P_A + P_B). \tag{8.27}$$

From Equation (8.9) it can be shown that

$$P_A + P_B = P_s - \frac{(A_A - A_B)}{2K_c \eta_{a_v}} \dot{y}. \tag{8.28}$$

Substituting Equation (8.28) into Equation (8.27) the following result may be written for the supplied volumetric flow rate to the valve:

$$Q_s = 2K_c P_s + \frac{(A_A - A_B)}{2\eta_{a_v}} \dot{y}. \tag{8.29}$$

Finally, substituting Equation (8.29) into Equation (8.26) it may be shown that the required volumetric displacement for the supply pump is given by

$$V_p = \frac{2K_c P_s}{\eta_{p_v} \omega} + \frac{(A_A - A_B)v_o}{2\eta_{p_v} \eta_{a_v} \omega}, \tag{8.30}$$

where $\dot{y}$ has been replaced with the no-load velocity requirement for the system v_o.

So far this discussion has been concerned with the appropriate sizing of the supply pump for the hydraulic control system shown in Figure 8-1. Though the pump size has been specified for the system based on certain system needs, nothing has been said about the physical construction of the pump. In practice this pump may be a gear pump, or it may be an axial-piston pump, or it may be any positive-displacement pump that satisfies the volumetric-displacement requirement shown in Equation (8.30). Usually the pump construction type is selected based on the required supply pressure of the system and the desired operating efficiency of the pump. As stated in Chapter 5 axial-piston pumps tend to operate at higher pressures and provide a higher operating efficiency as compared to gear and vane type pumps. On the other hand axial-piston pumps tend to increase the upfront cost of the control system. As the reader knows all of these things must be taken into account when selecting the pump construction type that will be used in the control system design.

Input Power Design.   The input power that is required to operate the hydraulic control system shown in Figure 8-1 is given by the standard power equation

$$\Pi = \frac{V_p \omega P_s}{\eta_p}, \tag{8.31}$$

where η_p is the overall efficiency of the pump given by the product of the volumetric and torque efficiency values $\eta_p = \eta_{p_t} \eta_{p_v}$. Obviously, by using an inefficient pump the input power requirement for this system goes up. This input power requirement is especially sensitive to the volumetric efficiency of the pump as the volumetric efficiency also tends to increase the required pump displacement V_p. Other parameters in Equation (8.31) are determined by the specified functional requirements of the hydraulic control system as noted in the previous discussion.

Case Study

To illustrate the design approach for a system like the one shown in Figure 8-1 we consider the following case study: A garbage truck is equipped with a trash compactor that is driven by a four-way valve-controlled, double rod linear actuator; that is, $A_A = A_B$. The trash compactor is expected to generate a compaction force that is equal to 10 kN. The no-load velocity of the actuator is specified to be 150 mm/s and the stroke of the actuator is ± 300 mm. The force and volumetric efficiencies for the linear actuator are 92% and 96%, respectively. The hydraulic pump that supplies fluid power to the control system is connected to the main engine of the garbage truck using a one-to-one speed ratio. During the trash compaction operation the truck engine operates at 2000 rpm. The pressure setting on the supply line relief valve is 20 MPa and the torque and volumetric efficiencies of the supply pump are 96% and 92%, respectively. You are to specify the actuator size, the valve coefficients, and the volumetric displacement of the pump for this control system. Also, identify the amount of power and torque that is required from the garbage truck engine to operate the control system.

The solution to this design problem begins by specifying the actuator size that meets the working load requirements of the trash compaction system. For the double rod actuator Equation (8.16) may be used to yield

$$A_A = A_B = \frac{2F_w}{\eta_{a_f} P_s} = \frac{2 \times 10 \text{ kN}}{0.92 \times 20 \text{ MPa}} = 1.087 \times 10^{-3} \text{ m}^2.$$

If it is assumed that the rod diameter is ¾ of the piston diameter it can be shown that the rod diameter is 42.18 mm and the piston diameter is 56.24 mm. The stroke of the actuator is given in the problem statement as ± 300 mm.

Now that the actuator has been designed we move toward specifying the valve coefficients for the four-way control valve shown in Figure 8-1. These coefficients are dependent on the porting geometry of the valve. For rectangular ports the underlapped dimension u and the port height h are given in Equation (8.25). If we assume that the discharge coefficient for the valve is given by $C_d = 0.62$ and that the fluid density is given by $\rho = 850 \, \text{kg/m}^3$ Equation (8.25) may be used to calculate the porting dimensions for the rectangular flow passage as $u = 0.94 \, \text{mm}$ and $h = 3.78 \, \text{mm}$. These results are based on the calculated actuator area $A_A = 1.087 \times 10^{-3} \, \text{m}^2$ and the no-load velocity requirement that has been stated in the problem $v_o = 0.15 \, \text{m/s}$. Using these results with Equation (8.22) the valve coefficients may be calculated as

$$K_q = \frac{2A_A v_o}{u \eta_{a_v}} = \frac{2 \times 1.087 \times 10^{-3} \, \text{m}^2 \times 0.15 \, \text{m/s}}{0.94 \, \text{mm} \times 0.96} = 0.359 \frac{\text{m}^2}{\text{s}}$$

$$K_c = \frac{2A_A v_o}{P_s \eta_{a_v}} = \frac{2 \times 1.087 \times 10^{-3} \, \text{m}^2 \times 0.15 \, \text{m/s}}{20 \, \text{MPa} \times 0.96} = 1.70 \times 10^{-11} \frac{\text{m}^3}{\text{Pa s}}$$

$$K_p = \frac{K_q}{K_c} = \frac{P_s}{u} = \frac{20 \, \text{MPa}}{0.94 \, \text{mm}} = 2.12 \times 10^{10} \frac{\text{Pa}}{\text{m}}.$$

These equations describe the required flow characteristics of the spool valve. Nothing has been said about the actuation mechanism of the valve; however, for a garbage truck application it is likely that the valve will be moved using a manual lever.

Now that the actuator and valve have been designed the volumetric displacement of the supply pump must be specified. This specification is governed by Equation (8.30). Using this result the specified volumetric displacement for the pump may be calculated as

$$V_p = \frac{2K_c P_s}{\eta_{p_v} \omega} = \frac{2 \times 1.70 \times 10^{-11} \, \text{m}^3/(\text{Pa s}) \times 20 \, \text{MPa}}{0.92 \times 2000 \, \text{rpm}} = 22 \frac{\text{cm}^3}{\text{rev}}.$$

Note: In this result the no-load velocity term vanishes due to the double rod actuator design.

Finally, the amount of power that is required from the engine of the garbage truck must be evaluated. This power is governed by Equation (8.31) and may be calculated as follows:

$$\Pi = \frac{V_p \omega P_s}{\eta_p} = \frac{22 \, \text{cm}^3/\text{rev} \times 2000 \, \text{rpm} \times 20 \, \text{MPa}}{0.96 \times 0.92} = 16.72 \, \text{kW}.$$

The reader will recall that the overall efficiency η_p is the product of the torque and volumetric efficiencies that have been specified in the problem statement. Recognizing that the input torque to the pump is equal to the input power divided by the shaft speed the input torque requirement may be calculated as

$$T = \frac{\Pi}{\omega} = \frac{16.72 \, \text{kW}}{2000 \, \text{rpm}} = 79.8 \, \text{N m}.$$

For this application the garbage truck engine must be capable of generating this amount of torque at the specified rotational speed of 2000 rpm.

Design Summary. In this subsection of the chapter the steady-state design characteristics of the four-way valve-controlled, single rod linear actuator shown in Figure 8-1 are considered. As previously noted this system is extremely common in its use throughout the hydraulic control industry. In various applications this system may be used to control the kinematic properties of the load or even the load force itself. In the following subsection standard control designs for these various applications are considered.

8.2.4 Control

Position Control.   As previously noted the four-way valve-controlled linear actuator shown in Figure 8-1 is traditionally used to accomplish a number of control objectives. One of the more common objectives of this control system is to accurately position the load at a prescribed location within the trajectory range of the actuator. Typically, this control function is carried out under slow-moving conditions of the actuator; therefore, the plant description for this control problem may safely neglect any transient contributions that would normally be significant during high-speed operations. With this being the case Equation (8.10) may be used to describe the controlled system by neglecting the transient terms of inertia and viscous damping. This result is given by

$$ky = \eta_{a_f}(A_A + A_B)\, \mathrm{K}_p\, x - F, \tag{8.32}$$

where K_p is the pressure sensitivity of the control valve, η_{a_f} is the force efficiency of the actuator, and the other parameters of Equation (8.32) are shown in Figure 8-1. Using a standard proportional-integral (PI) controller the control law for the displacement of the four-way spool valve may be written as

$$x = \mathrm{K}_e\,(y_d - y) + \mathrm{K}_i \int (y_d - y)\, dt, \tag{8.33}$$

where K_e is the proportional gain, K_i is the integral gain, and y_d is the desired position of the load. The most practical way to enforce the control law that has been defined in Equation (8.33) is to use an electrohydraulic position control for the four-way spool valve coupled with a microprocessor that is capable of reading feedback information and generating the appropriate output signal for the valve actuator. A typical electrohydraulic valve that may be used in this application is shown in Figure 4-15 of Chapter 4. For more information on using microprocessors in the control loop the reader is referred to practical texts that address this issue [3, 4]. Figure 8-2 shows the block diagram for the position control that is described by Equations (8.32) and (8.33).

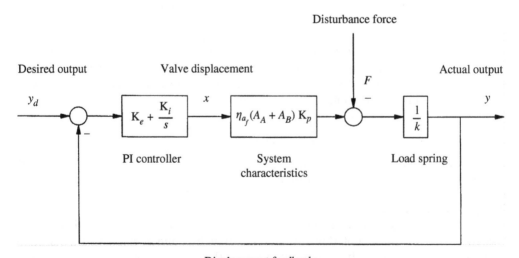

Figure 8-2. Position control of the four-way valve-controlled linear actuator.

If we assume that the position control of the system shown in Figure 8-1 is a regulation control problem and that the load force F and the desired position of the load y_d are constants Equations (8.32) and (8.33) may both be differentiated and combined to write the following equation of motion for the controlled system:

$$\tau \dot{y} + y = y_d. \tag{8.34}$$

In this equation the system time constant is given by

$$\tau = \frac{1}{K_i} \left(\frac{k}{\eta_{a_f}(A_A + A_B) K_p} + K_e \right). \tag{8.35}$$

The reader will recognize Equation (8.34) as a description of a first-order system that asymptotically approaches the desired position of the load in time without exhibiting any overshoot or oscillation. As time goes to infinity the steady-state error of this system vanishes, which is the primary advantage that is gained by using the integral component of the PI controller. The settling time for the first-order system is given by $t_s = 4\,\tau$.

The design objective for the position control problem is to shape the first-order response by designing the appropriate time constant for the system. By making a proper selection of the proportional and integral control gains this time constant can be either increased or diminished based on the desired settling time for the control system output. Though the first-order controlled system of Equation (8.34) shows a high degree of simplicity, which is a good thing, the reader must keep in mind that this system response is based on a slow-moving device and that if the desired settling time becomes too short higher-order dynamics will become significant and a more sophisticated analysis of the system will be necessary to ensure that the system response is acceptable.

Velocity Control.   Another common control objective for the system shown in Figure 8-1 is that of velocity control. This objective seeks to establish a specific output velocity for the load based on a desired velocity that is prescribed by the application. This objective is typically carried out for load systems that do not include a load spring. Therefore, by neglecting the spring that is shown in Figure 8-1; that is, setting $k = 0$, the equation of motion for the velocity-controlled system may be written using the general form of Equation (8.10). This result is given by

$$m\dot{v} + \left(c + \frac{A_A^2 + A_B^2}{2K_c} \right) v = \eta_{a_f}(A_A + A_B) K_p\, x - F, \tag{8.36}$$

where v is the instantaneous velocity of the load. Again, K_c and K_p are the pressure-flow coefficient and the pressure sensitivity of the four-way control valve respectively and η_{a_f} is the force efficiency of the actuator. All other parameters of Equation (8.36) are shown in Figure 8-1. For the velocity control objective we select a full proportional-integral-derivative (PID) controller with the following control law for the spool valve:

$$x = K_e (v_d - v) + K_i \int (v_d - v)\, dt + K_d \frac{d}{dt}(v_d - v), \tag{8.37}$$

where K_e is the proportional gain, K_i is the integral gain, K_d is the derivative gain, and v_d is the desired velocity of the load. Again, the most practical way to enforce the control law that has been defined in Equation (8.37) is to use an electrohydraulic position control for

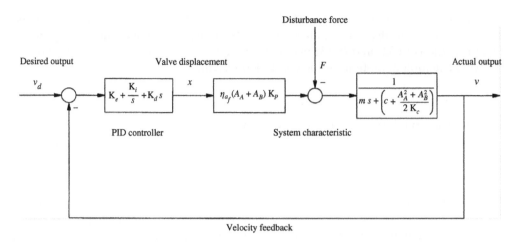

Figure 8-3. Velocity control of the four-way valve-controlled linear actuator.

the four-way spool valve coupled with a microprocessor that is capable of reading feedback information and generating the appropriate output signal for the valve actuator. Figure 8-3 shows the block diagram for the velocity control that is described by Equations (8.36) and (8.37).

If it is assumed that the velocity control of the system shown in Figure 8-1 is a regulation control problem and that the load force F and the desired velocity of the load v_d are constants Equations (8.36) and (8.37) may both be differentiated and combined to write the following equation of motion for the controlled system:

$$\ddot{v} + 2\zeta\omega_n\dot{v} + \omega_n^2 v = \omega_n^2 v_d, \tag{8.38}$$

where the undamped natural frequency and the damping ratio are given respectively as

$$\omega_n = \sqrt{\frac{K_i}{\dfrac{m}{\eta_{a_f}(A_A + A_B)K_p} + K_d}} \quad \text{and}$$

$$\zeta = \frac{\left(\dfrac{c}{\eta_{a_f}(A_A + A_B)K_p} + \dfrac{A_A^2 + A_B^2}{2\eta_{a_f}(A_A + A_B)K_q} + K_e\right)}{2\sqrt{K_i\left(\dfrac{m}{\eta_{a_f}(A_A + A_B)K_p} + K_d\right)}}. \tag{8.39}$$

The reader will recognize Equation (8.38) as a description of a second-order system that may exhibit overshoot and oscillation before reaching a steady-state condition in time. Whether overshoot and oscillation is observed depends on whether the damping ratio ζ is less than or greater than unity. If the damping ratio is less than unity the system is an underdamped system and the output will be characterized by overshoot and oscillation before reaching a steady-state condition. On the other hand, if the damping ratio is greater than unity the system is an overdamped system and the output will appear to be very much like a first-order system that asymptotically approaches a steady-state condition without any overshoot or oscillation. If it is assumed that the controlled system is of the underdamped variety, then the rise time, settling time, and maximum percent overshoot can be calculated according to the their definitions in Chapter 3. The settling time is given by $t_s = 4/(\omega_n \zeta)$.

The design objective for the velocity control problem is to shape the second-order response by designing the appropriate undamped natural frequency and damping ratio for the system. By making the proper selection of the proportional, integral, and derivative control gains the undamped natural frequency and the damping ratio for the control system may be adjusted accordingly. From Equation (8.39) it can be seen that the proportional gain provides an adjustment for the effective damping of the system while the integral and derivative gains are used to adjust the effective spring rate and mass of the system respectively. If the derivative gain is neglected, that is, a PI control is used instead of the full PID control, it can be seen from Equation (8.39) that the undamped natural frequency of the system may be adjusted by solely adjusting the integral gain K_i. The reader will recall that the undamped natural frequency is the primary descriptor for the rise time of the underdamped second-order system. Once the desired natural frequency has been obtained the damping ratio for the PI-controlled system may be adjusted by adjusting the proportional gain K_e. The damping ratio is the sole contributor to the maximum percent overshoot. This series approach to tuning the PI controller is very convenient and has been noted as one of its primary advantages in Chapter 3.

Force Control. Finally, another control objective that is occasionally used for the hydraulic control system shown in Figure 8-1 is that of force control. Force-controlled systems are usually configured very much like a velocity-controlled system without a load spring and when they are used they often switch between velocity and force control depending on the immediate needs of the application. Furthermore, force-controlled systems are typically operated in slow motion so as to gradually apply the load force to whatever the application is trying to resist. Under these slow motion conditions the load inertia and viscous damping effects in the system may be safely ignored without sacrificing accuracy in the modeling process. By setting the left-handside of Equation (8.10) equal to zero the following equation may be used to describe the applied load force for the force-controlled system:

$$F = \eta_{a_f}(A_A + A_B) K_p \, x. \tag{8.40}$$

Using a standard proportional-integral (PI) controller the control law for the displacement of the four-way spool valve may be written as

$$x = K_e (F_d - F) + K_i \int (F_d - F) \, dt, \tag{8.41}$$

where K_e is the proportional gain, K_i is the integral gain, and F_d is the desired force that is to be exerted on the load. In most applications, the instantaneous load force F is measured by sensing the fluid pressures in sides A and B of the linear actuator and calculating the load force according to Equation (8.11). Again, a microprocessor is necessary to carry out these functions and an electrohydraulic position control for the four-way spool valve is used to enforce the control law of Equation (8.41). Figure 8-4 shows the block diagram for the force control that is described by Equations (8.40) and (8.41).

If it is assumed that the force control of the system shown in Figure 8-1 is a regulation control problem and that the desired load force F_d is a constant Equations (8.40) and (8.41) may both be differentiated and combined to write the following dynamic equation for the force-controlled system:

$$\tau \dot{F} + F = F_d, \tag{8.42}$$

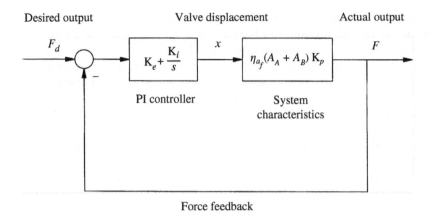

Desired output   Valve displacement   Actual output

Figure 8-4. Force control of the four-way valve-controlled linear actuator.

where the system time constant is given by

$$\tau = \frac{1}{K_i} \left(\frac{1}{\eta_{a_f}(A_A + A_B)\, K_p} + K_e \right). \tag{8.43}$$

Again, the reader will recognize Equation (8.42) as a description of a first-order system that asymptotically approaches the desired load force F_d in time without exhibiting any overshoot or oscillation. The dynamic similarities of this system should be compared to the position-controlled system that was previously discussed.

The design objective for the force control problem is to shape the first-order response by designing the appropriate time constant for the system. By making a proper selection of the proportional and integral control gains this time constant can be either increased or diminished based on the desired settling time for the control system output. Again, the reader is reminded that the first-order controlled system of Equation (8.42) has been designed based on the assumption that the linear actuator moves slowly. If in fact the desired response time for this system becomes too short higher-order dynamics that have been neglected in the modeling process will become more significant and the observed output may be unsatisfactory. Under such conditions a more sophisticated analysis of the system is needed to ensure satisfactory response characteristics.

Case Study

To illustrate the force control objective that has been described for the linear actuation system shown in Figure 8-1 we extend the case study that was presented in the previous subsection of this chapter. The reader will recall that this case study was concerned with a trash compaction system to be used on a mobile garbage truck with the functional requirement of generating a 10 kN compaction force. Based on this requirement and a specified no-load velocity requirement of 150 mm/s the hydraulic control system was designed so that the pressurized area of the actuator was given by $A_A = A_B = 1.087 \times 10^{-3}$ m^2. Furthermore, the pressure sensitivity K_p for the four-way control valve was designed to be 2.12×10^{10} Pa/m and the underlapped valve dimension was given by $u = 0.94$ mm. Based on these design values you are to select the appropriate control gains for the PI controller that will produce a settling time of 12 seconds for the compaction force control system.

The solution to this problem begins by recognizing that the settling time for the first-order control system is given by $t_s = 4\,\tau$, where the time constant τ is given in Equation (8.43). If we design the proportional gain such that

$$K_e = \frac{4}{5}\frac{x_{\max}}{F_{\max}} = \frac{4}{5}\frac{u/2}{F_w} = \frac{4}{5}\frac{0.94\text{ mm}/2}{10\text{ kN}} = 3.76 \times 10^{-8}\,\frac{\text{m}}{\text{N}},$$

the required integral gain for this system may be determined as

$$K_i = \frac{4}{t_s}\left(\frac{1}{\eta_{f_a}(A_A + A_B)\,K_p} + K_e\right)$$

$$= \frac{4}{12\text{ s}}\left(\frac{1}{0.92 \times 2.17 \times 10^{-3}\text{ m}^2 \times 2.12\text{E}+10\text{ Pa/m}} + 3.76 \times 10^{-8}\,\frac{\text{m}}{\text{N}}\right)$$

$$= 2.04 \times 10^{-8}\,\frac{\text{m}}{\text{N s}}.$$

As shown by this example the system design parameters have a significant impact on the dynamic response of the force control as these parameters produce dynamic contributions that are on the scale of the reasonable proportional gain that has been specified.

Control Summary. In this subsection of the chapter we consider the three control objectives that are typically carried out using the four-way valve-controlled linear actuator shown in Figure 8-1. These objectives include position control, velocity control, and force control all of which have employed a form of the standard PID controller. For the position control and the force control an important slow speed assumption is employed for the purposes of simplifying the control problem for realistic modes of operation. Due to the slow speed assumptions behind these control strategies the reader is cautioned about using these controllers for high-speed applications that are rarely encountered for these control objectives. Under such high-speed conditions higher-order dynamics may tend to manifest themselves and may produce a less than desirable system output.

8.2.5 Summary

In this section of the chapter the four-way valve-controlled linear actuator of Figure 8-1 is analyzed, designed, and controlled. It should be noted that throughout this study certain speed conditions have been assumed regarding the operation of the control system. For instance, during the development of the governing system equations in Subsection 8.2.2 the pressure transient effects were neglected due to the assumption that they occurred very quickly in relationship to the overall process time of the load dynamics. If indeed this is not the case a higher-order model of the control system will be necessary to capture the relevant pressure dynamics of the system. This situation, however, will only occur for very rare and specialized applications and the analysis presented here is deemed to be sufficient for traditional and more typical applications of the system. In Subsection 8.2.3 a design methodology for the valve-controlled linear actuator is presented. This design method begins by considering the steady output objectives of the system and then works backward to identify the hydraulic components that are necessary to achieve these objectives. As shown in this subsection the actuator, valve, and pump are closely linked in their design and the proper specification of these components is important when seeking to satisfy the overall performance objectives for the system. Finally, in Subsection 8.2.4 the dynamic control objectives of the actuation system are presented. These objectives vary depending on the application; however, they generally consist of either a position control, a velocity control, or a force control. In the consideration of these control objectives it has been assumed that the control law can be implemented by using an electrohydraulic control valve and a microprocessor with sufficient sensing of the controlled parameters. A typical electrohydraulic control valve that may be used for these applications has been presented in Subsection 4.5.4 of Chapter 4. For information on implementing the microprocessor within the control design the reader is referred to practical texts dealing with this subject [3, 4].

8.3 THREE-WAY VALVE CONTROL OF A SINGLE-ROD LINEAR ACTUATOR

8.3.1 Description

In this section of our chapter we introduce the three-way valve-controlled single-rod linear actuator as shown in Figure 8-5. This system has many things in common with the four-way valve-controlled linear actuator shown in Figure 8-1; however, there are some differences as well. While both actuators may be used to control position, velocity, and force they each use a slightly different mechanical approach for doing so. The reader will recall that the four-way valve-controlled linear actuator was controlled by adjusting the fluid pressure on both sides of the actuator; but, in the case of the three-way valve design shown in Figure 8-5 the fluid pressure is only controlled on side A while the fluid pressure on side B is held

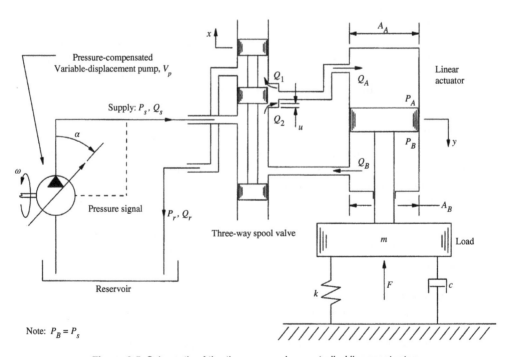

Figure 8-5. Schematic of the three-way valve-controlled linear actuator.

constant by the system supply pressure P_s. This mode of control requires that the three-way valve be used with a single-rod linear actuator as a double-rod linear actuator would not be capable of generating enough force to move the actuator against the load.

Again, the reader will recall that the linear actuator in Figure 8-1 was generalized to include the analysis of a double-rod actuator by simply setting the pressurized areas within the actuator equal to each other. This cannot be done for the three-way valve-controlled system; therefore, the system in Figure 8-5 cannot be generalized to include the use of a double-rod design.

The hydraulic system shown in Figure 8-5 utilizes a standard three-way spool valve to control the output characteristics of a single-rod linear actuator. In this figure the load to be moved by the actuator is shown as a single mass-spring-damper system with a load disturbance force given by F. The mass, spring rate, and viscous drag coefficient for the load are shown in Figure 8-5 by the symbols m, c, and k, respectively. The linear actuator is shown to be constructed using a single rod, which is connected to both the load and the actuator piston. Due to the single-rod design the pressurized area on the top of the piston A_A is greater than the pressurized area on the bottom of the piston A_B. The volumetric flow of hydraulic fluid into side A of the actuator is controlled by the three-way control valve shown to the left of the linear actuator. This valve is shown to be constructed as a symmetric open center design with an underlapped dimension on the spool given by the dimension u. The displacement of the spool valve x is shown to be positive in the upward direction. As the valve is moved in the positive x-direction flow is directed into side A of the actuator which then requires flow to exit the actuator from side B. These two volumetric flow rates are shown in Figure 8-5 by the symbols Q_A and Q_B, respectively. In order to provide a hydraulic power source for the valve-controlled system Figure 8-5 shows a variable-displacement pump that provides a sufficient volumetric flow rate into the supply line so as to maintain a prescribed supply pressure for the system. In other words, the displacement of the supply pump is controlled by a pressure-compensating design similar to what has been presented and discussed in Chapter 5. The maximum pump displacement per unit of rotation is given by the symbol V_p while the symbol α shown in Figure 8-5 is the swash plate angle of the pump that is used to decrease the volumetric displacement when a lower flow demand is required from the system. The pump is driven by an external power source not shown in Figure 8-1 at an angular velocity ω.

8.3.2   Analysis

Load Analysis.   The analysis of the three-way valve-controlled linear actuator begins by considering the output dynamics of the load. These dynamics are preferred above the dynamics of the actuator itself because the actuator dynamics are either very small due to insignificant inertial effects or very fast due to the nearly incompressible nature of the hydraulic fluid that is used to activate the actuator. A brief comparison of Figure 8-5 with Figure 8-1 shows that the basic construction of the load and linear actuator for both the three-way and four-way valve-controlled systems are identical. Therefore, Equation (8.1) may be used to describe the load dynamics for the three-way valve-controlled system shown in Figure 8-5. This equation is rewritten here for convenience:

$$m\ddot{y} + c\dot{y} + ky = \eta_{a_f}(A_A P_A - A_B P_B) - F - F_o, \tag{8.44}$$

where P_A and P_B are the fluid pressures on the A and B sides of the actuator, respectively, η_{a_f} is the force efficiency of the actuator, and where F_o is the nominal spring or bias load that is applied to the actuator when y equals zero. At the nominal steady-state operating conditions of the system the following design conditions apply:

$$y = 0, \qquad P_A = P_s/2, \qquad P_B = P_s, \qquad \text{and} \qquad F = 0. \tag{8.45}$$

Under these conditions it can be shown from the steady form of Equation (8.44) that the nominal force exerted on the actuator by the load is given as

$$F_o = \eta_{a_f} \left(A_A \frac{P_s}{2} - A_B P_s \right). \tag{8.46}$$

Substituting this result back into Equation (8.44) produces the following equation for describing the output dynamics of the load:

$$m\ddot{y} + c\dot{y} + ky = \eta_{a_f}[A_A (P_A - P_s/2) - A_B (P_B - P_s)] - F. \tag{8.47}$$

From Equation (8.47) it can be seen that the pressures P_A and P_B are the required inputs for adjusting the output dynamics of the load. These pressures result from the changing flow and volumetric conditions of the actuator itself and are considered in the next paragraph.

Pressure Analysis. As previously stated, for most applications the transient behavior of the fluid pressure within the actuator is extremely fast compared to the transient behavior of the load itself. Again, this may not be the case for systems that utilize long transmission lines between the control valve and the actuator, or for systems that operate using a fluid with a large amount of entrained air. For now we assume that this is not the case and that the more typical system conditions describe our problem. Using these assumptions we can say for the incompressible fluid that the volumetric flow rate into side A of the actuator is

$$Q_A = \frac{A_A}{\eta_{a_v}} \dot{y}, \tag{8.48}$$

where η_{a_v} is the volumetric efficiency of the actuator. From the volumetric flow rates that are shown at the three-way spool valve in Figure 8-5 it may be seen that

$$Q_A = Q_2 - Q_1. \tag{8.49}$$

Using Equation (8.7) with this equation it can be shown that the flow rate into side A of the actuator is given by

$$Q_A = 2K_q x - 2K_c (P_A - P_s/2), \tag{8.50}$$

where K_q and K_c are the flow gain and the pressure-flow coefficient for the valve, respectively. Using this result with Equation (8.48) the following equation for the operating pressures on side A of the linear actuator may be written as

$$P_A - P_s/2 = K_p x - \frac{A_A}{2K_c \eta_{a_v}} \dot{y}, \tag{8.51}$$

where K_p is the valve's pressure sensitivity given by the ratio of the flow gain to the pressure-flow coefficient K_q/K_c. This result is same as the one presented in Equation (8.9).

For the three-way valve-controlled system shown in Figure 8-5 the fluid pressure on side B of the linear actuator is given by

$$P_B = P_s. \tag{8.52}$$

Note: This pressure feature on side B is the primary difference between the three-way and four-way valve design that is shown in Figures 8-5 and 8-1, respectively. From Equation (8.51) it can be seen that the fluid pressure on side A of the linear actuator is increased by moving the spool valve in the positive x-direction. As is shown in the summary of the next paragraph this adjustment in fluid pressure by the motion of the spool valve is the mechanism for adjusting the output motion of the load shown in Figure 8-5. *Note:* Equation (8.51) shows a linear velocity dependence for the fluid pressure as well. As to be shown subsequently this feature results in favorable damping characteristics for the system.

Analysis Summary.   To summarize the analysis of the linear actuation system shown in Figure 8-5 we substitute Equation (8.51) and (8.52) into Equation (8.47) to produce the following equation of motion for the system:

$$m\ddot{y} + \left(c + \frac{A_A^2}{2K_c} \right) \dot{y} + ky = \eta_{a_f} A_A K_p\, x - F, \tag{8.53}$$

where it has been assumed that $\eta_{a_f}/\eta_{a_v} \approx 1$. From this equation it can be seen that the mechanical design of the linear actuator and the three-way spool valve have a decisive impact on the overall dynamics of the hydraulic control system. In particular, it may be seen from Equation (8.53) that these design parameters help to shape the effective damping of the system and to provide an adequate gain relationship between the input motion of the spool valve and the output motion of the load. It is also important to remember that this result assumes that the supply pressure P_s is controlled very well and that its dynamics have a negligible impact on the overall system performance. The following subsection of this chapter considers the mechanical design parameters that must be specified in order to achieve a satisfactory output of the system from a steady-state perspective.

8.3.3 Design

Actuator Design.   As with all hydraulic system design work it is necessary to consider the actuator design first since this is the device that performs the output function that is specified by the end user. In other words, the linear actuator that is shown in Figure 8-5 must be designed in accordance with the expected load requirements for the system. Usually, for a given application the working load force is known and can be used to determine the required pressurized areas that are needed within the actuator to develop this load for a specified working pressure. If we consider the right-hand side of Equation (8.47) it is clear that the steady working force generated by the single rod actuator is given by

$$F = \eta_{a_f}[A_A\,(P_A - P_s/2) - A_B\,(P_B - P_s)], \tag{8.54}$$

where A_A and A_B are the designed pressurized areas on sides A and B of the actuator, and P_A and P_B are the fluid pressures on the each side of the actuator. At the working force that is specified for the application it is common to use a fluid pressure on side A of the linear actuator that is less than the full supply pressure so as to provide a margin of excessive force capability for the system. Typical design specifications at the working force conditions are given by

$$F = F_w, \qquad P_A = \frac{3}{4} P_s, \qquad \text{and} \qquad P_B = P_s, \tag{8.55}$$

where F_w is the working force of the hydraulic control system and P_s is the supply pressure of the hydraulic system. Substituting these conditions into Equation (8.54) produces the following equation for sizing the pressurized area on side A of the linear actuator that is shown in Figure 8-5:

$$A_A = 4 \frac{F_w}{\eta_{a_f} P_s}. \tag{8.56}$$

Using this result with Equation (8.46) it can also be shown that the pressurized area on side B of the actuator is given by

$$A_B = \frac{A_A}{2} - \frac{F_o}{\eta_{a_f} P_s} = \frac{2F_w - F_o}{\eta_{a_f} P_s}, \tag{8.57}$$

where F_o is the nominal bias load when $y = 0$.

These results may be used to design or select a linear actuator that will provide a sufficient working force for a given supply pressure. Other design considerations must also be taken into account when designing the linear actuator. For example, the stroke of the actuator must also be sufficient for the application. Machine design considerations must also be taken into account to avoid premature failure of the actuator. Again, the reader is referred to standard machine design texts for this information [1, 2].

Valve Design.   As it turns out the valve design for the three-way valve-controlled linear actuator follows the same basic principles as that of the four-way valve-controlled linear actuator that was discussed in Subsection 8.2.3. For the valve design governing equations the reader is referred to Equations (8.22) and (8.25).

Pump Design.   Now that the linear actuator and the control valve have been designed it is time to specify the hydraulic pump that will be used to power the hydraulic control system. As shown in Figure 8-5 this pump is a variable-displacement pump that produces a volumetric flow rate that is proportional to the angular input speed of the pump shaft ω. The instantaneous displacement of the pump is also proportional to the swash-plate angle α; therefore, the supply flow to the three-way control valve may be described using the following equation:

$$Q_s = \eta_{p_v} V_p \frac{\alpha}{\alpha_{\max}} \omega, \tag{8.58}$$

where V_p is the maximum volumetric displacement of the pump per unit of rotation, $\alpha_{\max}$ is the maximum swash plate angle of the pump, and η_{p_v} is the pump volumetric efficiency. *Note:* The volumetric efficiency of the pump is known to change with the swash plate angle, the system pressure, and the input speed. From Figure 8-5 it can be seen that the supply flow must equal the volumetric flow rate that is crossing metering land 2 minus the volumetric flow rate that is exiting the linear actuator from side B. Recognizing that the flow rate from side B is given by $Q_B = A_B \dot{y}/\eta_{a_v}$, and using Equations (8.7) and (8.51), it may be shown that the supply flow is given by

$$Q_s = Q_2 - Q_B = K_c P_s + \frac{(A_A/2 - A_B)}{\eta_{a_v}} \dot{y}. \tag{8.59}$$

Finally, substituting Equation (8.59) into Equation (8.58) it may be shown that the maximum required volumetric displacement for the supply pump is given by

$$V_p = \frac{K_c P_s}{\eta_{p_v} \omega} + \frac{(A_A/2 - A_B) v_o}{\eta_{p_v} \eta_{a_v} \omega},$$ (8.60)

where $\dot{y}$ has been replaced with the no-load velocity requirement for the system v_o and where the swash-plate angle α has been set equal to α_{max}.

So far this discussion has been concerned with the appropriate sizing of the supply pump for the hydraulic control system shown in Figure 8-5. Throughout this discussion the parameter α has been used to describe the swash-plate angle of the pump assuming that the pump is a swash plate type pump; however, this parameter could be generalized to mean any displacement control variable that adjusts the volumetric flow rate of the pump with the objective to maintain a prescribed discharge pressure P_s. The reader will recall from Chapter 5 that vane pumps, radial piston pumps, and bent-axis type pumps can also be constructed to deliver a variable output flow based on the adjustment of a mechanical parameter. In this case the symbol α simply describes the adjustment of that mechanical parameter, which is used to adjust the output flow rate of the pump. As previously noted, the pump construction type is usually selected based on the required supply pressure of the system and the desired operating efficiency of the pump. Axial-piston pumps tend to operate at higher pressures and provide a higher operating efficiency as compared to gear and vane type pumps; however, these performance benefits must be weighed against the additional upfront cost of the control system that is usually accompanied by their use.

Input Power Design.   The equation for determining the maximum input power that is required to operate the hydraulic control system shown in Figure 8-5 is identical to that of the previous system. This expression is given in Equation (8.31).

Case Study

The figure shows a grain dispenser that is used to measure the amount of grain purchased by a customer from a grain supplier. The top portion of the figure shows the cart positioned directly below the grain dispenser during the filling process; however, the cart must be repositioned as shown in the bottom portion of the figure for delivering the grain to the customer. The cart works against a spring that biases the cart toward the filling position. A three-way valve-controlled single rod linear actuator is used to position the cart throughout the process. The cart must travel a distance of 15 feet between the filling and dumping positions. The supply pressure to the hydraulic control system is 1000 psi and the spring rate for the biasing spring is 50 lbf per foot. In the filling position the biasing spring is compressed 1 foot. The force and volumetric efficiencies of the actuator are given by 91% and 97%, respectively. The no-load velocity requirement for the cart is given at 20 feet per minute. In this problem the torque and volumetric efficiencies of the pump are given by 92% and 89%, respectively, and the shaft speed for the pump is 1750 rpm. You are to specify the design parameters for the actuator, the valve, and the pump that are used to build the hydraulic control system. What is the maximum horsepower that should be available for controlling this system?

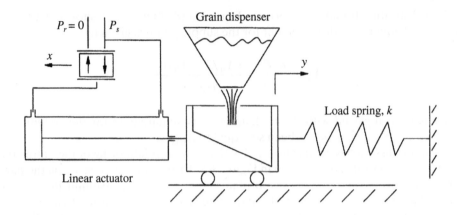

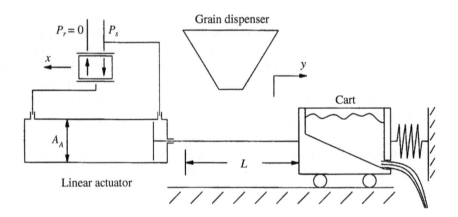

The solution to this problem begins by sizing the single rod linear actuator according to the spring preload and the working force requirement of the system. The spring preload is calculated from the spring load at the midpoint of the actuator stroke. This result is given by

$$F_o = k(L_o + L/2) = 50 \text{ lbf/ft} \times (1 \text{ ft} + 7.5 \text{ ft}) = 425 \text{ lbf}.$$

Where L_o is the initial spring compression length and L is the total stroke of the actuator. The working force for this system may be calculated similarly as

$$F_w = k(L_o + L) = 50 \text{ lbf/ft} \times (1 \text{ ft} + 15 \text{ ft}) = 800 \text{ lbf}.$$

Using these results with the available system pressure Equations (8.56) and (8.57) may be used to calculate the pressurized areas of the linear actuator:

$$A_A = 4\frac{F_w}{\eta_{af} P_s} = 4 \times \frac{800 \text{ lbf}}{0.91 \times 1000 \text{ psi}} = 3.517 \text{ in}^2$$

$$A_B = \frac{2F_w - F_o}{\eta_{af} P_s} = \frac{2 \times 800 \text{ lbf} - 425 \text{ lbf}}{0.91 \times 1000 \text{ psi}} = 1.291 \text{ in}^2.$$

From these results it can be shown that the rod diameter and piston diameter for the linear actuator are given by 1.68 in and 2.12 in, respectively. The actuator stroke is given in the problem statement as $\pm$ 7.5 ft.

Now that the actuator has been designed the valve coefficients must be specified to yield acceptable steady-state performance. These coefficients are dependent on the porting geometry of the valve. For rectangular ports the underlapped dimension u and the port height h are given in Equation (8.25). If we assume that the discharge coefficient for the valve is given by $C_d = 0.62$ and that the fluid density is given by $\rho = 0.0307$ lbm/in^3 Equation (8.25) may be used to calculate the porting dimensions for the rectangular flow passage as $u = 0.0574$ in and $h = 0.2297$ in. *Note:* In making these calculations one must be careful with the conversion from pound-mass (lbm) to pound-force (lbf). These results are based on the calculated actuator area $A_A = 3.517$ in^2 and the no-load velocity requirement that has been stated in the problem $v_o = 20$ ft/min. Using these results with Equation (8.22) the valve coefficients may be calculated as

$$K_q = \frac{2A_A v_o}{u \eta_{a_v}} = \frac{2 \times 3.517 \text{ in}^2 \times 20 \text{ ft/min}}{0.0574 \text{ in} \times 0.97} = 505.07 \frac{\text{in}^2}{\text{s}}$$

$$K_c = \frac{2A_A v_o}{P_s \eta_{a_v}} = \frac{2 \times 3.517 \text{ in}^2 \times 20 \text{ ft/min}}{1000 \text{ psi} \times 0.97} = 0.0290 \frac{\text{in}^3}{\text{psi s}}$$

$$K_p = \frac{K_q}{K_c} = \frac{P_s}{u} = \frac{1000 \text{ psi}}{0.0574 \text{ in}} = 17415 \frac{\text{psi}}{\text{in}}.$$

These equations describe the required flow characteristics of the spool valve. Nothing has been said about the actuation mechanism of the valve; however, for the grain dispensing system the displacement control is most likely an automated function. Therefore, an electrohydraulic spool valve is used with a microprocessor for the purposes of achieving the final cart position. The next subsection of this chapter deals with the control strategy of this system.

Now that the actuator and valve have been designed the maximum volumetric displacement of the supply pump must be specified. This specification is governed by Equation (8.60). Using this result the specified volumetric displacement for the pump may be calculated as

$$V_p = \frac{K_c P_s}{\eta_{p_v} \omega} + \frac{(A_A/2 - A_B)v_o}{\eta_{p_v} \eta_{a_v} \omega}$$

$$= \frac{0.0290 \text{ in}^3/(\text{psi s}) \times 1000 \text{ psi}}{0.89 \times 1750 \text{ rpm}} + \frac{(1.76 \text{ in}^2 - 1.29 \text{ in}^2) \times 20 \text{ ft/min}}{0.89 \times 0.97 \times 1750 \text{ rpm}} = 1.19 \frac{\text{in}^3}{\text{rev}}.$$

Finally, the maximum amount of power that is required to operate this system at any given time must be evaluated. This power is governed by Equation (8.31) and may be calculated as follows:

$$\Pi = \frac{V_p \omega P_s}{\eta_p} = \frac{1.19 \text{ in}^3/\text{rev} \times 1750 \text{ rpm} \times 1000 \text{ psi}}{0.92 \times 0.89}$$

$$= 6.43 \text{ hp}.$$

The reader will recall that the overall efficiency η_p is the product of the torque and volumetric efficiencies that have been specified in the problem statement.

Summary.   In this subsection of the chapter the steady-state design characteristics of the three-way valve-controlled single-rod linear actuator shown in Figure 8-5 have been considered. In the following subsection, standard control designs for various applications are presented and discussed.

8.3.4 Control

Position Control.   The three-way valve-controlled linear actuator shown in Figure 8-5 is traditionally used in position control applications such as those used in aircraft controls and tracer-controlled machine tools [5]. Typically this control function is carried out under slow-moving conditions of the actuator; therefore, the plant description for this control problem may safely neglect any transient contributions that would normally be significant during high-speed operations. With this being the case Equation (8.53) may be used to describe the controlled system by neglecting the transient terms of inertia and viscous damping. This result is given by

$$ky = \eta_{a_f} A_A K_p \, x - F, \tag{8.61}$$

where K_p is the pressure sensitivity of the control valve, η_{a_f} is the force efficiency of the actuator, and the other parameters of Equation (8.32) are shown in Figure 8-5. Using a standard proportional-integral (PI) controller the control law for the displacement of the three-way spool valve may be written as

$$x = K_e \, (y_d - y) + K_i \int (y_d - y) \, dt, \tag{8.62}$$

where K_e is the proportional gain, K_i is the integral gain, and y_d is the desired position of the load. As mentioned in the discussion of the four-way valve-controlled linear actuator, the most practical way to enforce the control law that has been defined in Equation (8.62) is to use an electrohydraulic position control for the three-way spool valve coupled with a microprocessor that is capable of reading feedback information and generating the appropriate output signal for the valve actuator. Again, typical electrohydraulic actuated valves have been presented in Chapter 4 and standard texts on the use of microprocessors are also available [3, 4]. Figure 8-6 shows the block diagram for the position control that is described by Equations (8.61) and (8.62).

If we assume that the position control of the system shown in Figure 8-5 is a regulation control problem and that the load force F and the desired position of the load y_d are constants, Equations (8.61) and (8.62) may both be differentiated and combined to write the following equation of motion for the controlled system:

$$\tau \dot{y} + y = y_d, \tag{8.63}$$

where the system time constant is given by

$$\tau = \frac{1}{K_i} \left(\frac{k}{\eta_{a_f} A_A K_p} + K_e \right). \tag{8.64}$$

Note: This result is very similar to the position control result of the four-way valve-controlled system shown in Equation (8.35). Again, the reader will recognize Equation (8.63) as a description of a first-order system that asymptotically approaches

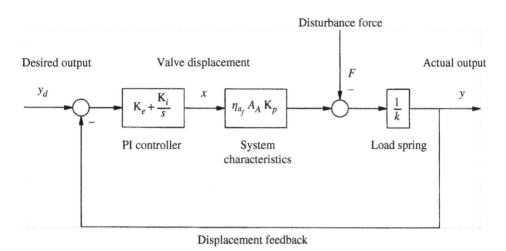

Figure 8-6. Position control of the three-way valve-controlled linear actuator.

the desired position of the load in time without exhibiting any overshoot or oscillation. Once again, the integral control has caused the steady-state error of this system to vanish as time approaches infinity. The settling time for the first-order system is given by $t_s = 4\,\tau$.

The design objective for the position control problem is to shape the first-order response by designing the appropriate time constant for the system. By making a proper selection of the proportional and integral control gains this time constant can be either increased or diminished based on the desired settling time for the control system output. It is important to remind the reader that the position control scheme that is described here is based on the assumption that the system moves at a slow enough rate as to make the effects of inertia and viscous drag negligible. If, however, the desired settling time becomes too short, thus decreasing the system time constant, higher-order dynamics will become significant and a more sophisticated analysis of the system will be necessary to ensure that the system response is acceptable.

Case Study

In the previous case study a three-way valve-controlled linear actuator was designed for controlling the position of a cart used in a grain dispensing process. In this process the no-load velocity requirement was specified as 20 feet per minute and the total distance to be traveled was given by 15 feet. The force efficiency of the actuator was 91%. The spring rate for the load spring was given by 50 lbf per foot. The pressure sensitivity for the control valve was calculated to be 17415 psi/in and the pressurized area on side A of the actuator was designed to be 3.517 in^2. Using a proportional gain that is equal to 0.13×10^{-3} (no units) calculate the integral gain that is necessary to achieve satisfactory performance of this system.

The solution to this problem begins by identifying what is meant by "satisfactory performance" for this system. From the control objective we may assume that satisfactory performance is achieved if the grain cart is moved to its new location within the time period that is expected by the system operator. In other words, we must design the controller so that it exhibits an appropriate settling time for the operation.

Using information from the problem statement we can calculate an appropriate settling time as

$$t_s = \frac{L}{v_o} = \frac{15 \text{ ft}}{20 \text{ ft/min}} = 45 \text{ s},$$

where L is the total distance to be traveled and v_o is the no-load velocity that was specified in the problem statement. Recalling that the settling time is equal to four time constants we can say that $\tau = 11.25$ s. Rearranging Equation (8.64) it may be shown that

$$K_i = \frac{1}{\tau}\left(\frac{k}{\eta_{a_f} A_A K_p} + K_e\right) = \frac{1}{11.25 \text{ s}}\left(\frac{50 \text{ lbf/ft}}{0.91 \times 3.517 \text{ in}^2 \times 17415 \text{ psi/in}} + 0.13 \times 10^{-3}\right)$$

$$= \frac{0.0187 \times 10^{-3}}{\text{s}}.$$

Thus the control gains that must be used in the microprocessor for sufficient displacement control are $K_e = 0.13 \times 10^{-3}$ and $K_i = 0.0187 \times 10^{-3}/\text{s}$.

Velocity Control.   Another control objective that is possible for the system shown in Figure 8-5 is that of velocity control. This objective seeks to establish a specific output velocity for the load based on a desired velocity that is prescribed by the application. This objective is typically carried out for load systems that do not include a load spring. Therefore, by neglecting the spring that is shown in Figure 8-5; that is, setting $k = 0$, the equation of motion for the velocity-controlled system may be written using the general form of Equation (8.53). This result is given by

$$m\dot{v} + \left(c + \frac{A_A^2}{2K_c}\right) v = \eta_{a_f} A_A K_p x - F, \tag{8.65}$$

where v is the instantaneous velocity of the load. Again, K_c and K_p are the pressure-flow coefficient and the pressure sensitivity of the three-way control valve respectively and η_{a_f} is the actuator force efficiency. All other parameters of Equation (8.65) are shown in Figure 8-5. For the velocity control objective we select a full proportional-integral-derivative (PID) controller with the following control law for the spool valve:

$$x = K_e (v_d - v) + K_i \int(v_d - v) \, dt + K_d \frac{d}{dt}(v_d - v), \tag{8.66}$$

where K_e is the proportional gain, K_i is the integral gain, K_d is the derivative gain, and v_d is the desired velocity of the load. Again, the most practical way to enforce the control law that has been defined in Equation (8.66) is to use an electrohydraulic position control for the three-way spool valve coupled with a microprocessor that is capable of reading feedback information and generating the appropriate output signal for the valve actuator. Figure 8-7 shows the block diagram for the velocity control that is described by Equations (8.65) and (8.66).

If it is assumed that the velocity control of the system shown in Figure 8-7 is a regulation control problem, and that the load force F and the desired velocity of the load v_d are constants, Equations (8.65) and (8.66) may both be differentiated and combined to write the following equation of motion for the controlled system:

$$\ddot{v} + 2\zeta\omega_n\dot{v} + \omega_n^2 v = \omega_n^2 v_d, \tag{8.67}$$

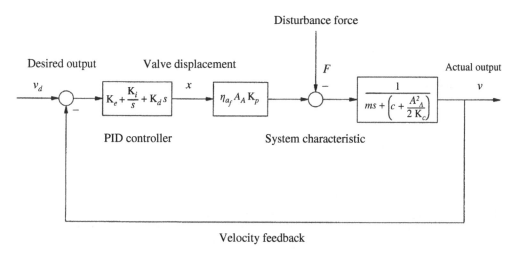

Figure 8-7. Velocity control of the three-way valve-controlled linear actuator.

where the undamped natural frequency and the damping ratio are given by

$$\omega_n = \sqrt{\dfrac{K_i}{\dfrac{m}{\eta_{a_f}A_A K_p} + K_d}} \quad \text{and} \quad \zeta = \dfrac{\left(\dfrac{c}{\eta_{a_f}A_A K_p} + \dfrac{A_A}{2\eta_{a_f}K_q} + K_e\right)}{2\sqrt{K_i\left(\dfrac{m}{\eta_{a_f}A_A K_p} + K_d\right)}}. \tag{8.68}$$

The reader will recognize Equation (8.67) as a description of a second-order system that may exhibit overshoot and oscillation before reaching a steady-state condition in time. Whether overshoot and oscillation is observed depends on whether the damping ratio ζ is less than or greater than unity. If the damping ratio is less than unity the system is an underdamped system and the output will be characterized by overshoot and oscillation before reaching a steady-state condition. On the other hand, if the damping ratio is greater than unity the system is an overdamped system and the output will appear to be very much like a first-order system that asymptotically approaches a steady-state condition without any overshoot or oscillation. If it is assumed that the controlled system is of the underdamped variety then the rise time, settling time, and maximum percent overshoot can be calculated according to the their definitions in Chapter 3. The settling time for the underdamped system is given by $t_s = 4/(\omega_n\zeta)$.

The design objective for the velocity control problem is to shape the second-order response by designing the appropriate undamped natural frequency and damping ratio for the system. By making the proper selection of the proportional, integral, and derivative control gains the undamped natural frequency and the damping ratio for the system may be adjusted accordingly. From Equation (8.68) it can be seen that the proportional gain provides an adjustment for the effective damping of the system while the integral and derivative gains are used to adjust the effective spring rate and mass of the system, respectively. If the derivative gain is neglected, that is, a PI control is used instead of the full PID control, it can be seen from Equation (8.68) that the undamped natural frequency of the system may be adjusted by solely adjusting the integral gain K_i. The reader will recall that the undamped natural frequency is the primary descriptor for the rise time of the underdamped second-order system. Once the desired natural frequency has been obtained the damping ratio for the PI-controlled system may be adjusted by adjusting the proportional gain K_e. The damping ratio is the sole contributor to the maximum percent overshoot. This series approach to tuning the PI controller is very convenient and has been noted as one of its primary advantages over the PID control in Chapter 3.

Force Control.   A third control objective that is occasionally used for the hydraulic control system shown in Figure 8-5 is that of force control. As previously noted force-controlled systems are usually configured very much like a velocity-controlled system without a load spring, and when they are used they often switch between velocity and force control depending on the immediate needs of the application. Furthermore, force-controlled systems are typically operated in slow motion so as to gradually apply the load force to whatever the application is trying to resist. Under these slow motion conditions the load inertia and viscous damping effects in the system may be safely ignored without sacrificing accuracy in the modeling process. By setting the left-hand side of Equation (8.53) equal to zero the following equation may be used to describe the applied load force for the force-controlled system:

$$F = \eta_{a_f} A_A K_p\, x. \tag{8.69}$$

Using a standard proportional-integral (PI) controller the control law for the displacement of the three-way spool valve may be written as

$$x = K_e\,(F_d - F) + K_i \int (F_d - F)\, dt, \tag{8.70}$$

where K_e is the proportional gain, K_i is the integral gain, and F_d is the desired force that is to be exerted on the load. In most applications the instantaneous load force F is measured by sensing the fluid pressure on side A of the linear actuator and calculating the load force according to Equation (8.54). Again, a microprocessor is necessary to carry out these functions and an electrohydraulic position control for the three-way spool valve will be used to enforce the control law of Equation (8.70). Figure 8-8 shows the block diagram for the force control that is described by Equations (8.69) and (8.70).

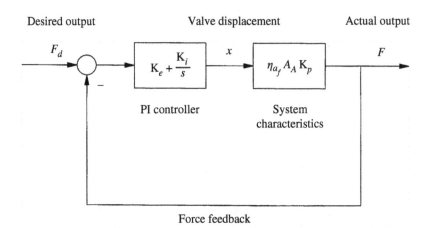

Figure 8-8. Force control of the three-way valve-controlled linear actuator.

If it is assumed that the force control of the system shown in Figure 8-5 is a regulation control problem and that the desired load force F_d is a constant, Equations (8.69) and (8.70) may both be differentiated and combined to write the following dynamic equation for the force-controlled system:

$$\tau \dot{F} + F = F_d, \tag{8.71}$$

where the system time constant is given by

$$\tau = \frac{1}{K_i}\left(\frac{1}{\eta_{a_f} A_A K_p} + K_e\right). \tag{8.72}$$

Again, the reader will recognize Equation (8.71) as a description of a first-order system that asymptotically approaches the desired load force F_d in time without exhibiting any overshoot or oscillation. The dynamic similarities of this system should be compared to the position-controlled system that was previously discussed.

The design objective for the force control problem is to shape the first-order response by designing the appropriate time constant for the system. By making a proper selection of the proportional and integral control gains this time constant can be either increased or diminished based on the desired settling time for the control system output. Again, the reader is reminded that the first-order controlled system of Equation (8.71) has been designed based on the assumption that the linear actuator moves slowly. If in fact the desired response time for this system becomes too short, higher-order dynamics that have been neglected in the modeling process will become more significant and the observed output may be unsatisfactory. Under such conditions a more sophisticated analysis of the system is needed to ensure satisfactory response characteristics.

Control Summary. In this subsection of the chapter we considered the three control objectives that are typically carried out using the three-way valve-controlled linear actuator shown in Figure 8-5. These objectives include position control, velocity control, and force control all of which have employed a form of the standard PID controller. It is also noted that this control system has been predominately used as a position control as opposed to a velocity or force control [5]. For the position control and the force control an important slow-speed assumption has been employed for the purposes of simplifying the control problem for realistic modes of operation. Due to the slow-speed assumptions behind these control strategies the reader is cautioned about using these controllers for high-speed applications that may be encountered for these control objectives. Under such high-speed conditions higher-order dynamics may tend to manifest themselves and may produce a less than desirable system output.

8.3.5 Summary

In this section of the chapter the three-way valve-controlled linear actuator of Figure 8-5 is analyzed, designed, and controlled. It should be noted that throughout this study certain speed conditions have been assumed regarding the operation of the control system. For instance, during the development of the governing system equations in Subsection 8.3.2 the pressure transient effects were neglected due to the assumption that they occurred very quickly in relationship to the overall process time of the load dynamics. If indeed this is not the case, a higher-order model of the control system will be necessary to capture the relevant pressure dynamics of the system. This situation, however, only occurs for very rare and specialized applications and the analysis presented here is deemed to be sufficient for traditional and more typical applications of the system. Note also that this system depends on a very stable and robust pump control that is able to provide a constant supply pressure to the system. Any overshoot or oscillation in this pump control is felt directly by the actuator and the load system. In Subsection 8.3.3 a design methodology for the three-way valve-controlled linear actuator is presented. Similar to the previous system, this design method begins by considering the steady output objectives of the system and then works backward to identify the hydraulic components that are necessary to achieve these objectives. As shown in this subsection the actuator, valve, and pump are closely linked in their design and the proper specification of these components is important when seeking to satisfy the overall performance objectives for the system. Finally, in Subsection 8.3.4 the dynamic control objectives of the actuation system are presented. These objectives vary depending on the application; however, they generally consist of either a position control, a velocity control, or a force control with the position control being the most common. In consideration of these control objectives it is assumed that the control law can be implemented by using an electrohydraulic control valve and a microprocessor with sufficient sensing of the controlled parameters. Typical electrohydraulic actuated control valves are presented in Chapter 4. For information on implementing the microprocessor within the control design the reader is once again referred to practical texts that exist on this subject [3, 4].

8.4 FOUR-WAY VALVE CONTROL OF A ROTARY ACTUATOR

8.4.1 Description

In this section of our chapter we introduce the four-way valve-controlled rotary actuator. As shown in Figure 8-9 this system utilizes a standard four-way valve to control the output characteristics of a fixed-displacement rotary actuator. In this figure the system load to be

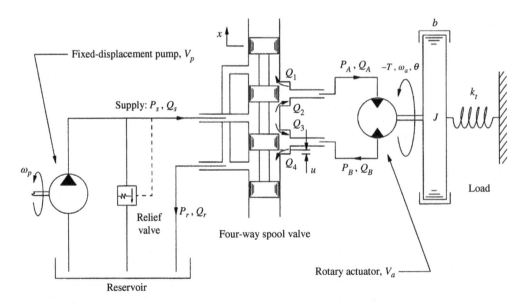

Figure 8-9. Schematic of the four-way valve-controlled rotary actuator.

moved by the actuator is shown as a rotary mass-spring-damper system with a load disturbance torque given by T. The mass moment of inertia, torsional spring rate, and viscous
drag coefficient for the load are shown in Figure 8-9 by the symbols J, b, and k_t respectively.
The rotary actuator is shown with the fixed volumetric displacement per unit of rotation V_a
and is connected to the rotating mass at the output shaft.

The volumetric flow of hydraulic fluid into and out of the actuator is controlled by the
four-way control valve shown to the left of the rotary actuator. This valve is shown to be
constructed as a symmetric open center design with an underlapped dimension on the spool
given by the dimension u. The displacement of the spool valve x is shown to be positive in
the upward direction. As the valve is moved in the positive x-direction flow is directed into
side A of the actuator, which then requires flow to exit the actuator from side B. These
two volumetric flow rates are shown in Figure 8-9 by the symbols Q_A and Q_B, respectively.
In order to provide a hydraulic power source for the valve-controlled system Figure 8-9
shows a fixed-displacement pump that provides a constant volumetric flow rate into the
supply line leading to the four-way control valve. This pump is sized according to its volumetric displacement V_p and is driven by an external power source (not shown in Figure 8-9)
at an angular velocity ω_p. The pressure in the discharge line of the pump is controlled
using a high-pressure relief valve that is set at the desired supply pressure P_s. As noted
in the introduction of this chapter this standby pressure source results in a disadvantage for
valve-controlled systems in that it creates a continual energy drain on the overall system
even when the system is doing no useful work.

8.4.2 Analysis

Load Analysis. The analysis of the system shown in Figure 8-9 begins with the load
dynamics. When we develop the equations of motion for the load we will neglect the inertia
of the rotary actuator itself since it is much smaller than the actual forces that are generated
by the actuator. Again, the reader will recall that this high torque to inertia ratio is one of
the principle advantages of using a hydraulic system as opposed to an electric system. By
lumping the viscous drag effects of both the actuator and the load into the rotary dashpot
shown in Figure 8-9 the equation of motion for the load may be written as

$$J\ddot{\theta} + b\dot{\theta} + k_t\theta = \eta_{a_t} V_a (P_A - P_B) - T, \tag{8.73}$$

where P_A and P_B are the fluid pressures on the A and B sides of the actuator respectively. In this equation the torque exerted on the load by the rotary actuator is given by
$\eta_{a_t} V_a (P_A - P_B)$ where η_{a_t} is the torque efficiency of the actuator. From this result it can be
seen that the actuator pressures P_A and P_B are the required inputs for adjusting the position
of the load. These pressures result from the changing flow and volume conditions within the
actuator itself and are considered in our next paragraph.

Pressure Analysis. In order to analyze the fluid pressures on sides A and B of the rotary
actuator it is once again assumed that the pressure transients that result from fluid compressibility are negligible. This assumption is valid for a system design in which the transmission
lines between the valve and the actuator are short and in which small volumes of fluid exist
on either side of the actuator. The omission of pressure transient effects is also valid for
systems in which the load dynamics are much slower than the pressure dynamics themselves. Since the load dynamics typically occur over a range of seconds and the pressure
dynamics typically occur over a range of milliseconds, it is usually safe to neglect the time

variation of the pressure in favor of the time variation of the overall system. For design and operating conditions in which long transmissions lines between the valve and the actuator are used, or in which a large amount of entrained air is captured within the fluid causing the fluid bulk-modulus to be reduced, it may be necessary to conduct a transient analysis of the pressure conditions on both sides of the actuator. This may also be the case if the actuator dynamics are very fast. For now we assume that the more typical system conditions exist and the pressure transients within the transmission lines are neglected. Using these assumptions we can say for the incompressible fluid that the volumetric flow rates in and out of the actuator are given by

$$Q_A = Q_B = \frac{V_a}{\eta_{a_v}} \dot{\theta},$$ (8.74)

where Q_A and Q_B are the volumetric flow rates into and out of the actuator respectively, η_{a_v} is the volumetric efficiency, and V_a is the volumetric displacement of the actuator per unit of rotation. It should be noted that the volumetric efficiency in Equation (8.74) only takes into account the cross-port leakage that occurs within the actuator from side A to side B. Internal leakage back to the reservoir from either side of the actuator has been neglected to simplify the analysis that follows. Since the four-way valve shown in Figure 8-9 is identical to the valve that was analyzed in Subsection 8.2.2 it may be seen that the volumetric flow rates into and out of the actuator are given by Equation (8.8). Using this result with Equation (8.74) it may be observed that the pressures on sides A and B of the rotary actuator are

$$P_A = \frac{P_s}{2} + K_p x - \frac{V_a}{2 K_c \eta_{a_v}} \dot{\theta},$$

$$P_B = \frac{P_s}{2} - K_p x + \frac{V_a}{2 K_c \eta_{a_v}} \dot{\theta},$$ (8.75)

where K_p is the valve's pressure sensitivity given by the ratio of the flow gain to the pressure-flow coefficient K_q / K_c. Subtracting the two results of Equation (8.75) from each other produces the following expression for the pressure drop across the rotary actuator:

$$P_A - P_B = 2 K_p x - \frac{V_a}{K_c \eta_{a_v}} \dot{\theta}.$$ (8.76)

From Equation (8.75) it can be seen that the fluid pressure on side A of the linear actuator is increased by moving the spool valve in the positive x-direction and that the fluid pressure on side B of the actuator is decreased by the same spool valve motion. As is shown in the summary of the next paragraph this adjustment in fluid pressure by the motion of the spool valve is the mechanism for adjusting the output motion of the load shown in Figure 8-9. *Note:* Equation (8.75) shows a rotary velocity dependence for the fluid pressure as well. As is shown subsequently this feature results in favorable damping characteristics for the system.

Analysis Summary.   To summarize the analysis of the rotary actuation system shown in Figure 8-9 we substitute Equation (8.76) into Equation (8.73) to produce the following equation of motion for the system:

$$J\ddot{\theta} + \left(b + \frac{V_a^2}{K_c} \right) \dot{\theta} + k_t \theta = 2\eta_{a_t} V_a K_p x - T,$$ (8.77)

where it has been assumed that $\eta_{a_t}/\eta_{a_v} \approx 1$. Once again, from this equation it can be seen that the mechanical design of the rotary actuator and the four-way spool valve have a decisive impact on the overall dynamics of the hydraulic control system. In particular, it may be seen from Equation (8.77) that these design parameters help to shape the effective damping of the system and to provide an adequate gain relationship between the input motion of the spool valve and the output motion of the load. The following subsection of this chapter is used to consider the mechanical design parameters that must be specified in order to achieve a satisfactory output of the system from a steady-state perspective.

8.4.3   Design

Actuator Design.   We begin our design consideration of the system shown in Figure 8-9 by first considering the fixed-displacement rotary actuator. Usually, for a given application, the working load torque is specified along with the working actuator pressures and can be used to determine the required volumetric displacement of the actuator V_a. If we consider the right-hand side of Equation (8.73) it is clear that the steady working torque generated by rotary actuator is

$$T = \eta_{a_t} V_a(P_A - P_B), \tag{8.78}$$

where V_a is the volumetric displacement of the actuator, η_{a_t} is the torque efficiency of the actuator, and P_A and P_B are the fluid pressures on each side of the actuator. At the working torque that is specified for the application it is common to use fluid pressures in the rotary actuator that are less than the full supply pressure so as to provide a margin of excessive torque-capability for the system. Typical design specifications at the working torque conditions are given by

$$T = T_w, \qquad P_A = \frac{3}{4}P_s, \qquad \text{and} \qquad P_B = \frac{1}{4}P_s, \tag{8.79}$$

where T_w is the working torque of the hydraulic control system and P_s is the supply pressure to the hydraulic control valve. Substituting these conditions into Equation (8.78) produces the following equation for sizing the required volumetric displacement of the rotary actuator that is shown in Figure 8-9:

$$V_a = 2\frac{T_w}{\eta_{a_t}P_s}. \tag{8.80}$$

This result may be used to design or select a rotary actuator that provides a sufficient working torque for a given supply pressure. Since the rotary actuator is simply a hydraulic pump operating in a reverse direction the type of actuator to be used may range from a gear type machine to an axial-piston machine. Usually the choice of the actuator is based on the performance that is needed from the actuator itself. For instance, if a high operating efficiency is needed with a high starting torque, the best actuator selection would be a bent-axis axial-piston machine. *Note:* Axial-piston machines (bent-axis or swash-plate type) must generally be used for operating pressures that exceed 20 MPa. On the other hand, if efficiency is not a significant factor in the application and if the pressures are below 20 MPa a less-expensive option would be to select either a gear or vane type actuator.

Valve Design.   Now that the rotary actuator has been designed to generate the necessary working torque for the control application, the next step is to design a control valve that will provide sufficient flow and pressure characteristics for the actuator. These characteristics are designed by specifying the appropriate flow gain K_q and pressure-flow coefficient K_c for the valve. The first part of Equation (8.75) is useful for specifying the needed valve coefficients for our design. This equation is given explicitly in terms of the flow gain and the pressure-flow coefficient as

$$K_c(P_A - P_s/2) = K_q x - \frac{V_a}{2\eta_{a_v}}\dot{\theta},$$
(8.81)

where $\dot{\theta}$ is the angular velocity of the load. For specifying the appropriate valve coefficients there are usually two operating conditions that are considered: (1) the steady working torque condition, and (2) the steady no-load velocity of the actuator. For both of these conditions the displacement of the open center valve is usually taken to be 1/4 of the underlapped valve dimension u that is shown in Figure 8-9. Once again, a specified working valve displacement of this magnitude keeps the valve operating near the vicinity of the null position about which the valve flow equations have been linearized and provides a margin of excessive capability for the system design. By using the following parameters for the steady working conditions of the hydraulic control system:

$$x = \frac{1}{4}u, \qquad P_A = \frac{3}{4}P_s, \qquad \text{and} \qquad \dot{\theta} = 0,$$
(8.82)

Equation (8.81) may be used to write the following equation that will satisfy the working specifications of the design:

$$K_c P_s = K_q u.$$
(8.83)

Similarly, by using the following parameters to identify the no-load velocity conditions of the hydraulic control system:

$$x = \frac{1}{4}u, \qquad P_A = \frac{1}{2}P_s, \qquad \text{and} \qquad \dot{\theta} = \omega_o,$$
(8.84)

where ω_o is the no-load velocity that is specified when the load disturbance torque T is zero, Equation (8.81) may be used to write the following equation that will satisfy the no-load velocity requirements of the design:

$$0 = K_q u - \frac{2V_a}{\eta_{a_v}}\omega_o.$$
(8.85)

A simultaneous solution for Equations (8.83) and (8.85) produces the following specifications for the valve flow gain, the pressure-flow coefficient, and the pressure sensitivity respectively:

$$K_q = \frac{2V_a\omega_o}{u\eta_{a_v}}, \qquad K_c = \frac{2V_a\omega_o}{P_s\eta_{a_v}}, \qquad \text{and} \qquad K_p = \frac{K_q}{K_c} = \frac{P_s}{u}.$$
(8.86)

A comparison of Equation (8.86) with Equation (8.22) shows that by simply replacing A_A with V_a and by replacing v_o with ω_o the governing equations for the rectangular flow passage geometry may be written as

$$u = \sqrt{\frac{V_a\omega_o}{2C_d \sqrt{P_s/\rho\eta_{a_v}}}} \qquad \text{and} \qquad h = 4\,u.$$
(8.87)

From these results it can be seen that as the volumetric displacement of the actuator V_a and the no-load velocity ω_o becomes large the valve size must also grow accordingly. Thus, the valve must be matched with the load requirements in order to satisfy the overall design objectives of the hydraulic control system.

So far in our consideration of the valve design we have only addressed issues regarding the standard flow coefficients and the nominal flow passage size that is required for adequate performance of the actuator. Issues regarding the physical design of the valve and the means of actuation for the valve have not been considered in this discussion. For consideration of these design aspects the reader is referred to the more in depth valve analysis that is presented in Chapter 4.

Pump Design. Now that the linear actuator and the control valve have been designed we must specify the hydraulic pump that will be used to power the hydraulic control system. As shown in Figure 8-9 this pump is a fixed-displacement pump that produces a volumetric flow rate that is proportional to the angular input speed of the pump shaft ω_p. Within the pump itself there is internal leakage that results in a reduction of the pump volumetric efficiency η_{p_v}. If we assume for now that the relief valve shown in Figure 8-9 is closed we may see that the supply flow to the four-way control valve is given by

$$Q_s = \eta_{p_v} V_p \omega_p, \tag{8.88}$$

where V_p is the volumetric displacement of the pump per unit of rotation and η_{p_v} is the volumetric efficiency of the pump. From Figure 8-9 it can be seen that the supply flow must equal the sum of the volumetric flow rates that are crossing metering lands 2 and 3 on the four-way spool valve. Using Equation (8.7) it may be shown that this supply flow is given by

$$Q_s = Q_2 + Q_3 = 3K_c P_s - K_c(P_A + P_B). \tag{8.89}$$

From Equation (8.75) it can be shown that

$$P_A + P_B = P_s, \tag{8.90}$$

which indicates no dependency on actuator speed whatsoever. Substituting this result into Equation (8.89) it may be shown that the supply flow to the four-way valve is given by

$$Q_s = 2K_c P_s. \tag{8.91}$$

Finally, substituting Equation (8.91) into Equation (8.88) it may be shown that the required volumetric displacement for the supply pump is given by

$$V_p = \frac{2K_c P_s}{\eta_{p_v} \omega_p}. \tag{8.92}$$

The reader should notice that this is the simplest expression that we have generated so far for sizing the supply pump, which results from the symmetric operation of the rotary actuator. In fact, since there is no actuator velocity dependence in Equation (8.92) the supply pump can theoretically be used without a parallel relief valve in the circuit since peak flows will not need to be clipped in order to avoid over pressurization of the system. In other words, the constant flow resistance offered by the open center four-way valve is sufficient for generating the needed supply pressure P_s provided that Equation (8.92) is satisfied.

Once again, this discussion is concerned with the appropriate sizing of the supply pump for the hydraulic control system shown in Figure 8-9. Though the pump size is specified for the system based on certain system needs, nothing is said about the physical construction of the pump. In practice this pump may be a gear pump, or it may be an axial-piston pump, or it may be any positive-displacement pump that satisfies the volumetric-displacement requirement shown in Equation (8.92). Usually the pump construction type is selected based on the required supply pressure of the system and the desired operating efficiency of the pump. As stated in Chapter 5 axial-piston pumps tend to operate at higher pressures and provide a higher operating efficiency as compared to gear and vane type pumps. On the other hand, axial-piston pumps tend to increase the upfront cost of the control system, which is occasionally a deterrent for their use.

Input Power Design. The input power that is required to operate the hydraulic control system shown in Figure 8-9 is given by the standard power equation

$$\Pi = \frac{V_p \omega_p P_s}{\eta_p},$$ (8.93)

where η_p is the overall efficiency of the pump given by the product of the volumetric and torque efficiency values $\eta_p = \eta_{p_t} \eta_{p_v}$. This result is essentially identical with that of Equation (8.31).

Case Study

To illustrate the design process associated with the system shown in Figure 8-9 consider the following case study: A cement mixing truck uses a mixing drum with an average drum radius of 1 meter. It takes 10000 Nm of torque to drive the drum when it is fully loaded. The no-load velocity of the drum is 5 rpm. The drum is driven by a four-way valve-controlled rotary actuator similar to what is shown in Figure 8-9 through a speed reducing gear set which has a gear ratio given by 100:1. The torque and volumetric efficiencies for the rotary actuator are 95% and 97%, respectively. The supply pump provides a supply pressure of 20 MPa while being driven by the truck engine at 1800 rpm. The torque and volumetric efficiencies of the pump are given by 95% and 96%, respectively. You are to design and specify the components of this hydraulic system.

The solution to this problem begins by considering the size of the rotary actuator that must be used to achieve the load objectives of the application. In order to determine the working torque that must be delivered by the rotary actuator we utilize the gear ratio:

$$T_w = \frac{10000 \, \text{Nm}}{100} = 100 \, \text{Nm}.$$

Using this result with Equation (8.80) it can be shown that the volumetric displacement of the rotary actuator should be

$$V_a = 2\frac{T_w}{\eta_{a_t} P_s} = 2\frac{100 \, \text{Nm}}{0.95 \times 20 \, \text{MPa}} = 66 \, \frac{\text{cm}^3}{\text{rev}}.$$

Now that the actuator has been designed we move toward specifying the valve coefficients for the four-way control valve shown in Figure 8-9. These coefficients are dependent on the porting geometry of the valve. For rectangular ports the underlapped dimension u and the port height h are given in Equation (8.87). If we assume that the discharge coefficient for the valve is given by $C_d = 0.62$ and that the fluid density is given by $\rho = 850 \, \text{kg/m}^3$ this equation may be used to calculate the porting dimensions for the rectangular flow passage as $u = 1.73$ mm and $h = 6.91$ mm. These results are based on the calculated volumetric displacement of the actuator $V_a = 66$ cm^3/rev and the no-load velocity requirement that has been stated in the problem $\omega_o = 100 \times 5$ rpm $= 500$ rpm, where 100 is the gear ratio between the actuator and the cement mixer drum. Using these results with Equation (8.86) the valve coefficients may be calculated as

$$K_q = \frac{2V_a\omega_o}{u\eta_{a_v}} = \frac{2 \times 66 \text{ cm}^3/\text{rev} \times 500 \text{ rpm}}{1.73 \text{ mm} \times 0.97} = 0.657 \frac{\text{m}^2}{\text{s}}$$

$$K_c = \frac{2V_a\omega_o}{P_s\eta_{a_v}} = \frac{2 \times 66 \text{ cm}^3/\text{rev} \times 500 \text{ rpm}}{20 \text{ MPa} \times 0.97} = 5.68 \times 10^{-11} \frac{\text{m}^3}{\text{Pa s}}$$

$$K_p = \frac{K_q}{K_c} = \frac{P_s}{u} = \frac{20 \text{ MPa}}{1.73 \text{ mm}} = 1.16 \times 10^{10} \frac{\text{Pa}}{\text{m}}.$$

These equations describe the required flow characteristics of the spool valve. Nothing has been said about the actuation mechanism of the valve; however, for a speed-controlled cement mixing application it is likely that the valve will be moved using a feedback controller and an electrohydraulic actuation system.

Now that the actuator and valve have been designed the volumetric displacement of the supply pump must be specified. This specification is governed by Equation (8.92). Using this result the specified volumetric displacement for the pump may be calculated as

$$V_p = \frac{2K_cP_s}{\eta_{p_v}\omega_p} = \frac{2 \times 5.68 \times 10^{-11} \text{ m}^3/(\text{Pa s}) \times 20 \text{ MPa}}{0.96 \times 1800 \text{ rpm}} = 79 \frac{\text{cm}^3}{\text{rev}}.$$

Finally, the amount of power that is required from the engine of the garbage truck must be evaluated. This power is governed by Equation (8.93) and may be calculated as follows:

$$\Pi = \frac{V_p\omega_pP_s}{\eta_p} = \frac{79 \text{ cm}^3/\text{rev} \times 1800 \text{ rpm} \times 20 \text{ MPa}}{0.95 \times 0.96} = 45.46 \text{ kW}.$$

The reader will recall that η_p is the overall efficiency of the pump.

8.4.4 Control

Position Control. The four-way valve-controlled rotary actuator shown in Figure 8-9 may be used to accomplish a number of control objectives. One of the possible objectives of this control system is to accurately position the load at a prescribed angular location within the trajectory range of the actuator. Typically this control function is carried out under slow moving conditions of the actuator; therefore, the plant description for this control problem may safely neglect any transient contributions that would normally be significant during high-speed operations. With this being the case Equation (8.77) may be used to describe the controlled system by neglecting the transient terms of inertia and viscous damping. This result is given by

$$k_t\theta = 2\eta_{a_t}V_aK_px - T, \tag{8.94}$$

where K_p is the pressure sensitivity of the control valve, η_{a_t} is the torque efficiency of the actuator, and the other parameters of Equation (8.94) are shown in Figure 8-9. Using a standard proportional-integral (PI) controller the control law for the displacement of the four-way spool valve may be written as

$$x = K_e(\theta_d - \theta) + K_i \int (\theta_d - \theta)\, dt, \tag{8.95}$$

where K_e is the proportional gain, K_i is the integral gain, and θ_d is the desired position of the load. The most practical way to enforce the control law that has been defined in Equation (8.95) is to use an electrohydraulic position control for the four-way spool valve coupled with a microprocessor that is capable of reading feedback information and generating the appropriate output signal for the valve actuator. A typical electrohydraulic valve that may be used in this application is shown in Figure 4-15 of Chapter 4. For more information on using microprocessors in the control loop the reader is referred to practical texts that address this issue [3, 4]. Figure 8-10 shows the block diagram for the position control that is described by Equations (8.94) and (8.95).

If we assume that the position control of the system shown in Figure 8-9 is a regulation control problem and that the load torque T and the desired position of the load θ_d are constants, Equations (8.94) and (8.95) may both be differentiated and combined to write the following equation of motion for the controlled system:

$$\tau \dot{\theta} + \theta = \theta_d. \tag{8.96}$$

In this equation the system time constant is given by

$$\tau = \frac{1}{K_i}\left(\frac{k_t}{2\, \eta_{a_t} V_a K_p} + K_e \right). \tag{8.97}$$

The reader will recognize Equation (8.96) as a description of a first-order system that asymptotically approaches the desired position of the load in time without exhibiting any overshoot or oscillation. As time goes to infinity the steady-state error of this system vanishes, which is the primary advantage that is gained by using the integral component of the PI controller. The settling time for the first-order system is given by $t_s = 4\,\tau$.

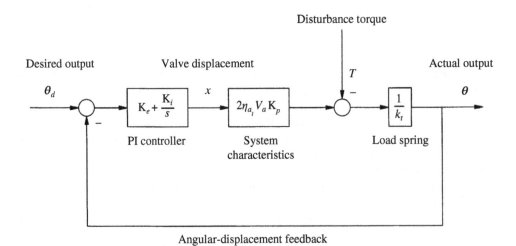

Figure 8-10. Position control of the four-way valve-controlled rotary actuator.

The design objective for the position control problem is to shape the first-order response by designing the appropriate time constant for the system. By making a proper selection of the proportional and integral control gains this time constant can be either increased or diminished based on the desired settling time for the control system output. Though the first-order controlled system of Equation (8.96) shows a high degree of simplicity, the reader must keep in mind that this system response is based on a slow moving device and that if the desired settling time becomes too short higher-order dynamics will become significant and a more sophisticated analysis of the system will be necessary to ensure that the system response is acceptable.

Velocity Control. Perhaps the most common control objective for the system shown in Figure 8-9 is that of velocity control. This objective seeks to establish a specific output angular velocity for the load based on a desired angular velocity that is prescribed by the application. This objective is typically carried out for load systems that do not include a load spring. Therefore, by neglecting the spring that is shown in Figure 8-9; that is, setting $k_t = 0$, the equation of motion for the velocity-controlled system may be written using the general form of Equation (8.77). This result is given by

$$J\dot{\omega}_a + \left(b + \frac{V_a^2}{K_c}\right)\omega_a = 2\eta_{a_t} V_a K_p x - T, \tag{8.98}$$

where ω_a is the instantaneous angular velocity of the load. Again, K_c and K_p are the pressure-flow coefficient and the pressure sensitivity of the four-way control valve respectively and η_{a_t} is the torque efficiency of the actuator. All other parameters of Equation (8.98) are shown in Figure 8-9. For the velocity control objective we will select a full proportional-integral-derivative (PID) controller with the following control law for the spool valve:

$$x = K_e(\omega_d - \omega_a) + K_i \int (\omega_d - \omega_a)\, dt + K_d \frac{d}{dt}(\omega_d - \omega_a), \tag{8.99}$$

where K_e is the proportional gain, K_i is the integral gain, K_d is the derivative gain, and ω_d is the desired angular velocity of the load. Again, the most practical way to enforce the control law that has been defined in Equation (8.99) is to use an electrohydraulic position control for the four-way spool valve, coupled with a microprocessor that is capable of reading feedback information and generating the appropriate output signal for the valve actuator. Figure 8-11 shows the block diagram for the velocity control that is described by Equations (8.98) and (8.99).

If it is assumed that the velocity control of the system shown in Figure 8-9 is a regulation control problem and that the load torque T and the desired angular velocity of the load ω_d are constants, Equations (8.98) and (8.99) may both be differentiated and combined to write the following equation of motion for the controlled system:

$$\ddot{\omega}_a + 2\zeta\omega_n\dot{\omega}_a + \omega_n^2\omega_a = \omega_n^2\omega_d, \tag{8.100}$$

where the undamped natural frequency and the damping ratio are given respectively as

$$\omega_n = \sqrt{\frac{K_i}{\dfrac{J}{2\eta_{a_t} V_a K_p} + K_d}} \quad \text{and} \quad \zeta = \frac{\left(\dfrac{b}{2\eta_{a_t} V_a K_p} + \dfrac{V_a}{2\eta_{a_t} K_q} + K_e\right)}{2\sqrt{K_i\left(\dfrac{J}{2\eta_{a_t} V_a K_p} + K_d\right)}}. \tag{8.101}$$

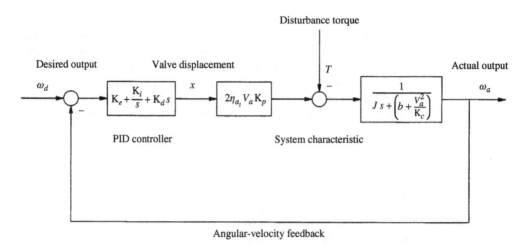

Figure 8-11. Velocity control of the four-way valve-controlled rotary actuator.

The reader will recognize Equation (8.100) as a description of a second-order system that may exhibit overshoot and oscillation before reaching a steady-state condition in time. Whether overshoot and oscillation is observed depends on whether the damping ratio ζ is less than or greater than unity. If the damping ratio is less than unity the system is an under-damped system and the output will be characterized by overshoot and oscillation before reaching a steady-state condition. On the other hand, if the damping ratio is greater than unity the system is an overdamped system and the output will appear to be very much like a first-order system that asymptotically approaches a steady-state condition without any overshoot or oscillation. If it is assumed that the controlled system is of the underdamped variety then the rise time, settling time, and maximum percent overshoot can be calculated according to the their definitions in Chapter 3.

The design objective for the angular velocity control problem is to shape the second-order response by designing the appropriate undamped natural frequency and damping ratio for the system. By making the proper selection of the proportional, integral, and derivative control gains the undamped natural frequency and damping ratio for the system may be adjusted accordingly. From Equation (8.101) it can be seen that the proportional gain provides an adjustment for the effective damping of the system while the integral and derivative gains are used to adjust the effective spring rate and mass of the system respectively. If the derivative gain is neglected; that is, a PI control is used instead of the full PID control, it can be seen from Equation (8.101) that the undamped natural frequency of the system may be adjusted by solely adjusting the integral gain K_i. The reader will recall that the undamped natural frequency is the primary descriptor for the rise time of the underdamped second-order system. Once the desired natural frequency has been obtained, the damping ratio for the PI-controlled system may be adjusted by adjusting the proportional gain K_e. The damping ratio is the sole contributor to the maximum percent overshoot. This series approach to tuning the PI controller is very convenient and has been noted as one of its primary advantages in Chapter 3. Nevertheless, reasonable control gains must be chosen so as to avoid saturating the mechanical limits of the control valve.

Case Study

In the previous case study a four-way valve-controlled rotary actuator was designed to control the output dynamics of a cement mixer drum. This drum had an average radius of 1 meter and was driven through a speed reducing gearset with a gear ratio of 100:1. The design of this system involved a rotary actuator with a volumetric displacement of 66 cm³/rev and torque efficiency given by 95%. The four-way valve was designed with a flow gain of 0.657 m²/s and a pressure sensitivity of 1.16×10^{10} Pa/m. The mass moment of inertia for the loaded rotating drum is given by 50000 kg m² and the viscous damping of the system is negligible. By neglecting the derivative gain in the PID control structure calculate the natural frequency and damping ratio of the system utilizing a proportional gain of 1.4×10^{-4} meter-seconds an integral gain of 9.33×10^{-6} meters. These gains are reasonable based on the maximum travel distance of the control valve and the expected errors in the velocity feedback signal. What is the maximum percent overshoot and the settling time associated with this response?

The solution to this problem begins by calculating the effective mass moment of inertia that is observed by the rotary actuator. By equating the rotating kinetic energy of the rotary actuator with the rotating kinetic energy of the mixing drum it can be shown that the mass moment of inertia that is felt by the rotary actuator is given by

$$J = \frac{50000 \text{ kg m}^2}{(100)^2} = 5 \text{ kg m}^2,$$

where 50000 kg m² is the mass moment of inertia of the rotating drum and 100 is the gear ratio between the rotary actuator and rotating drum. From Equation (8.101) the expression for the natural frequency may be rearranged and calculated as

$$\omega_n = \sqrt{\frac{2K_i \eta_{a_t} V_a K_p}{J}}$$

$$= \sqrt{\frac{2 \times 9.33 \times 10^{-6} \text{ m} \times 0.95 \times 66 \text{ cm}^3/\text{rev} \times 1.16 \times 10^{10} \text{ Pa/m}}{5 \text{ kg m}^2}}$$

$$= 0.657 \frac{\text{rad}}{\text{s}} = 0.105 \text{ Hz}.$$

Similarly, the expression for the damping ratio may be used to yield:

$$\varsigma = \frac{\left\{ \dfrac{V_a}{2\eta_{a_t} K_q} + K_e \right\}}{2\sqrt{\dfrac{JK_i}{2\eta_{a_t} V_a K_p}}} = \frac{\left\{ \dfrac{66 \text{ cm}^3/\text{rev}}{2 \times 0.95 \times 0.657 \text{m}^2/\text{s}} + 1.4 \times 10^{-5} \text{ m s} \right\}}{2\sqrt{\dfrac{5 \text{ kg m}^2 \times 9.33 \times 10^{-6} \text{ m}}{2 \times 0.95 \times 66 \text{ cm}^3/\text{rev} \times 1.16 \times 10^{10} \text{ Pa/m}}}}$$

$$= 0.7895.$$

Using the definition of the rise time and maximum percent overshoot for an under-damped second-order system as presented in Chapter 3 it may be shown that the settling time is 7.71 seconds and the maximum percent overshoot is 1.758%.

Torque Control.   Finally, another control objective that is occasionally used for the hydraulic control system shown in Figure 8-9 is that of torque control. Torque-controlled systems are usually configured very much like a velocity-controlled system without a load spring and when they are used they often switch between velocity and force control depending on the immediate needs of the application. Furthermore, torque-controlled systems are typically operated in slow motion so as to gradually apply the load force to whatever the application is trying to resist. Under these slow-motion conditions the load inertia and viscous damping effects in the system may be safely ignored without sacrificing accuracy in the modeling process. By setting the left-hand side of Equation (8.77) equal to zero the following equation may be used to describe the applied load torque for the torque-controlled system:

$$T = 2\eta_{a_t} V_a K_p x. \tag{8.102}$$

Using a standard proportional-integral (PI) controller the control law for the displacement of the four-way spool valve may be written as

$$x = K_e(T_d - T) + K_i \int (T_d - T) \, dt, \tag{8.103}$$

where K_e is the proportional gain, K_i is the integral gain, and T_d is the desired torque that is to be exerted on the load. In most applications the instantaneous load torque T is measured by sensing the fluid pressures in sides A and B of the rotary actuator and calculating the load torque according to Equation (8.78). Again, a microprocessor is necessary to carry out these functions and an electrohydraulic position control for the four-way spool valve is used to enforce the control law of Equation (8.103). Figure 8-12 shows the block diagram for the torque control that is described by Equations (8.102) and (8.103).

If it is assumed that the torque control of the system shown in Figure 8-9 is a regulation control problem, and that the desired load torque T_d is a constant, Equations (8.102) and (8.103) may both be differentiated and combined to write the following dynamic equation for the torque-controlled system:

$$\tau \dot{T} + T = T_d, \tag{8.104}$$

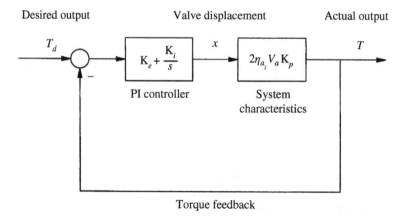

Figure 8-12. Torque control of the four-way valve-controlled rotary actuator.

where the system time constant is given by

$$\tau = \frac{1}{K_i} \left(\frac{1}{2\eta_{a_t} V_a K_p} + K_e \right). \tag{8.105}$$

Again, the reader will recognize Equation (8.104) as a description of a first-order system that asymptotically approaches the desired load torque T in time without exhibiting any overshoot or oscillation. The dynamic similarities of this system should be compared to the position-controlled system that was previously discussed.

The design objective for the torque control problem is to shape the first-order response by designing the appropriate time constant for the system. By making a proper selection of the proportional and integral control gains this time constant can be either increased or diminished based on the desired settling time for the control system output. Again, the reader is reminded that the first-order controlled system of Equation (8.104) has been designed based on the assumption that the linear actuator moves slowly. If in fact the desired response time for this system becomes too short higher-order dynamics that have been neglected in the modeling process will become more significant and the observed output may be unsatisfactory. Under such conditions a more sophisticated analysis of the system will be needed to ensure satisfactory response characteristics.

Control Summary. In this subsection of the chapter we considered the three control objectives that are typically carried out using the four-way valve-controlled rotary actuator shown in Figure 8-9. These objectives include position control, velocity control, and torque control all of which have employed a form of the standard PID controller. For the position control and the torque control an important slow-speed assumption has been employed for the purposes of simplifying the control problem for realistic modes of operation. Due to the slow-speed assumptions behind these control strategies the reader is cautioned about using these controllers for high-speed applications that are rarely encountered for these control objectives. Under such high-speed conditions higher-order dynamics may tend to manifest themselves and may produce a less than desirable system output.

8.4.5 Summary

In this section of the chapter the four-way valve-controlled rotary actuator of Figure 8-9 has been analyzed, designed, and controlled. It should be noted that throughout this study certain speed conditions have been assumed regarding the operation of the control system. For instance, during the development of the governing system equations in Subsection 8.4.2 the pressure transient effects in the transmission lines were neglected due to the assumption that they occurred very quickly in relationship to the overall process time of the load dynamics. If indeed this is not the case a higher-order model of the control system is necessary to capture the relevant pressure dynamics of the system. This situation, however, only occurs for very rare applications and the analysis presented here is deemed to be sufficient for traditional and more typical applications of the system. In Subsection 8.4.3 a design methodology for the valve-controlled rotary actuator has been presented. This design method begins by considering the steady output objectives of the system and then works backward to identify the hydraulic components that are necessary to achieve these objectives. As shown in this subsection the actuator, valve, and pump are closely linked in their design and the proper specification of these components is important when seeking to satisfy the overall performance objectives for the system. Finally, in Subsection 8.4.4 the dynamic control objectives of the actuation system have been presented. These objectives vary depending on the application; however, they generally consist of either a position control, a velocity control, or a torque control. In the consideration of these control objectives it has been assumed that the control law may be implemented by using an electrohydraulic control valve and a microprocessor with sufficient sensing of the controlled parameters. A typical electrohydraulic control valve that may be used for these applications has been presented in Subsection 4.5.4 of Chapter 4. For information on implementing the microprocessor within the control design the reader is referred to standard texts that address this issue [3, 4].

8.5 CONCLUSION

This chapter describes a few standard valve-controlled hydraulic systems that exist in practice today. This chapter has by no means exhausted the subject of valve-controlled hydraulic systems; however, the systems that have been presented are representative of other hydraulic control systems that either currently exist or could be designed to satisfy various control objectives. Variations from the presented hydraulic control systems of this chapter include applications that often seek to improve the overall efficiency of the system. Such applications typically employ a load-sensing variable-displacement pump in which the downstream pressures of the control valve are sensed and the supply pressure is then adjusted to be approximately 2 MPa higher than the sensed working pressure in the actuator. This online adjustment of the supply pressure causes the valve coefficients to continuously vary and thereby complicates the analysis of the system significantly. Nevertheless, the gain that is made in the overall system efficiency is substantial and therefore the added complexity of the system analysis, design, and control is generally worthwhile.

The control sections of this chapter are written to describe various feedback structures that may be used to automate the hydraulic control system. These feedback structures have depended on simplified models of the controlled plant and have also utilized classical PID control methods for achieving the desired output of the system. While PID control methods have usually proven successful in hydraulic control systems it should be noted that PID controllers are linear controllers that work well for systems that can be described linearly over a wide range of operating conditions. Unfortunately, this class of systems does not usually include hydraulic systems, which generally exhibit more nonlinearity than most control systems, and therefore the PID gains often need to be readjusted and scheduled within the microprocessor to achieve adequate control at various operating conditions of the system. Other methods of nonlinear control, which have not been addressed in this chapter, are also becoming more available to the control engineer.

Though it has not been mentioned explicitly it should be noted that the control systems presented in this chapter are often employed without using automatic feedback methods for achieving the end objective of the system. These open loop control systems depend on a great deal of repeatability within the system, which can only be guaranteed as long as the operating environment (e.g., temperature, operating load) does not change drastically. In other cases of control the control loop is closed by the feedback of a human operator who is watching the system output and adjusting the control valve position manually to achieve the end objective. This mode of control is especially common for systems that are used to conduct modular type work such as that of a backhoe loader or a robot manipulator in which the output objectives change frequently.

The three valve-controlled hydraulic systems presented in this chapter are: the four-way valve-controlled linear actuator, the three-way valve-controlled linear actuator, and the four-way valve-controlled rotary actuator. In the first system the analysis is generalized to include the design of both a single-rod and a double-rod linear actuator; however, the reader is reminded that the three-way valve control is only applicable for the single rod linear actuation design. Since rotary actuators are symmetric in their design and operation they too are only controllable using a four-way valve. Variations from these three basic systems can be used to design unique hydraulic control systems that may be used to achieve specific control objectives for new applications.

8.6 REFERENCES

[1] Norton R. L. 2000. *Machine Design: An Integrated Approach*. Prentice-Hall, Upper Saddle River, NJ.

[2] Hamrock B. J., B. Jacobson, and S. R. Schmid. 1999. *Fundamentals of Machine Elements*. WCB/McGraw-Hill, New York.

[3] Kleman A. 1989. *Interfacing Microprocessors in Hydraulic Systems*. Marcel Dekker, New York and Basel.

[4] Bolton W. 1996. *Mechatronics: Electronic Control Systems in Mechanical Engineering*. Addison Wesley Longman Ltd., England.

[5] Merritt H. E. 1967. *Hydraulic Control Systems*. John Wiley & Sons, New York.

8.7 HOMEWORK PROBLEMS

8.7.1 Four-Way Valve Control of a Linear Actuator

8.1 A four-way valve controls a single rod linear actuator that is used to operate a cargo elevator at a distribution warehouse. The control system is similar to Figure 8-1. The maximum load capacity of the elevator is 10000 lbf including its own weight, which is 1000 lbf. The supply pressure for the system is 500 psi. Assuming that the actuator force efficiency is 93% design the piston and rod diameters for the actuator.

8.2 For the elevator design in Problem 8.1 the no-load velocity specification is 4 in/s and the volumetric efficiency of the actuator is 97%. Using rectangular porting geometry for the control valve determine the nominal port geometry required to operate the elevator. From this geometry and the operating conditions of the system calculate the flow gain, the pressure-flow coefficient, and the pressure sensitivity of the valve. Assume that the fluid density is 0.0307 lbm/in^3 and that the discharge coefficient of the valve is 0.62.

8.3 For the elevator design in Problems 8.1 and 8.2 specify the required volumetric displacement of the supply pump assuming that the input speed to the pump is 1800 rpm. The volumetric efficiency of the pump is 89%. How much power is required to operate this pump assuming that the torque efficiency is 92%?

8.4 The velocity of the elevator in Problems 8.1 through 8.3 is controlled using a PI controller, which actuates the four-way spool valve using a proportional gain given by 1.22×10^{-2} seconds and an integral gain of 2.04×10^{-3} (no units). If the elevator is fully loaded with a mass of 10,000 lbm, calculate the undamped natural frequency and the damping ratio for the controlled system. Do you expect the elevator to exhibit overshoot and oscillation? Explain your answer.

8.7.2 Three-Way Valve Control of a Single Rod Linear Actuator

8.5 A three-way valve-controlled linear actuator similar to what is shown in Figure 8-5 is used to operate a household trash compactor. The actuator must be capable of compacting the trash with a maximum force of 2225 N using a supply pressure of 2 MPa. The nominal bias load on the actuator is zero and the force efficiency is 90%. Specify the required pressurized areas on sides A and B of the actuator. What is the piston and the rod diameter required for this device?

8.6 For the compactor design in Problem 8.5 the no-load velocity specification is 5 mm/s and the volumetric efficiency of the actuator is 87%. Using rectangular porting geometry for the three-way control valve, determine the nominal port geometry required to operate the compactor. From this geometry and the operating conditions of the system calculate the flow gain, the pressure-flow coefficient, and the pressure sensitivity of the valve. Assume that the fluid density is 850 kg/m³ and that the discharge coefficient of the valve is 0.62.

8.7 For the compactor design in Problems 8.5 and 8.6 specify the maximum volumetric displacement of the supply pump assuming that the input speed to the pump is 500 rpm. The volumetric efficiency of the pump is 89%. How much power is required to operate this pump assuming that the torque efficiency is 92%?

8.8 The applied force of the trash compactor in Problems 8.5 through 8.7 is controlled using a PI controller, which actuates the three-way spool valve using a proportional gain of 6.18×10^{-8} m/N and an integral gain of 1.24×10^{-8} m/(N s). Using these gains calculate the system time constant and settling time. Is this an acceptable response for a trash compactor? Explain your answer.

8.7.3 Four-Way Valve Control of a Rotary Actuator

8.9 A valve-controlled rotary actuator similar to Figure 8-9 is used to operate a machining lathe with a torque resistance of 100 in-lbf. The actuator is connected to the lathe through a gearbox with a 10:1 speed ratio and the supply pressure to the control system is 5000 psi. Assuming that the torque efficiency of the actuator is 95% calculate the required volumetric displacement of the actuator per revolution.

8.10 For the lathe design in Problem 8.9 the no-load angular velocity of the lathe is specified as 8500 rpm and the volumetric efficiency of the actuator is 96%. Using rectangular porting geometry for the three-way control valve, determine the nominal port geometry required to operate the lathe. From this geometry and the operating conditions of the system calculate the flow gain, the pressure-flow coefficient, and the pressure sensitivity of the valve. Assume that the fluid density is 0.0307 lbm/in³ and that the discharge coefficient of the valve is 0.61.

8.11 For the lathe design in Problems 8.9 and 8.10 specify the required volumetric displacement of the supply pump assuming that the input speed to the pump is 1800 rpm. The volumetric efficiency of the pump is 95%. How much power is required to operate this pump assuming that the torque efficiency is 92%? Is the relief valve in Figure 8-9 a required part of the system design? Explain your answer.

8.12 The angular velocity of the lathe in Problems 8.9 through 8.11 is controlled using a PID controller, which actuates the four-way spool valve using a proportional gain of 1.42×10^{-4} in-s, an integral gain of 2.83×10^{-5} in, and a derivative gain 9.44×10^{-4} in-s^2. Assuming that the rotating inertia and viscous drag of the lathe are negligible, calculate the undamped natural frequency and the damping ratio for the controlled system. From these results calculate the rise time, settling time, and maximum percent overshoot of the controlled system. What is the bandwidth frequency of the controlled system?

9

PUMP-CONTROLLED HYDRAULIC SYSTEMS

9.1 INTRODUCTION

In this chapter we consider the analysis, design, and control of pump-controlled hydraulic systems. These systems utilize a pump as opposed to a control valve for directing available hydraulic power to and from an actuator that is used to generate useful output. It is well known that pump-controlled hydraulic systems exhibit an efficiency advantage over valve-controlled systems due to the fact that the control valve introduces a pressure drop, which results in significant heat dissipation. The pump-controlled system does not utilize this valve; therefore, the immediate power needs of the output are met directly by the power source, which increases the overall operating efficiency of the system. This being said, there are four disadvantages of the pump-controlled system that may be listed as:

1. The response characteristics of pump-controlled systems can be slower than that of a valve-controlled system due to the longer transmission lines that are usually used for reaching the output actuator. This is an effect of fluid compressibility, which is not modeled in this chapter.
2. If the disadvantage just mentioned is solved by trying to locate the pump closer to the actuator the working unit, that is, the pump including the actuator and power source, it is generally too bulky to be acceptable for most applications.
3. Closed circuit pump-controlled systems generally require auxiliary hydraulic systems to replenish the low-pressure line of the circuit. This requires another pump, two check valves, and a relief valve. This additional hardware adds additional cost the hydraulic control system.
4. Pump-controlled systems are comprised of a single pump that operates a single actuator. Multiple actuators cannot share the power that is generated from one pump, which means that the total pump cost must be included with the overall cost of a single actuator.

In short, the efficiency advantage of the pump-controlled system must be weighed against the several disadvantages that have just been mentioned before one chooses to use a pump-controlled hydraulic system as opposed to a valve-controlled system. In the discussion that follows it is assumed that the pump-controlled system has been selected based on the efficiency advantage that it provides.

Like valve-controlled systems the subject of pump-controlled hydraulic systems is most easily divided between the control of linear actuators (hydraulic pistons and cylinders) and rotary actuators (hydraulic motors). The linear actuator system is commonly used to implement devices that utilize translational motion and provide a linear force output. Since the pump operates symmetrically as it sends flow to and receives flow from the output actuator, the pump-controlled linear actuation system is only suitable for double-rod linear actuators. Typical example applications for these systems include industrial robots and flight-surface controls that are used in the aerospace industry. The rotary actuation system is commonly used for driving a rotating shaft, which then produces a rotary torque output. Pump-controlled rotary actuators are often called hydrostatic transmissions and are frequently used within the power train of off-highway earthmoving equipment or lawn and garden tractors. They are also used within the aerospace industry to provide a constant speed drive for various flight applications. These applications illustrate the wide variation of use for hydraulic control systems, which find their niche in high-power applications that require a large effort to inertia ratios for achieving a very stiff dynamic response.

In the following discussion two pump-controlled hydraulic systems are presented: the fixed-displacement, pump-controlled linear actuator, and the variable-displacement pump-controlled rotary actuator. Both of these hydraulic control systems are analyzed to determine the equation of motion that governs the dynamic behavior of the output load. Once the governing equation is established, the design of the hydraulic system is evaluated using steady-state forms of the governing equation. Finally, both sections of this chapter conclude with a presentation of the various control schemes that may be used for each hydraulic system. These control schemes use classical PID controllers that are implemented for achieving position, velocity, and force or torque control.

9.2   FIXED-DISPLACEMENT PUMP CONTROL OF A LINEAR ACTUATOR

9.2.1   Description

In this section of our chapter we introduce the fixed-displacement, pump-controlled linear actuator. This system is frequently employed in the aerospace industry for controlling fight surfaces. As shown in Figure 9-1 this system utilizes a speed-controlled, fixed-displacement pump to control the output characteristics of a double-rod linear actuator. In this figure, the load to be moved by the actuator is shown as a single mass-spring-damper system with a load disturbance force given by F. The mass, spring rate, and viscous drag coefficient for the load are shown in Figure 9-1 by the symbols m, c, and k, respectively. The linear actuator is shown to be constructed with a double rod that facilitates symmetric action of the actuator. The rods are connected to both the load and the actuator piston. Since the actuator is a double-rod design, the pressurized areas are identical on both sides of the actuator.

The volumetric flow of hydraulic fluid into the actuator is controlled by the output flow of the pump shown to the far left of Figure 9-1. This pump is shown to be constructed as a fixed-displacement pump with a volumetric displacement per unit of rotation given by

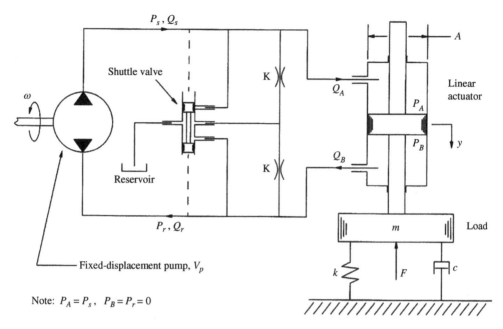

Figure 9-1. Schematic of the fixed-displacement, pump-controlled linear actuator.

the symbol V_p. For a positive angular velocity of the pump shaft pump flow is directed into side A of the actuator, which then requires the load to move downward and flow to exit the actuator from side B. For a negative angular velocity of the pump shaft pump flow is directed into side B of the actuator, which then requires the load to move upward and flow to exit the actuator from side A. These two volumetric flow rates into and out of the actuator are shown in Figure 9-1 by the symbols Q_A and Q_B respectively. In Figure 9-1 a shuttle valve is used to connect the low pressure side of the hydraulic control system to the reservoir. This feature does the following things: (a) it keeps the low pressure side of the circuit at a constant reservoir pressure; that is, zero gauge pressure, (b) it keeps the fixed-displacement pump from drawing a vacuum and causing fluid cavitation, and (c) it allows for the return flow to be cooled by a low-pressure radiator not shown in Figure 9-1. The shuttle valve shifts up or down depending on which side of the circuit is at high pressure. The dashed lines in Figure 9-1 indicate pressure signals that are used to move the shuttle valve. The schematic in Figure 9-1 shows a leak path on both sides of the hydraulic circuit that is characterized by the leakage coefficient K. This low Reynolds number flow occurs naturally due to the inherent internal leakage of the system; or, it may be designed intentionally to enhance the "flushing" characteristics of the circuit. The pump in Figure 9-1 is shown to be driven by an input shaft that rotates at a variable angular velocity ω. This rotation is achieved by a prime mover (not shown in Figure 9-1) that is capable of altering the input speed to the pump based on the control demands of the application. This prime mover may be an internal combustion engine that varies its speed based on throttle position or it may be an electric motor that is equipped with a variable frequency drive. The selection of the prime mover is often based on the application itself. For terrestrial mobile applications an internal combustion engine is often used while aerospace or industrial applications typically utilize electric motors as the power source for the hydraulic system.

9.2.2 Analysis

Load Analysis.   The analysis of the system shown in Figure 9-1 begins with the load. When developing the equations of motion for the load we neglect the inertia of the actuator itself since it is much smaller than the actual forces that are generated by the actuator. By lumping the viscous drag effects of both the actuator and the load into the dashpot shown in Figure 9-1 the equation of motion for the load may be written as

$$m\ddot{y} + c\dot{y} + ky = \eta_{a_f} A(P_A - P_B) - F, \tag{9.1}$$

where P_A and P_B are the fluid pressures on the A and B sides of the actuator respectively, η_{a_f} is the force efficiency of the actuator, and F is the load force exerted on the actuator. From this result it can be seen that the actuator pressures P_A and P_B are the required inputs for adjusting the position of the load. These pressures result from the changing flow and volume conditions within the actuator itself and are considered in our next paragraph.

Pressure Analysis.   In order to analyze the fluid pressures on sides A and B of the actuator, it is assumed that the pressure transients that result from fluid compressibility in the transmission lines are negligible. This assumption is especially valid for a system design in which the transmission lines between the valve and the actuator are very short; that is, small volumes of fluid exist on either side of the actuator. The omission of pressure transient effects is also valid for systems in which the load dynamics are much slower than the pressure dynamics themselves. Since the load dynamics typically occur over a range of seconds and the pressure dynamics typically occur over a range of milliseconds, it is usually safe to neglect the time variation of the pressure in favor of the time variation of the overall system. For design and operating conditions in which long transmissions lines between the valve and the actuator are used, or in which a large amount of entrained air is captured within the fluid causing the fluid bulk-modulus to be reduced, it may be necessary to conduct a transient analysis of the pressure conditions on both sides of the actuator. This may also be the case if the actuator dynamics are very fast. For now, we assume that this is not the case and that the more typical system conditions describe our problem in which the pressure transients may be safely neglected. Using these assumptions we can use the volumetric efficiency results of Chapter 6 to describe the volumetric flow rate into side A of the linear actuator as

$$Q_A = \frac{A}{\eta_{a_v}}\dot{y}, \tag{9.2}$$

where A is the pressurized area of the actuator and η_{a_v} is the volumetric efficiency of the actuator. From the flow rates that are shown in Figure 9-1 it may be seen that the actuator flow rate is given by

$$Q_A = Q_s - KP_A \tag{9.3}$$

for an incompressible fluid. In this equation, K is the low Reynolds number leakage coefficient and Q_s is the supply flow delivered from the pump. This supply flow is mathematically described as

$$Q_s = \eta_{p_v} V_p \omega, \tag{9.4}$$

where η_{p_v} is the volumetric efficiency of the pump, V_p is the volumetric displacement of the pump per unit of rotation, and ω is the instantaneous shaft speed of the pump. Using the results of Equations (9.2) through (9.4) the fluid pressure on side A of the actuator may be described as

$$P_A = \frac{\eta_{p_v} V_p}{K} \omega - \frac{A}{\eta_{a_v} K} \dot{y}. \tag{9.5}$$

As previously mentioned the shuttle valve in Figure 9-1 is used to connect the low-pressure side of the hydraulic circuit to the reservoir. Therefore, the pressure on side B of the linear actuator is given by

$$P_B = P_r = 0, \tag{9.6}$$

where P_r is the fluid pressure in the reservoir, which is taken to be zero gauge pressure. *Note*: Equation (9.5) shows a pump velocity, and an actuator velocity dependence for the fluid pressure on side A. As to be shown subsequently an adjustment of the pump velocity term is used to provide a control input to the dynamic load equation while the linear velocity term is useful in providing favorable damping characteristics for the system.

Analysis Summary. To summarize the analysis of the linear actuation system shown in Figure 9-1 we substitute Equations (9.5) and (9.6) into Equation (9.1) to produce the following equation of motion for the system:

$$m\ddot{y} + \left(c + \frac{\eta_{a_f}}{\eta_{a_v}} \frac{A^2}{K}\right) \dot{y} + ky = \frac{\eta_{a_f} \eta_{p_v} V_p A}{K} \omega - F. \tag{9.7}$$

From Equation (9.7) it can be seen that the mechanical design of the linear actuator and the volumetric displacement of the pump have a decisive impact on the overall dynamics of the hydraulic control system. In particular it may be seen from Equation (9.7) that these design parameters help to shape the effective damping of the system and to provide an adequate gain relationship between the input velocity of the pump and the output motion of the load. The following subsection of this chapter considers the mechanical design parameters that must be specified in order to achieve a satisfactory output of the system from a steady-state perspective.

9.2.3 Design

Actuator Design. When considering the mechanical design of the hydraulic control system that is shown in Figure 9-1 it is necessary to start by sizing the linear actuator in accordance with the expected load requirements. Usually, for a given application the maximum working load is known along with the maximum working pressure in the system. These two quantities can be used together with the steady form of Equation (9.1) to design the pressurized area of the linear actuator. The steady form of Equation (9.1) is given by

$$F = \eta_{a_f} A (P_A - P_B), \tag{9.8}$$

where A is the designed pressurized area of the actuator, η_{a_f} is the force efficiency of the actuator, and $P_A - P_B$ is the fluid pressure drop across the actuator. At the working force that is specified for the application the fluid pressure on side A of the linear actuator is given by the full supply pressure since there is no intermediate valve that creates a pressure drop between the pump and the actuator. In this case the design specifications at the working force conditions are given by

$$F = F_w, \quad P_A = P_s, \quad \text{and} \quad P_B = 0, \tag{9.9}$$

where F_w is the working force of the hydraulic control system and P_s is the supply pressure that is specified at the working condition. Substituting these conditions into Equation (9.8) produces the following equation for sizing the pressurized area of the linear actuator shown in Figure 9-1:

$$A = \frac{F_w}{\eta_{a_f} P_s}. \tag{9.10}$$

These results may be used to design or select a linear actuator that will provide a sufficient working force for a given supply pressure. Other design considerations must also be taken into account when designing the linear actuator. For example, the stroke of the actuator or distance of piston travel must be sufficient for the application. Other machine design aspects must also be considered regarding acceptable pressure vessel stresses that will exist within the actuator itself and the sealing mechanisms that must be used to minimize both internal and external leakage. These machine design topics are beyond the scope of this current text. For further information on these topics the reader is referred to other texts that provide a more thorough address of these subjects [1, 2].

Pump Design.   Now that the linear actuator has been designed it is time to specify the hydraulic pump that will be used to power the hydraulic control system. As shown in Figure 9-1 this pump is a fixed-displacement pump that produces a volumetric flow rate that is proportional to the angular input speed of the pump shaft ω. In order to size the required volumetric displacement of the pump it is common to specify a no-load velocity requirement for the linear actuator and to size the pump in such a way as to achieve this velocity requirement. By setting P_A equal to zero for the no-load case Equation (9.5) may be used to show that the volumetric displacement of the pump must be designed so that

$$V_p = \frac{A v_o}{\omega}, \tag{9.11}$$

where v_o is the specified no-load velocity for the system and where the volumetric efficiencies for the pump and actuator η_{p_v} and η_{a_v} have been set equal to unity for the no-load condition of the system. In Equation (9.11) A is the designed area of the actuator and ω is the angular velocity of the pump at the no-load condition specified in the problem. *Note*: This is most likely the maximum pump speed.

So far, this discussion has been concerned with the appropriate sizing of the supply pump for the hydraulic control system shown in Figure 9-1. Though the pump size has been specified for the system based on certain system needs nothing has been said about the physical construction of the pump. In practice this pump may be a gear pump, or it may be an axial-piston pump, or it may be any positive-displacement pump that satisfies the volumetric-displacement requirement shown in Equation (9.11). Usually, the pump construction type is selected based on the required supply pressure of the system and the desired operating efficiency of the pump. As stated in Chapter 5, axial-piston pumps tend to operate at higher pressures and provide a higher operating efficiency as compared to gear and vane type pumps. On the other hand, axial-piston pumps tend to increase the upfront cost of the control system. As the reader knows all of these things must be taken into account when selecting the pump construction type that will be used in the control system design shown in Figure 9-1.

Input Power Design.   The input power that is required to operate the hydraulic control system shown in Figure 9-1 is given by the standard power equation

$$\Pi = \frac{V_p \omega P_s}{\eta_p}, \tag{9.12}$$

where η_p is the overall efficiency of the pump given by the product of the volumetric and torque efficiency values $\eta_p = \eta_{p_t}\eta_{p_v}$, and ω and P_s are the instantaneous operating speeds and pressures of the pump. To determine the maximum power that is required to operate the hydraulic control system shown in Figure 9-1 Equation (9.12) should be evaluated at the maximum combination of operating speeds and pressures that will be encountered in the application. Obviously, by using an inefficient pump the input power requirement for this system goes up. Other parameters in Equation (9.12) are determined by the specified functional requirements of the hydraulic control system as noted in the previous discussion.

Case Study

To illustrate the design of the hydraulic control system shown in Figure 9-1 we consider the following case study: An aircraft application utilizes an electrohydraulic actuator similar to what is shown in Figure 9-1 for positioning the control flaps on an aircraft wing. The wind resistance on the flap linearly increases the load from 0 to 20000 lbf as the actuator extends itself from −6 in to +6 in for a total travel distance of 12 in. The no-load velocity requirement for the actuator is given by 6 in/s and the actuator exhibits a force and volumetric efficiency of 98% and 97%, respectively. The maximum working pressure of the hydraulic system is 3400 psi and the maximum pump speed is given by 16000 rpm. The torque and volumetric efficiency of the pump is 95% and 94%, respectively. You are to design the hydraulic components of this system to achieve satisfactory performance of the system.

The design of this system begins with the actuator. Using the result of Equation (9.10) the pressurized area of the actuator is given by

$$A = \frac{F_w}{\eta_{a_f} P_s} = \frac{20000 \text{ lbf}}{0.98 \times 3400 \text{ psi}} = 6.0 \text{ in}^2.$$

If the rod diameter is assumed to be 1 in then the piston diameter may be calculated to be 2.94 in. Again, the stroke of this actuator is given by ±6 in.

Now that the actuator has been designed to handle the maximum load requirements of the system the volumetric displacement of the pump must be specified to ensure that the actuator responds fast enough to satisfy the no-load velocity requirement of the system. Equation (9.11) may be used to calculate the necessary pump displacement as

$$V_p = \frac{A v_o}{\omega} = \frac{6 \text{ in}^2 \times 6 \text{ in/s}}{16000 \text{ rpm}} = 0.135 \frac{\text{in}^3}{\text{rev}},$$

which is a fairly small pump made possible only by the high rotating speed of 16000 rpm. At the working pressure of 3400 psi this pump is most likely a piston type pump.

The peak power requirement for this application may be determined by using Equation (9.12). This result is given by

$$\Pi = \frac{V_p \omega P_s}{\eta_p} = \frac{0.135 \text{ in}^3/\text{rev} \times 16000 \text{ rpm} \times 3400 \text{ psi}}{0.95 \times 0.94} = 20.7 \text{ hp}.$$

The reader will recall that the total pump efficiency η_p is given by the product of the volumetric and torque efficiencies of the pump.

Design Summary.   In this subsection of the chapter the steady-state design characteristics of the fixed-displacement, pump-controlled double-rod linear actuator shown in Figure 9-1 are considered. As previously noted, this system is commonly used in the aerospace industry for the position control of flight surfaces; however, in various applications this system may also be used to control the output velocity of the load or even the load force itself. In the following subsection standard control designs for these various applications are considered using classical PID control structures.

9.2.4 Control

Position Control.   As previously noted, the fixed-displacement pump-controlled linear actuator shown in Figure 9-1 is traditionally used to accomplish a number of control objectives. One of the more common objectives of this control system is to accurately position the load at a prescribed location within the trajectory range of the actuator. Typically, this control function is carried out under slow-moving conditions of the actuator; therefore, the plant description for this control problem may safely neglect any transient contributions that would normally be significant during high-speed operations. With this being the case Equation (9.7) may be used to describe the controlled system by neglecting the transient terms of inertia and viscous damping. This result is given by

$$ky = \frac{\eta_{a_f}\eta_{p_v}V_pA}{K}\,\omega - F, \tag{9.13}$$

where η_{a_f} is the force efficiency of the actuator, η_{p_v} is the volumetric efficiency of the pump, and the other parameters of Equation (9.13) are shown in Figure 9-1. Using a standard proportional-integral (PI) controller, the control law for the angular velocity of the fixed-displacement pump may be written as

$$\omega = K_e(y_d - y) + K_i \int (y_d - y)\,dt, \tag{9.14}$$

where K_e is the proportional gain, K_i is the integral gain, and y_d is the desired position of the load. Equation (9.14) must be enforced using a speed-controlled prime mover for the power source of this system coupled with a microprocessor that is capable of reading feedback information and generating the appropriate output signal for the speed control. Practical texts that address the interface of the mechanical system with the microprocessor are given in the References section of this chapter. Figure 9-2 shows the block diagram for the position control that is described by Equations (9.13) and (9.14).

If we assume that the position control of the system shown in Figure 9-1 is a regulation control problem and that the load force F and the desired position of the load y_d are constants, Equations (9.13) and (9.14) may both be differentiated and combined to write the following equation of motion for the controlled system:

$$\tau \dot{y} + y = y_d. \tag{9.15}$$

In this equation the system time constant is given by

$$\tau = \frac{1}{K_i}\left(\frac{kK}{\eta_{a_f}\eta_{p_v}AV_p} + K_e\right). \tag{9.16}$$

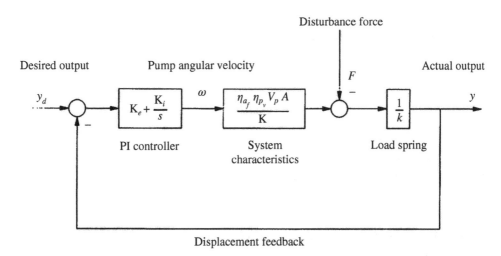

Figure 9-2. Position control of the fixed-displacement, pump-controlled linear actuator.

The reader will recognize Equation (9.15) as a description of a first-order system that asymptotically approaches the desired position of the load-in time without exhibiting any overshoot or oscillation. As time goes to infinity the steady-state error of this system vanishes, which is the primary advantage that is gained by using the integral component of the PI controller. The settling time for the first-order system is given by $t_s = 4\tau$.

The design objective for the position control problem is to shape the first-order response by designing the appropriate time constant for the system. By making a proper selection of the proportional and integral control gains this time constant can be either increased or diminished based on the desired settling time for the control system output. Though the first-order controlled system of Equation (9.15) shows a high degree of simplicity, the reader must keep in mind that this system response is based on a slow-moving device and that if the desired settling time becomes too short higher-order dynamics will become significant and a more sophisticated analysis of the system will be necessary to ensure that the system response is acceptable.

Case Study

In the previous case study an aircraft application was presented in which an electrohydraulic actuator similar to what is shown in Figure 9-1 was used for positioning the control flaps on an aircraft wing. In this application the wind resistance on the flap increased linearly from 0 to 20000 lbf as the actuator extends itself from −6 in to +6 in for a total travel distance of 12 in. The no-load velocity requirement for the actuator was given by 6 in/s and the actuator exhibited a force efficiency of 98%. The maximum pump speed was given by 16000 rpm. The volumetric efficiency of the pump was given by 94%. In the design of this system, the pressurized area of the actuator was determined to be 6.0 in^2 and the volumetric displacement of the pump was specified as 0.135 in^3/rev. The leakage coefficient for each side of the hydraulic loop is given by 1.283×10^{-3} in^3/(psi s). Using a standard PI controller to achieve position control of the actuator calculate the settling time of the control system using realistic proportional and integral control gains.

The solution to this problem begins by recognizing that the settling time for the first-order control system is given by $t_s = 4\tau$ where the time constant τ is given in

Equation (9.16). From Chapter 3 we learned that realistic proportional and integral control gains may be obtained as:

$$K_e = \frac{4}{5}\frac{\omega_{\max}}{y_{\max}} = \frac{4}{5}\frac{16000\ \text{rpm}}{12\ \text{in}} = 17.78\ \frac{\text{rev}}{\text{ins}},$$

$$K_i = \frac{2K_e}{t_s} = \frac{2 \times (1067\ \text{rpm/in})}{2\text{s}} = 17.78\ \frac{\text{rev}}{\text{ins}^2},$$

where at this point the settling time t_s is only an estimate of the actual settling time. Another important feature of the control system that must be determined is the load spring rate. From the problem statement we can see that this quantity is determined by

$$k = \frac{\Delta F}{\Delta y} = \frac{20000\ \text{lbf}}{12\ \text{in}} = 1667\ \frac{\text{lbf}}{\text{in}}.$$

From Equation (9.16), we see that the system time constant is given by

$$\tau = \frac{1}{K_i}\left(\frac{kK}{\eta_{a_f}\eta_{p_v}AV_p} + K_e\right).$$

$$= \frac{1}{17.78\ \text{rev/(in s}^2)}\left(\frac{1667\ \text{lbf/in} \times 1.283 \times 10^{-3}\ \text{in}^3/(\text{psi s})}{0.98 \times 0.94 \times 6.0\ \text{in}^2 \times 0.135\ \text{in}^3/\text{rev}} + 17.78\ \frac{\text{rev}}{\text{in s}}\right) = 1.15\ \text{s}.$$

The actual settling time will be equal to four time constants; therefore, $t_s = 4.6$ s. As shown by this example the dynamic response of the control system is dominated by the control gain selection as the system design parameters have only a 15% impact on the time constant. This dominance of the control gain selection provides for a more robust control design since uncertainties and variations in the system parameters will have only a minor impact on the overall system response.

Velocity Control.   Another possible control objective for the system shown in Figure 9-1 is that of velocity control. This objective seeks to establish a specific output velocity for the load based on a desired velocity that is prescribed by the application. This objective is typically carried out for load systems that do not include a load spring. Therefore, by neglecting the spring that is shown in Figure 9-1; that is, setting $k = 0$, the equation of motion for the velocity-controlled system may be written using the general form of Equation (9.7). This result is given by

$$m\dot{v} + \left(c + \frac{\eta_{a_f}}{\eta_{a_v}}\frac{A^2}{K}\right)v = \frac{\eta_{a_f}\eta_{p_v}V_p}{K}\omega - F, \tag{9.17}$$

where v is the instantaneous velocity of the load. Again, η_{a_f} is the force efficiency of the actuator and η_{p_v} is the volumetric efficiency of the pump. All other parameters of Equation (9.17) are shown in Figure 9-1. For the velocity control objective we select a full proportional-integral-derivative (PID) controller with the following control law for the angular velocity input to the fixed-displacement pump:

$$\omega = K_e(v_d - v) + K_i\int(v_d - v)\ dt + K_d\frac{d}{dt}(v_d - v), \tag{9.18}$$

where K_e is the proportional gain, K_i is the integral gain, K_d is the derivative gain, and v_d is the desired velocity of the load. Again, the most practical way to enforce the control law

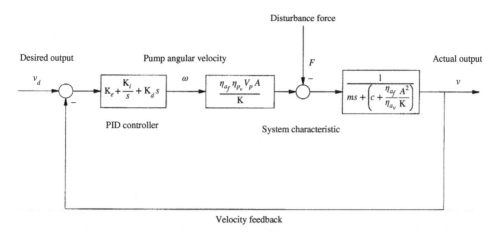

Figure 9-3. Velocity control of the fixed-displacement, pump-controlled linear actuator.

of Equation (9.18) is to use a speed-controlled prime mover for the power source of this system coupled with a microprocessor that is capable of reading feedback information and generating the appropriate output signal for the speed control. Figure 9-3 shows the block diagram for the velocity control that is described by Equations (9.17) and (9.18).

If it is assumed that the velocity control of the system shown in Figure 9-1 is a regulation control problem and that the load force F and the desired velocity of the load v_d are constants, Equations (9.17) and (9.18) may both be differentiated and combined to write the following equation of motion for the controlled system:

$$\ddot{v} + 2\zeta\omega_n\dot{v} + \omega_n^2 v = \omega_n^2 v_d, \tag{9.19}$$

where the undamped natural frequency and the damping ratio are given respectively as

$$\omega_n = \sqrt{\frac{K_i}{\dfrac{mK}{\eta_{a_f}\eta_{p_v}AV_p} + K_d}} \quad \text{and} \quad \zeta = \frac{\left(c + \dfrac{\eta_{a_f}}{\eta_{a_v}}\dfrac{A^2}{K} + \dfrac{\eta_{a_f}\eta_{p_v}AV_p}{K}K_e\right)}{2\left(m + \dfrac{\eta_{a_f}\eta_{p_v}AV_p}{K}K_d\right)\omega_n}. \tag{9.20}$$

The reader will recognize Equation (9.19) as a description of a second-order system that may exhibit overshoot and oscillation before reaching a steady-state condition in time. Whether overshoot and oscillation is observed depends on whether the damping ratio ζ is less than or greater than unity. If the damping ratio is less than unity the system is an underdamped system and the output will be characterized by overshoot and oscillation before reaching a steady-state condition. On the other hand, if the damping ratio is greater than unity the system is an overdamped system and the output will appear to be very much like a first-order system that asymptotically approaches a steady-state condition without any overshoot or oscillation. If it is assumed that the controlled system is of the underdamped variety then the rise time, settling time, and maximum percent overshoot may be calculated according to their definitions in Chapter 3.

The design objective for the velocity control problem is to shape the second-order response by designing the appropriate undamped natural frequency and damping ratio for the system. By making the proper selection of the proportional, integral, and derivative control gains the undamped natural frequency and damping ratio for the system may be adjusted accordingly. From Equation (9.20) it can be seen that the proportional gain provides an adjustment for the effective damping of the system while the integral and derivative gains are used to adjust the effective spring rate and mass of the system, respectively. If the derivative gain is neglected, that is, a PI control is used instead of the full PID control, it can be seen from Equation (9.20) that the undamped natural frequency of the system may be adjusted by solely adjusting the integral gain K_i. The reader will recall that the undamped natural frequency is the primary descriptor for the rise time of the underdamped second-order system. Once the desired natural frequency has been obtained the damping ratio for the PI-controlled system may be adjusted by adjusting the proportional gain K_e. The damping ratio is the sole contributor to the maximum percent overshoot. This series approach to tuning the PI controller is very convenient and has been noted as one of its primary advantages of this controller in Chapter 3.

Force Control.   Finally, another control objective that is occasionally used for the hydraulic control system shown in Figure 9-1 is that of force control. Force-controlled systems are usually configured very much like a velocity-controlled system without a load spring and when they are used they often switch between velocity and force control depending on the immediate needs of the application. Furthermore, force-controlled systems are typically operated in slow motion so as to gradually apply the load force to the object that the application is trying to resist. Under these slow-motion conditions the load inertia and viscous damping effects in the system may be safely ignored without sacrificing accuracy in the modeling process. By setting the left-hand side of Equation (9.7) equal to zero the following equation may be used to describe the applied load force for the force-controlled system:

$$F = \frac{\eta_{a_f} \eta_{p_v} V_p A}{K} \omega. \tag{9.21}$$

Using a standard proportional-integral (PI) controller, the control law for the angular velocity of the pump shaft may be written as

$$\omega = K_e(F_d - F) + K_i \int (F_d - F)\, dt, \tag{9.22}$$

where K_e is the proportional gain, K_i is the integral gain, and F_d is the desired force that is to be exerted on the load. In most applications the instantaneous load force F is measured by sensing the fluid pressures on sides A and B of the linear actuator and calculating the load force according to Equation (9.8). Again, a microprocessor is necessary to carry out these functions along with a variable speed drive for the pump that is used to enforce the control law of Equation (9.22). Figure 9-4 shows the block diagram for the force control that is described by Equations (9.21) and (9.22).

If it is assumed that the force control of the system shown in Figure 9-1 is a regulation control problem and that the desired load force F_d is a constant, Equations (9.21) and (9.22) may both be differentiated and combined to write the following dynamic equation for the force-controlled system:

$$\tau \dot{F} + F = F_d, \tag{9.23}$$

where the system time constant is given by

$$\tau = \frac{1}{K_i}\left(\frac{K}{\eta_{a_f}\eta_{p_v}AV_p} + K_e\right). \tag{9.24}$$

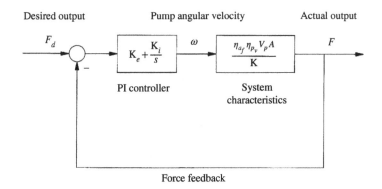

Figure 9-4. Force control of the fixed-displacement, pump-controlled linear actuator.

Again, the reader will recognize Equation (9.23) as a description of a first-order system that asymptotically approaches the desired load force F in time without exhibiting any overshoot or oscillation. The dynamic similarities of this system should be compared to the position-controlled system that was previously discussed.

The design objective for the force control problem is to shape the first-order response by designing the appropriate time constant for the system. By making a proper selection of the proportional and integral control gains this time constant can be either increased or diminished based on the desired settling time for the control system output. Again, the reader is reminded that the first-order controlled system of Equation (9.23) has been designed based on the assumption that the linear actuator moves slowly. If in fact the desired response time for this system becomes too short higher-order dynamics that have been neglected in the modeling process will become more significant and the observed output may be unsatisfactory. Under such conditions a more sophisticated analysis of the system is needed to ensure satisfactory response characteristics.

Control Summary. In this subsection of the chapter we have considered the three control objectives that may be carried out using the fixed-displacement pump-controlled linear actuator shown in Figure 9-1. These objectives include position control, velocity control, and force control all of which have employed a form of the standard PID controller. For the position control and the force control an important slow speed assumption has been employed for the purposes of simplifying the control problem for realistic modes of operation. Due to the slow-speed assumptions behind these control strategies the reader is cautioned about using these controllers for high-speed applications that may be encountered for these control objectives. Under such high-speed conditions higher-order dynamics may tend to manifest themselves and may produce a less than desirable system output.

9.2.5 Summary

In this section of the chapter the fixed-displacement pump-controlled, linear actuator of Figure 9-1 is analyzed, designed, and controlled. Note that throughout this study certain speed conditions have been assumed regarding the operation of the control system. For instance, during the development of the governing system equations in Subsection 9.2.2 the pressure transient effects were neglected due to the assumption that they occurred very quickly in relationship to the overall process time of the load dynamics. If indeed this is not the case a higher-order model of the control system will be necessary to capture the relevant pressure dynamics of the system. This situation, however, will only occur for very rare and specialized applications and the analysis presented here is deemed to be sufficient for traditional and more typical applications of the system. In Subsection 9.2.3 a design methodology for the pump-controlled linear actuator are presented. This design method begins by considering the steady output objectives of the system and then works backward to identify the system components that are necessary to achieve these objectives. As shown in this subsection the design of the actuator and pump are closely linked and the proper specification of these components is important when seeking to satisfy the overall perfor- mance objectives for the system. Finally, in Subsection 9.2.4 the dynamic control objectives of the actuation system are presented. These objectives vary depending on the application; however, they generally consist of either a position control, a velocity control, or a force control. In the consideration of these control objectives it has been assumed that the con- trol law may be implemented by using a speed-controlled input to the fixed-displacement pump and a microprocessor with sufficient sensing of the controlled parameters. For infor- mation on implementing the microprocessor within the control design the reader is referred to practical texts within this field [3, 4].

9.3   VARIABLE-DISPLACEMENT PUMP CONTROL OF A ROTARY ACTUATOR

9.3.1   Description

In this section of our chapter we introduce the variable-displacement, pump-controlled rotary actuator. This system is frequently employed in the off-highway mobile equipment industry for propelling work vehicles without the use of standard transmissions. Within such applications this control system is often referred to as a continuously variable hydrostatic transmission; or, in more common language this system is often referred to simply as a "hystat." As shown in Figure 9-5 this system utilizes a variable-displacement pump to control the output characteristics of a fixed-displacement rotary actuator. In this figure the load to be moved by the actuator is shown as a single mass-spring-damper system with a load disturbance torque given by T.

The mass, spring rate, and viscous drag coefficient for the load are shown in Figure 9-5 by the symbols J, b, and k_t, respectively. The rotary actuator is shown to be a fixed-displacement type with a volumetric displacement of fluid per unit of rotation given by the symbol V_a. The volumetric flow of hydraulic fluid into the actuator is controlled by the output flow of the variable-displacement pump shown to the far left of Figure 9-5. This pump is shown to be constructed as a variable-displacement pump with a maximum volumetric displacement per unit of rotation given by the symbol V_p and is typically operated at a fixed input speed given by ω_p. The variable displacement of the pump is achieved by altering the swash plate angle α. In this design the swash plate angle may be positive or negative depending on which direction flow is being directed through the hydraulic circuit. When the swash plate angle changes sign it is said to have gone "over center" and the direction of flow has been reversed. For a positive swash plate angle pump flow is directed into side A of the actuator, which then requires the load to turn in a positive angular direction and flow to exit the actuator from side B. For a negative swash plate angle pump flow is directed into side B of the actuator, which then requires the load to move in a negative angular direction and flow to exit the actuator from side A. These two volumetric flow rates into and out of the

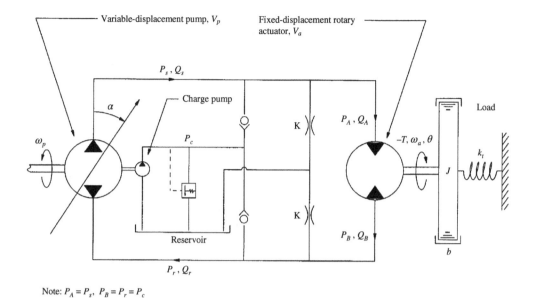

Note: $P_A = P_s$, $P_B = P_r = P_c$

Figure 9-5. Schematic of the variable-displacement, pump-controlled rotary actuator.

actuator are shown in Figure 9-5 by the symbols Q_A and Q_B, respectively. In Figure 9-5 orifice flow passages from both sides of the circuit are shown with flow coefficients given by the symbol K. These flow passages exist inherently within the system; however, they may also be designed intentionally for the purposes of directing fluid flow into the reservoir for cooling and filtration. *Note*: Coolers and filters are not shown in Figure 9-5. What is shown in Figure 9-5 is a charge pump circuit that is used to pressurize the low-pressure side of the control system and to sometimes provide an auxiliary pressure source for the variable-displacement pump control. The charge pump itself is a fixed-displacement pump (usually a gear pump or a gerotor pump) that is used in conjunction with a relief valve for the purposes of providing makeup flow to the main pump inlet. The pressure setting on the relief valve P_c is typically about 2 MPa while the high-pressure side of the hydraulic circuit often operates at a pressure over 10 times this amount. The check valves that are shown in Figure 9-5 by the ball and cone symbols are used to block the high-pressure side from the charge circuit and to open the low-pressure side to the charge circuit. This feature ensures that the low-pressure side is always maintained at a constant pressure given by P_c and prevents pump cavitation during fast alterations of the swash plate angle α.

9.3.2 Analysis

Load Analysis. As usual, the analysis of the system shown in Figure 9-5 begins with the load. When developing the equations of motion for the load we neglect the angular inertia of the actuator itself since it is much smaller than the actual torque that is generated by the actuator. By lumping the viscous drag effects of both the actuator and the load into the rotational dashpot shown in Figure 9-5 the equation of motion for the load may be written as

$$J\ddot{\theta} + b\dot{\theta} + k_t\theta = \eta_{a_t} V_a (P_A - P_B) - T, \qquad (9.25)$$

where P_A and P_B are the fluid pressures on the A and B sides of the actuator respectively, η_{a_t} is the torque efficiency of the actuator, and where T is the load force exerted on the actuator. From this result it can be seen that the actuator pressures P_A and P_B are the required inputs for adjusting the position of the load. These pressures result from the changing flow and volume conditions within the actuator itself and are considered in our next paragraph.

Pressure Analysis.   In order to analyze the fluid pressures on sides A and B of the actuator it is assumed that the pressure transients that result from fluid compressibility in the transmission lines are negligible. This assumption is especially valid for a system design in which the transmission lines between the pump and the actuator are very short; that is, small volumes of fluid exist on either side of the actuator. The omission of pressure transient effects is also valid for systems in which the load dynamics are much slower than the pressure dynamics themselves. Since the load dynamics typically occur over a range of seconds and the pressure dynamics typically occur over a range of milliseconds, it is usually safe to neglect the time variation of the pressure in favor of the time variation of the overall system. For design and operating conditions in which long transmissions lines between the pump and the actuator are used, or in which a large amount of entrained air is captured within the fluid causing the fluid bulk-modulus to be reduced, it may be necessary to conduct a transient analysis of the pressure conditions on both sides of the actuator. This may also be the case if the actuator dynamics are very fast. For now we assume that this is not the case and that the more typical system conditions describe our problem in which the pressure transients may be safely neglected. Using these assumptions we can use the volumetric efficiency results of Chapter 6 to describe the volumetric flow rate into side A of the rotary actuator as

$$Q_A = \frac{V_a}{\eta_{a_v}}\dot{\theta}, \tag{9.26}$$

where V_a is the volumetric displacement of the actuator and η_{a_v} is the volumetric efficiency of the actuator. From the flow rates that are shown in Figure 9-5 it may be seen that the actuator flow rate is given by

$$Q_A = Q_s - KP_A, \tag{9.27}$$

for an incompressible fluid. In this equation K is the low Reynolds number leakage coefficient and Q_s is the supply flow delivered from the pump. This supply flow is mathematically described as

$$Q_s = \frac{\eta_{p_v} V_p \omega_p}{\alpha_{\max}}\alpha, \tag{9.28}$$

where η_{p_v} is the volumetric efficiency of the pump, V_p is the maximum volumetric displacement of the pump per unit of rotation, ω_p is the instantaneous shaft speed of the pump, and $\alpha_{\max}$ is the maximum swash plate angle of the pump. Using the results of Equations (9.26) through (9.28) the fluid pressure on side A of the actuator may be described mathematically as

$$P_A = \frac{\eta_{p_v} V_p \omega_p}{K\alpha_{\max}}\alpha - \frac{V_a}{\eta_{a_v} K}\dot{\theta}. \tag{9.29}$$

As previously mentioned the check valves in Figure 9-5 are used to connect the low-pressure side of the hydraulic circuit to the charge circuit. Therefore, the pressure on side B of the linear actuator is given by

$$P_B = P_r = P_c, \tag{9.30}$$

where P_c is the fluid pressure in the charge circuit, which is normally about 2 MPa. *Note:* Equation (9.29) shows a pump swash plate angle and an actuator velocity dependence for the fluid pressure on side A. As to be shown subsequently an adjustment of the pump swash plate angle will be used to provide a control input to the dynamic load equation while the angular velocity term will be useful in providing favorable damping characteristics for the system.

Analysis Summary.   To summarize the analysis of the linear actuation system shown in Figure 9-5 we substitute Equations (9.29) and (9.30) into Equation (9.25) to produce the following equation of motion for the system:

$$J\ddot{\theta} + \left(b + \frac{\eta_{a_t}}{\eta_{a_v}} \frac{V_a^2}{K} \right) \dot{\theta} + k_t\theta = \frac{\eta_{a_t}\eta_{p_v} V_a V_p \omega_p}{K\alpha_{max}}\alpha - \eta_{a_t} V_a P_c - T. \tag{9.31}$$

From Equation (9.31) it can be seen that the mechanical design of the rotary actuator and the volumetric displacement of the pump have a decisive impact on the overall dynamics of the hydraulic control system. In particular it may be seen from Equation (9.31) that these design parameters help to shape the effective damping of the system and to provide an adequate gain relationship between the input velocity of the pump and the output motion of the load. The following subsection of this chapter considers the mechanical design parameters that must be specified in order to achieve a satisfactory output of the system from a steady-state perspective.

9.3.3 Design

Actuator Design.   When considering the mechanical design of the hydraulic control system that is shown in Figure 9-5 it is necessary to start by sizing the rotary actuator in accordance with the expected load requirements. Usually, for a given application the maximum working torque for the system is known along with the maximum working pressure. These two quantities can be used together with the steady form of Equation (9.25) to design the fixed volumetric displacement of the rotary actuator. The steady form of Equation (9.25) is given by

$$T = \eta_{a_t} V_a(P_A - P_B), \tag{9.32}$$

where V_a is the designed volumetric displacement of the actuator, η_{a_t} is the torque efficiency of the actuator, and $P_A - P_B$ is the fluid pressure drop across the actuator. At the working torque that is specified for the application the fluid pressure on side A of the rotary actuator is given by the full supply pressure since there is no intermediate valve that creates a pressure drop between the pump and the actuator. In this case the design specifications at the working torque conditions are given by

$$T = T_w, \quad P_A = P_s, \quad \text{and} \quad P_B = P_c, \tag{9.33}$$

where T_w is the working torque of the hydraulic control system, P_s is the supply pressure that is specified at the working condition, and P_c is the charge pressure connected to side B by the open check valve. Substituting these conditions into Equation (9.32) produces the following equation for sizing the volumetric displacement of the rotary actuator shown in Figure 9-5:

$$V_a = \frac{T_w}{\eta_{a_t}(P_s - P_c)}. \tag{9.34}$$

These results may be used to design or select a rotary actuator that provides a sufficient working torque for a given supply pressure. Since the rotary actuator is simply a hydraulic pump operating in a reverse direction the type of actuator to be used may range from a gear type machine to an axial-piston machine. Usually, the choice of the actuator is based on the performance that is needed from the actuator itself. For instance, if a high operating efficiency is needed with a high starting torque the best actuator selection would be a bent-axis axial-piston machine. *Note*: Axial-piston machines (bent-axis or swash-plate type) must generally be used for operating pressures that exceed 20 MPa. On the other hand, if efficiency is not a significant factor in the application and if the pressures are below 20 MPa a less-expensive option would be to select either a gear or a vane type actuator.

Pump Design. Now that the rotary actuator has been designed it is time to specify the hydraulic pump that will be used to power the hydraulic control system. As shown in Figure 9-5 this pump is a variable-displacement pump that produces a volumetric flow rate that is proportional to the swash plate angle of the pump α. For these systems the input speed to the pump is maintained at a constant value ω_p. *Note*: The swash plate angle in this problem is used to describe any type of adjustment that is made to the displacement of a variable-displacement machine and is not generally constrained to describe a swash plate type machine. Since swash plate type machines dominate the applications in which variable-displacement pumps are used the adjustment parameter α is referred to here as the swash plate angle to capture the majority of applications. In order to determine the maximum volumetric displacement of the pump it is common to specify a no-load velocity requirement for the rotary actuator and to size the pump in such a way as to achieve this velocity requirement. By setting P_A equal to zero for the no-load case, Equation (9.29) may be used to show that the volumetric displacement of the pump must be designed so that

$$V_p = \frac{\omega_o}{\omega_p} V_a, \tag{9.35}$$

where ω_o is the specified no-load velocity for the system and where the volumetric efficiencies for the pump and actuator η_{p_v} and η_{a_v} have been set equal to unity for the no-load condition of the system. In Equation (9.35) V_a is the designed volumetric displacement of the actuator given in Equation (9.34) and ω_p is the angular velocity of the pump at the no-load condition specified in the problem. From Equation (9.35) it can be seen that the ratio of the pump volumetric displacement to the actuator volumetric displacement $V_p : V_a$ is equal to the no-load speed ratio or "gear" ratio of the transmission ω_o/ω_p.

So far, this discussion has been concerned with the appropriate sizing of the supply pump for the hydraulic control system shown in Figure 9-5. Though the pump size has been specified for the system based on certain system needs nothing has been said about the physical construction of the pump. Since the pump is of the variable-displacement type it is most likely an axial-piston swash plate type pump. Nevertheless it may be any positive-displacement pump that is constructed to vary its displacement based on the adjustment of some internal parameter of the machine. For instance, a vane pump or a radial piston pump may be used with a variable eccentricity dimension. Usually, the pump construction type is selected based on the required supply pressure of the system and the desired operating efficiency of the pump. As stated in Chapter 5 axial-piston pumps tend to operate at higher pressures and provide a higher operating efficiency as compared to gear and vane type pumps. On the other hand, axial-piston pumps tend to increase the upfront cost of the control system. As the reader knows all of these things must be taken into account when selecting the pump construction type that will be used in the control system design.

Input Power Design.   The input power that is required to operate the hydraulic control system shown in Figure 9-5 is given by the standard power equation

$$\Pi = \frac{V_p \omega_p (P_s - P_c)}{\eta_p} + \frac{V_c \omega_p P_c}{\eta_c}, \tag{9.36}$$

where η_p is the overall efficiency of the main pump given by the product of the volumetric and torque efficiency values $\eta_p = \eta_{p_t} \eta_{p_v}$ and ω_p and P_s are the instantaneous operating speeds and pressures of the pump. In Equation (9.36) the charge pressure of the system is given by P_c and the second term on the right-hand side describes the power that is required to operate the charge circuit. As shown in Figure 9-5 the charge pump is directly coupled to the main pump using a through drive shaft. This arrangement requires the charge pump to operate at the same angular velocity as the main pump. In Equation (9.36) η_c is the overall efficiency of the charge pump. To determine the maximum power that is required to operate the hydraulic control system shown in Figure 9-5. Equation (9.36) should be evaluated at the maximum combination of operating speeds and pressures that will be encountered in the application. Obviously, by using an inefficient pump the input power requirement for this system goes up. Other parameters in Equation (9.36) are determined by the specified functional requirements of the hydraulic control system as noted in the previous discussion.

Case Study

It is common for track type vehicles to be equipped with two independent pump-controlled rotary actuators as shown in Figure 9-5. Together, these two independent hydraulic systems are known as a dual path hydrostatic transmission. The figure shows this arrangement for a typical track type tractor. Steering for this vehicle is achieved by causing one track to rotate at a different angular velocity relative to the other track. For straight line driving both tracks rotate at the same velocity.

The track type vehicle shown in the figure has a total mass of 24100 kg. During its most severe operation the torque that must be delivered to each track is given by 72000 Nm. This torque is generated by the rotary actuator working through a final drive gear ratio of 60:1. This gear ratio is shown in the figure by the symbol m_{FD}. The maximum no-load angular velocity for the motor is given by 1800 rpm which is also the angular velocity of the engine that drives each pump through a 1:1 gear ratio.

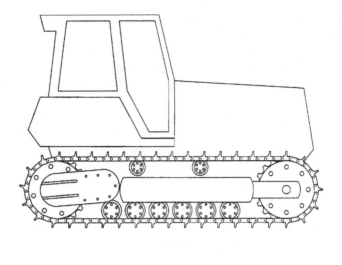

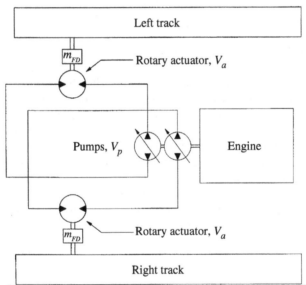

The radius of the track hub is given by 0.381 m. The torque and volumetric efficiencies for the rotary actuators are given by 95% and 96%, respectively, and these are the same efficiency numbers to be used for each pump. The maximum working pressure for the hydraulic circuit is 42 MPa and the charge pressure is 2 MPa. Determine the pump and actuator size that is needed for this application. How much power will be demanded from the engine during peak operating conditions (ignore the power loss due to the charge circuit)? What will be the maximum ground velocity of the vehicle at these conditions? Assume that the coefficient of leakage for each rotary actuator circuit is given by 5×10^{-11} m^3/(Pa s).

The solution to this problem begins by sizing the actuator to handle the required torque capacity for each track. From the problem statement we can see that this torque is given by

$$T_w = \frac{72000 \, \text{Nm}}{m_{FD}} = \frac{72000 \, \text{Nm}}{60} = 1200 \, \text{Nm}.$$

Using this result Equation (9.34) may be used to size the actuator as follows:

$$V_a = \frac{T_w}{\eta_{a_t}(P_s - P_c)} = \frac{1200 \text{ Nm}}{0.95 \times (42 \text{ MPa} - 2 \text{ MPa})}$$

$$= 3.158 \times 10^{-5} \frac{\text{m}^3}{\text{rad}} \approx 200 \frac{\text{cm}^3}{\text{rev}}.$$

Since the no-load velocity requirement of the actuator is equivalent to the angular velocity of the pump and engine Equation (9.35) may be used to show that the maximum volumetric displacement of the pump is given by the same displacement as that of the actuator. In other words, $V_p = V_a = 200 \text{ cm}^3/\text{rev}$. The power required to operate two rotary actuation systems without considering the charge circuit requirements may be determined from Equation (9.36) as:

$$\Pi = 2 \times \frac{V_p \omega_p (P_s - P_c)}{\eta_p} = 2 \times \frac{200 \text{ cm}^3/\text{rev} \times 1800 \text{ rpm} \times 40 \text{ MPa}}{0.95 \times 0.96}$$

$$= 526.3 \text{ kW,}$$

which is a little more than 700 hp. The maximum ground velocity under fully loaded conditions may be determined by evaluating the angular velocity of the actuator when the fluid pressure in the circuit is at 42 MPa. Rearranging Equation (9.29) this angular velocity is given by

$$\omega_a = \frac{\eta_{a_v} \eta_{p_v} V_p}{V_a} \omega_p - \frac{\eta_{a_v} K}{V_a} P_A$$

$$= \frac{0.96 \times 0.96 \times 200 \text{ cm}^3/\text{rev}}{200 \text{ cm}^3/\text{rev}} \times 1800 \text{ rpm} - \frac{0.96 \times 5 \times 10^{-11} \text{ m}^3/(\text{Pa s})}{200 \text{ cm}^3/\text{rev}} \times 42 \text{ MPa}$$

$$= 1{,}659 \text{ rpm} - 605 \text{ rpm} = 1054 \text{ rpm.}$$

In this result the swash plate angle of the pump has been set equal to its maximum value to achieve a maximum velocity output of the actuator. The ground velocity under these conditions will be

$$v = \frac{\omega_a}{m_{FD}} R = \frac{1054 \text{ rpm}}{60} \times 0.381 \text{ m} = 2.52 \frac{\text{km}}{\text{hr}},$$

where R is the radius of the track hub. It should be noted that the high-pressure conditions of operation have reduced the maximum achievable ground speed by almost 35%.

Summary.   In this subsection of the chapter the steady-state design characteristics of the variable-displacement, pump-controlled rotary actuator shown in Figure 9-5 are considered. As previously noted this system is commonly used in the propel function of many off highway mobile applications; however, in various situations this system may also be used to control the output position of the load or even the load torque itself. In the following subsection standard control designs for these various applications are considered using classical PID control structures.

9.3.4 Control

Position Control. As previously noted the variable-displacement, pump-controlled rotary actuator shown in Figure 9-5 may be used to accomplish a number of control objectives. One control objective of this system may be to accurately position the load at a prescribed location within the trajectory range of the actuator. Typically this control function is carried out under slow moving conditions of the actuator; therefore, the plant description for this control problem may safely neglect any transient contributions that would normally be significant during high-speed operations. With this being the case Equation (9.31) may be used to describe the controlled system by neglecting the transient terms of inertia and viscous damping. This result is given by

$$k_t\theta = \frac{\eta_{a_t}\eta_{p_v}V_aV_p\omega_p}{\mathrm{K}\alpha_{\max}}\alpha - \eta_{a_t}V_aP_c - T,\tag{9.37}$$

where η_{a_t} is the force efficiency of the actuator, η_{p_v} is the volumetric efficiency of the pump, and the other parameters of Equation (9.37) are shown in Figure 9-5. Using a standard proportional-integral (PI) controller the control law for the swash plate angle of the variable-displacement pump may be written as

$$\alpha = \mathrm{K}_e(\theta_d - \theta) + \mathrm{K}_i\int(\theta_d - \theta)\,dt,\tag{9.38}$$

where K_e is the proportional gain, K_i is the integral gain, and θ_d is the desired angular position of the load. Equation (9.38) must be enforced using an electrohydraulic displacement-controlled pump coupled with a microprocessor that is capable of reading feedback information and generating the appropriate output signal for the displacement control. A typical displacement-controlled pump has been presented in Section 5.4.4 of Chapter 5. Practical texts that address the interface of the mechanical system with the microprocessor are given in the Reference section of this chapter. Figure 9-6 shows the block diagram for the position control that is described by Equations (9.37) and (9.38).

If we assume that the position control of the system shown in Figure 9-5 is a regulation control problem and that the load torque T and the desired position of the load θ_d are

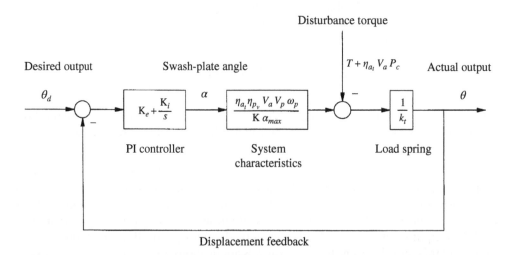

Figure 9-6. Position control of the variable-displacement, pump-controlled rotary actuator.

constants, Equations (9.37) and (9.38) may both be differentiated and combined to write the following equation of motion for the controlled system:

$$\tau \dot{y} + y = y_d. \tag{9.39}$$

In this equation the system time constant is given by

$$\tau = \frac{1}{K_i} \left(\frac{k_t K \alpha_{max}}{\eta_{a_t} \eta_{p_v} V_a V_p \omega_p} + K_e \right). \tag{9.40}$$

The reader will recognize Equation (9.39) as a description of a first-order system that asymptotically approaches the desired position of the load in time without exhibiting any overshoot or oscillation. As time goes to infinity the steady-state error of this system vanishes, which is the primary advantage that is gained by using the integral component of the PI controller. The settling time for the first-order system is given by $t_s = 4\tau$.

The design objective for the position control problem is to shape the first-order response by designing the appropriate time constant for the system. By making a proper selection of the proportional and integral control gains this time constant can be either increased or diminished based on the desired settling time for the control system output. Though the first-order controlled system of Equation (9.39) shows a high degree of simplicity the reader must keep in mind that this system response is based on a slow moving device and that if the desired settling time becomes too short higher-order dynamics will become significant and a more sophisticated analysis of the system will be necessary to ensure that the system response is acceptable.

Velocity Control. A more common control objective for the system shown in Figure 9-5 is that of angular velocity control. This objective seeks to establish a specific output angular velocity for the load based on a desired angular velocity that is prescribed by the application. This objective is typically carried out for load systems that do not include a load spring. Therefore, by neglecting the spring that is shown in Figure 9-5; that is, setting $k_t = 0$, the equation of motion for the velocity-controlled system may be written using the general form of Equation (9.31). This result is given by

$$J \dot{\omega}_a + \left(b + \frac{\eta_{a_t}}{\eta_{a_v}} \frac{V_a^2}{K} \right) \omega_a = \frac{\eta_{a_t} \eta_{p_v} V_a V_p \omega_p}{K \alpha_{max}} \alpha - \eta_{a_t} V_a P_c - T, \tag{9.41}$$

where ω_a is the instantaneous angular velocity of the load. Again, η_{a_t} is the torque efficiency of the actuator, η_{a_v} is the volumetric efficiency of the actuator, and η_{p_v} is the volumetric efficiency of the pump. All other parameters of Equation (9.41) are shown in Figure 9-5. For the velocity control objective we select a full proportional-integral-derivative (PID) controller with the following control law for the swash plate angle of the variable-displacement pump:

$$\alpha = K_e(\omega_d - \omega_a) + K_i \int (\omega_d - \omega_a) \, dt + K_d \frac{d}{dt}(\omega_d - \omega_a), \tag{9.42}$$

where K_e is the proportional gain, K_i is the integral gain, K_d is the derivative gain, and ω_d is the desired velocity of the load. Again, the most practical way to enforce the control law of Equation (9.18) is to use an electrohydraulic displacement-controlled pump coupled with a microprocessor that is capable of reading feedback information and generating the appropriate output signal for the displacement control. A typical displacement-controlled

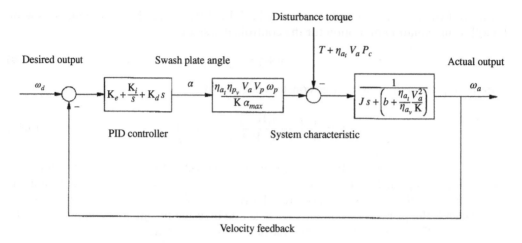

Figure 9-7. Velocity control of the variable-displacement, pump-controlled rotary actuator.

pump has been presented in Section 5.4.4 of Chapter 5. Figure 9-7 shows the block diagram for the angular velocity control that is described by Equations (9.41) and (9.42).

If it is assumed that the velocity control of the system shown in Figure 9-5 is a regulation control problem and that the load torque T and the desired velocity of the load ω_d are constants, Equations (9.41) and (9.42) may both be differentiated and combined to write the following equation of motion for the velocity-controlled system:

$$\ddot{\omega}_a + 2\zeta\omega_n\dot{\omega}_a + \omega_n^2\omega_a = \omega_n^2\omega_d, \tag{9.43}$$

where the undamped natural frequency and the damping ratio are given respectively as

$$\omega_n = \sqrt{\dfrac{K_i}{\dfrac{JK\alpha_{max}}{\eta_{a_t}\eta_{p_v}V_aV_p\omega_p} + K_d}} \quad \text{and}$$

$$\zeta = \dfrac{\left(\dfrac{bK\alpha_{max}}{\eta_{a_t}\eta_{p_v}V_aV_p\omega_p} + \dfrac{V_a\alpha_{max}}{\eta_{a_v}\eta_{p_v}V_p\omega_p} + K_e\right)}{2\sqrt{K_i\left(\dfrac{JK\alpha_{max}}{\eta_{a_t}\eta_{p_v}V_aV_p\omega_p} + K_d\right)}}. \tag{9.44}$$

The reader will recognize Equation (9.43) as a description of a second-order system that may exhibit overshoot and oscillation before reaching a steady-state condition in time. Whether overshoot and oscillation is observed depends on whether the damping ratio ζ is less than or greater than unity. If the damping ratio is less than unity the system is an under-damped system and the output will be characterized by overshoot and oscillation before reaching a steady-state condition. On the other hand, if the damping ratio is greater than unity the system is an overdamped system and the output will appear to be very much like a first-order system that asymptotically approaches a steady-state condition without any overshoot or oscillation. If it is assumed that the controlled system is of the underdamped variety then the rise time, settling time, and maximum percent overshoot may be calculated according to their definitions in Chapter 3. The design objective for the velocity control problem is to shape the second-order response by designing the appropriate undamped natural frequency and damping ratio for the system. By making the proper selection of the proportional, integral, and derivative control gains the undamped natural frequency and damping ratio for the system may be adjusted accordingly. From Equation (9.44) it can be seen that the proportional gain provides an adjustment for the effective damping of the system while the integral and derivative gains are used to adjust the effective spring rate and mass of the system respectively. If the derivative gain is neglected; that is, a PI control is used instead of the full PID control, it can be seen from Equation (9.44) that the undamped natural frequency of the system may be adjusted by solely adjusting the integral gain K_i. The reader will recall that the undamped natural frequency is the primary descriptor for the rise time of the underdamped second-order system. Once the desired natural frequency has been obtained the damping ratio for the PI-controlled system may be adjusted by adjusting the proportional gain K_e. The damping ratio is the sole contributor to the maximum percent overshoot. This series approach to tuning the PI controller is very convenient and has been noted as one of this controller's primary advantages in Chapter 3.

Case Study

In the previous case study a hydrostatic transmission similar to the control system of Figure 9-5 was designed to accomplish the propel function of 24100 kg track type vehicle. In this design the pump and rotary actuator were specified with volumetric displacements given by 200 cm³/rev and the constant shaft speed of the pump was 1800 rpm. The volumetric and torque efficiencies for both the pump and actuator were specified as 96% and 95%, respectively. The leakage coefficient in the circuit was given by $5 \times 10^{-11} \text{m}^3/(\text{Pa s})$. The gear ratio of the final drive between the actuator and the track was specified as 60:1 and the radius of the track hub was given by 0.381 m. The maximum swash plate angle for the pump is given by 18 degrees. Using a PI controller with the following gains: $K_e = 1.33 \times 10^{-3}\text{s}$ and $K_i = 0.67 \times 10^{-3}$ (no units), calculate the undamped natural frequency and the damping ratio of the velocity-controlled system. How do you expect this system to behave?

The solution to this problem begins by recognizing that the velocity control depends on the effective mass moment of inertia for the system. This quantity may be roughly determined by neglecting the rotating kinetic energy of the tracks themselves and equating the rotating kinetic energy of the transmission to the translating kinetic energy of the track type vehicle. This expression is given as:

$$\frac{1}{2}J\omega_a^2 = \frac{1}{2}\left(\frac{M}{2}\right)v^2,$$

where M is the total mass of the vehicle (only half of this mass is used for the calculations of a single track) and v is the translational ground velocity. Since the ground velocity is given by the following expression for a no-slip travel velocity:

$$v = \frac{\omega_a}{m_{FD}}R,$$

where m_{FD} is the gear ratio of the final drive and R is the track hub radius, the mass moment of inertia that is felt by the rotary actuator is given by

$$J = \frac{M}{2}\left(\frac{R}{m_{FD}}\right)^2 = \frac{24100 \text{ kg}}{2}\left(\frac{0.381 \text{ m}}{60}\right)^2 = 0.486 \text{ kg m}^2.$$

Using Equation (9.44) it can be shown that the natural frequency for the velocity-controlled system using a PI control is given by

$$
\omega_n = \sqrt{\frac{\eta_{a_t}\eta_{p_v}V_aV_p\omega_p K_i}{JK\alpha_{max}}}
$$

$$
= \sqrt{\frac{0.95 \times 0.96 \times (200 \text{ cm}^3/\text{rev})^2 \times 1800 \text{ rpm} \times 0.67 \times 10^{-3}}{0.486 \text{ kg m}^2 \times 5 \times 10^{-11} \text{ m}^3/(\text{Pa s}) \times 18 \text{ deg}}}
$$

$$
= 3.9 \frac{\text{rad}}{\text{s}} = 0.622 \text{ Hz}.
$$

Similarly, the damping ratio for the system is given by

$$
\zeta = \frac{\left(\dfrac{V_a\alpha_{max}}{\eta_{a_v}\eta_{p_v}V_p\omega_p} + K_e\right)}{2\sqrt{\dfrac{K_i JK\alpha_{max}}{\eta_{a_t}\eta_{p_v}V_aV_p\omega_p}}}
$$

$$
= \frac{\left(\dfrac{200 \text{ cm}^3/\text{rev} \times 18 \text{ deg}}{0.95 \times 0.96 \times 200 \text{ cm}^3/\text{rev} \times 1800 \text{ rpm}} + 1.33 \times 10^{-3}\text{ s}\right)}{2\sqrt{\dfrac{0.67 \times 10^{-3} \times 0.486 \text{ kg m}^2 \times 5 \times 10^{-11}\text{m}^3/(\text{Pa s}) \times 18 \text{ deg}}{0.95 \times 0.96 \times (200 \text{ cm}^3/\text{rev})^2 \times 1800 \text{ rpm}}}} = 9.22.
$$

Since this damping ratio is greater than unity the expected response is one that asymptotically approaches the desired steady-state velocity without exhibiting any overshoot or oscillation.

Torque Control.   A third control objective that may occasionally be used for the hydraulic control system shown in Figure 9-5 is that of torque control. As previously mentioned torque-controlled systems are usually configured very much like a velocity-controlled system without a load spring and when they are used they often switch between velocity and torque control depending on the immediate needs of the application. Furthermore, torque-controlled systems are typically operated in slow motion so as to gradually apply the load torque to the object that the application is trying to resist. Under these slow-motion conditions the load inertia and viscous damping effects in the system may be safely ignored without sacrificing accuracy in the modeling process. By setting the left-hand side of Equation (9.31) equal to zero the following equation may be used to describe the applied load torque for the torque-controlled system:

$$
T = \frac{\eta_{a_t}\eta_{p_v}V_aV_p\omega_p}{K\alpha_{max}}\alpha - \eta_{a_t}V_aP_c. \tag{9.45}
$$

Using a standard proportional-integral (PI) controller the control law for the swash plate angle of the pump may be written as

$$
\alpha = K_e(T_d - T) + K_i\int (T_d - T)\, dt, \tag{9.46}
$$

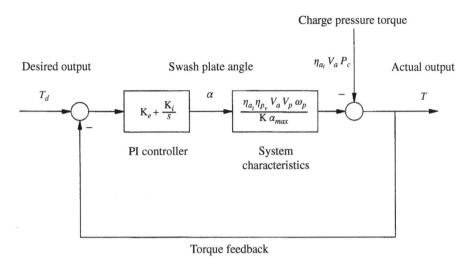

Figure 9-8. Torque control of the variable-displacement, pump-controlled rotary actuator.

where K_e is the proportional gain, K_i is the integral gain, and T_d is the desired torque that is to be exerted on the load. In most applications the instantaneous load torque T is measured by sensing the fluid pressures on sides A and B of the rotary actuator and calculating the load force according to Equation (9.32). Again, a microprocessor is necessary to carry out these functions along with an electrohydraulic displacement-controlled pump that is used to enforce the control law of Equation (9.46). Figure 9-8 shows the block diagram for the torque control that is described by Equations (9.45) and (9.46).

If it is assumed that the torque control of the system shown in Figure 9-5 is a regulation control problem and that the desired load torque T_d is a constant, Equations (9.45) and (9.46) may both be differentiated and combined to write the following dynamic equation for the torque-controlled system:

$$\tau \dot{T} + T = T_d, \tag{9.47}$$

where the system time constant is given by

$$\tau = \frac{1}{K_i} \left(\frac{K\alpha_{\max}}{\eta_{a_t} \eta_{p_v} V_a V_p \omega_p} + K_e \right). \tag{9.48}$$

Again, the reader will recognize Equation (9.47) as a description of a first-order system that asymptotically approaches the desired load torque T in time without exhibiting any overshoot or oscillation. The dynamic similarities of this system should be compared to the position-controlled system that was previously discussed.

The design objective for the torque control problem is to shape the first-order response by designing the appropriate time constant for the system. By making a proper selection of the proportional and integral control gains this time constant can be either increased or diminished based on the desired settling time for the control system output. Again, the reader is reminded that the first-order controlled system of Equation (9.47) has been designed based on the assumption that the rotary actuator moves slowly. If in fact the desired response time for this system becomes too short higher-order dynamics that have been neglected in the modeling process will become more significant and the observed output may be unsatisfactory. Under such conditions a more sophisticated analysis of the system will be needed to ensure satisfactory response characteristics.

Control Summary.   In this subsection of the chapter we have considered the three control objectives that may be carried out using the variable-displacement pump-controlled, rotary actuator shown in Figure 9-5. These objectives include position control, velocity control, and torque control all of which have employed a form of the standard PID controller. For the position control and the torque control an important slow-speed assumption has been employed for the purposes of simplifying the control problem for realistic modes of operation. Due to the slow-speed assumptions behind these control strategies the reader is cautioned about using these controllers for high-speed applications that may be encountered for these control objectives.

9.3.5 Summary

In this section of the chapter the variable-displacement pump-controlled rotary actuator of Figure 9-5 has been analyzed, designed, and controlled. It should be noted that throughout this study certain speed conditions have been assumed regarding the operation of the control system. For instance, during the development of the governing system equations in Subsection 9.3.2 the pressure transient effects were neglected due to the assumption that they occurred very quickly in relationship to the overall process time of the load dynamics. If indeed this is not the case a higher-order model of the control system will be necessary to capture the relevant pressure dynamics of the system. This situation, however, will only occur for very rare and specialized applications and the analysis presented here is deemed to be sufficient for traditional and more typical applications of the system. In Subsection 9.3.3 a design methodology for the pump-controlled rotary actuator has been presented. This design method begins by considering the steady output objectives of the system and then works backward to identify the system components that are necessary to achieve these objectives. As shown in this subsection the design of the actuator and pump are closely linked and the proper specification of these components is important when seeking to satisfy the overall performance objectives for the system. Finally, in Subsection 9.3.4 the dynamic control objectives of the actuation system have been presented. These objectives vary depending on the application; however, they generally consist of either a position control, a velocity control, or a torque control. In the consideration of these control objectives it has been assumed that the control law may be implemented by using an electrohydraulic displacement-controlled pump and a microprocessor with sufficient sensing of the controlled parameters.

9.4 CONCLUSION

This chapter has been written to describe two standard pump-controlled hydraulic systems that exist in practice today. In presenting these two systems this chapter has certainly not exhausted the subject of pump-controlled hydraulic systems; however, the systems that have been presented are representative of other hydraulic control systems that either currently exist or may be designed to satisfy various control objectives. Variations from the presented hydraulic control systems of this chapter include a crossbreed of components in which a linear actuator is controlled by a variable-displacement pump, or a rotary actuator is controlled by a fixed-displacement pump. These systems do exist and can be analyzed, designed, and controlled using similar methods that have been presented in this chapter.

The control sections of this chapter have been written to describe various feedback structures that may be used to automate the hydraulic control system. These feedback structures have depended on simplified models of the controlled plant and have also utilized classical PID control methods for achieving the desired output of the system. While PID control methods have usually proven successful in hydraulic control systems, it should be noted that PID controllers are linear controllers that work well for systems that can be described linearly over a wide range of operating conditions. Unfortunately, this class of systems does not usually include hydraulic systems, which generally exhibit more nonlinearity than most control systems and therefore the PID gains often need to be readjusted and scheduled within the microprocessor to achieve adequate control at various operating conditions of the system. Other methods of nonlinear control, which have not been addressed in this chapter, are also becoming more available to the control engineer.

Though it has not been mentioned explicitly it should be noted that the control systems presented in this chapter are often employed without using automatic feedback methods for achieving the end objective of the system. These open loop control systems depend on a great deal of repeatability within the system, which can only be guaranteed as long as the operating environment (e.g., temperature, operating load) does not change drastically. In other cases of control the control loop is closed by the feedback of a human operator who is watching the system output and adjusting the pump input manually to achieve the end objective.

9.5 REFERENCES

[1] Norton R. L. 2000. *Machine Design: An Integrated Approach*. Prentice-Hall, Upper Saddle River, NJ.

[2] Hamrock B. J., B. Jacobson, and S. R. Schmid. 1999. *Fundamentals of Machine Elements*. WCB/McGraw-Hill, New York.

[3] Kleman A. 1989. *Interfacing Microprocessors in Hydraulic Systems*. Marcel Dekker, New York and Basel.

[4] Bolton W. 1996. *Mechatronics: Electronic Control Systems in Mechanical Engineering*. Addison Wesley Longman Limited, England.

9.6 HOMEWORK PROBLEMS

9.6.1 Fixed-displacement Pump Control of a Linear Actuator

9.1   A fixed-displacement pump is used to control the position of pallet in a manufacturing process using a linear actuator. The system is similar to Figure 9-1. The actuation system is biased into a neutral position by a captured spring mechanism with a preload of 110 N and a spring rate of 1.2 N/mm. The pallet is displaced by the actuator a maximum distance of 300 mm in one direction using a maximum pressure of 1.75 MPa. The force efficiency of the actuator is 91%. Calculate the required pressurized area of the actuator. Assuming that the actuator rod diameter is 12.7 mm calculate the diameter of the actuator piston.

9.2   The no-load velocity specification for the actuation system of Problem 9.1 is 50 mm/s and the maximum pump speed is 500 rpm. Calculate the volumetric displacement of the pump required for this system.

9.3   What is the maximum power required to operate the actuation system described in Problems 9.1 and 9.2?

9.4   The position control system in Problem 9.1 and 9.2 utilizes a PI controller to adjust the input speed of the pump. The proportional gain is given by 0.14 per mm-s and the integral gain is 0.05 per mm-s^2. Calculate the settling time for this system assuming that the coefficient of leakage for the circuit is 1.5×10^{-12} m^3/(Pa s). In your opinion, is this a robust control design? Explain your answer.

9.6.2   Variable-displacement Pump Control of a Rotary Actuator

9.5   A hydrostatic transmission similar to Figure 9-5 is used to control a conveyor belt in a warehouse. The actuator must be capable of generating 240 in-lbf of torque using a supply pressure of 1000 psi. The charge pressure for the circuit is 50 psi and the actuator torque efficiency is 85%. Calculate the required volumetric displacement of the actuator per revolution.

9.6   For the hydrostatic transmission in Problem 9.5 the maximum no-load angular velocity of the actuator is 1600 rpm. Assuming that the pump speed is 1500 rpm, calculate the required volumetric displacement of the pump. How much power is required to operate the pump assuming that the torque efficiency is 89% and the volumetric efficiency is 82%? Neglect the charge circuit power for this calculation.

9.7   In order to keep the conveyor system of Problems 9.5 and 9.6 from failing the output torque of the actuator is limited using a torque control. For this system, the maximum swash plate angle is 14° and the circuit leakage coefficient is 8.9×10^{-3} in^3/(psi s). The proportional gain for the controller is 10^{-3} per in-lbf and the integral gain is 3×10^{-4} per in-lbf-s. Calculate the time constant and settling time for this controlled system. Comment on the robustness of the controller.

UNIT CONVERSIONS

LENGTH

1 m = 3.281 ft = 39.37 in = 100 cm = 1000 mm = 1×10^6 μm (a micron)
1 in = 2.54 cm = 25.4 mm

AREA

$1 m^2 = 10.764 ft^2 = 1550 in^2 = 10 \times 10^3 cm^2 = 1 \times 10^6 mm^2$
$1 in^2 = 6.4516 cm^2 = 645.16 mm^2$

MASS

1 kg = 2.2046 lbm
1 lbm = 1 lbf/g (g = gravitational constant, lbm = lb-mass, lbf = lb-force)

VOLUME

$1 m^3 = 35.315 ft^3$
$1 gal = 0.13368 ft^3 = 231 in^3 = 3.785 lit$
$1 lit = 1000 cm^3$
$1 in^3 = 16.387 cm^3$

DENSITY

$1 lbm/ft^3 = 16.019 kg/m^3$

TEMPERATURE

T, K = T, R (5/9) = (T, F + 459.67) (5/9) = T, C + 273.15
T, R = T, K (9/5) = (T, C + 273.15) (9/5) = T, F + 459.67
T, F = T, C (9/5) + 32
T, C = (T, F - 32) (5/9)

PRESSURE

$1 \text{ Pa} = 1 \text{ N/m}^2 = 1 \times 10^{-5} \text{ bar} = 1.4504 \times 10^{-4} \text{ psi}$
$1 \text{ psi} = 1 \text{ lbf/in}^2 = 6895 \text{ Pa}$
$1 \text{ atm} = 14.7 \text{ psi} = 101.325 \text{ kPa}$

FLOW

$1 \text{ lpm} = 1.6667 \times 10^{-5} \text{ m}^3/\text{s} = 16.667 \text{ cm}^3/\text{s} = 0.2642 \text{ gpm}$
$1 \text{ gpm} = 231 \text{ in}^3/\text{min} = 3.785 \text{ lpm}$

TORQUE

$1 \text{ in-lbf} = 0.1130 \text{ N-m}$
$1 \text{ N-m} = 8.8508 \text{ in-lbf}$

ANGULAR SPEED

$1 \text{ rpm} = 2\pi \text{ rad/min} = 0.1047 \text{ rad/s}$

FORCE

$1 \text{ N} = 1 \text{ kg-m/s}^2 = 0.2248 \text{ lbf}$
$1 \text{ lbf} = 1 \text{ lbm g}$ (g = gravitational constant, lbm = lb-mass, lbf = lb-force)

LINEAR VELOCITY

$1 \text{ m/s} = 3.281 \text{ ft/s} = 39.37 \text{ in/s} = 100 \text{ cm/s} = 1000 \text{ mm/s}$
$1 \text{ in/s} = 2.54 \text{ cm/s} = 25.4 \text{ mm/s}$

POWER

$1 \text{ hp} = 550 \text{ ft-lbf/s} = 745.7 \text{ N-m/s} = 745.7 \text{ W}$

INDEX

Note: Page references in *italics* refer to figures and tables.